# ROMANTIC ROCKS,
# AESTHETIC GEOLOGY

# Romantic Rocks, Aesthetic Geology

Noah Heringman

Cornell University Press

*Ithaca and London*

First published 2004 by Cornell University Press
First printing, Cornell Paperbacks, 2010
Printed in the United States of America

*Library of Congress Cataloging-in-Publication Data*
Heringman, Noah.
   Romantic rocks, aesthetic geology / Noah Heringman.
      p. cm.
Includes bibliographical references (p.  ) and index.
   ISBN 0-8014-4127-7 (cloth : alk. paper)
   1. English poetry—18th century—History and criticism. 2. English
poetry—19th century—History and criticism. 3. Geology in literature.
4. Geology—Great Britain—History—18th century. 5. Geology—Great
Britain—History—19th century. 6. Aesthetics, British—19th century.
7. Romanticism—Great Britain. 8. Sublime, The, in literature. 9.
Landscape in literature. I. Title.
PR575.G46H47 2004
821'.70936—dc22                               2003023065

Cornell University Press strives to use environmentally responsible
suppliers and materials to the fullest extent possible in the publishing
of its books. Such materials include vegetable-based, low-VOC inks
and acid-free papers that are recycled, totally chlorine-free, or partly
composed of nonwood fibers. For further information, visit our website
at www.cornellpress.cornell.edu.

Cloth printing    10  9  8  7  6  5  4  3  2  1

In Memoriam
Bernard Heringman
1923–1988

All the dry Land, and every Continent, is but a kind of
Mountain. Neither is there any Original Island upon the
Earth, but is either all a Rock, or hath Rocks and
Mountains in it.
—THOMAS BURNET, *The Sacred Theory of the Earth (1684)*


Das Erhabne wirkt versteinernd, und so dürften wir uns
nicht über das Erhabne der Natur und seine Wirkungen
wundern, oder nicht wissen, wo es zu suchen sei.

The sublime has power to petrify; hence we should not
be surprised at the sublime in nature or its influence, or
fail to know where to seek it.
—NOVALIS, *Die Lehrlinge zu Sais (1798–99)*, translated by Ralph Manheim

# Contents

# *Illustrations*

# *Preface*

      This book sets out to answer a simple yet puzzling question about Romantic poetry: Why are rocks and landforms so prominent in the literature of this period? To take a few of the most famous examples: Why does Shelley choose a mountain as the locus of a "voice . . . to repeal / Large codes of fraud and woe"? Why is it a cliff, in the boat-stealing episode of Wordsworth's *Prelude*, that chastises the young thief? Why is petrifaction, or "stonifying," in Blake's coinage, the ultimate figure of dehumanization? In the following chapters I argue that the literary culture producing this poetry was fundamentally shaped by many of the same cultural practices that formed geology as a science in the period 1770–1820. In reconstructing the complex historical basis of Romanticism's ambitious claims to reconfigure and explain the natural world, *Romantic Rocks, Aesthetic Geology* offers a paradigm for the divergence of arts and sciences and the formation of modern disciplines.

My route to these questions began with a study of writings on the sublime. Eighteenth-century discourse on the sublime is dominated by catalogues of aesthetic objects that typically begin with mountains (as in Joseph Addison's *Pleasures of the Imagination*, John Baillie's *Essay on the Sublime*, and Alexander Gerard's *Essay on Taste*, among many others). This emphasis continues in the later body of theory following Edmund Burke and culminating with Immanuel Kant, although theorists increasingly try to dismiss these objects as inessential triggers to the sublime experience. My curiosity about this dismissal was only heightened by the fact that the considerable recent literature on the sublime follows Kant in this if not in other respects. I felt that a clue to the real importance of the eighteenth-century examples must lie in the general ideas about rocks and landforms that were circulating in the period, in its geology—or, as it turned out, in travel narratives, theories of the earth, landscape architecture, and the other raw materials that would form the science called geology and Romanticism as well. An opportunity offered itself not only to do better justice to the lived aesthetic experience formalized in accounts of the sublime, but also to connect the practices of print culture with other practices surrounding Romantic literary culture.

What I ultimately found was a pervasive connection between scientific and literary culture that bears less on the sublime than on Romanticism and the history of the disciplines: writing about landscape in the Romantic period became a forum for the discussion of changing attitudes toward the earth's material and toward materiality itself. The result is not only a body of poetry obsessed with mountains, but also a geology steeped in aesthetics. This dialectic is still apparent in Charles Lyell's *Principles of Geology* (1830), a work generally recognized as marking the emergence of geology as a modern science. Numerous histories of geology detail the theoretical paradigm around which geology coalesced, a paradigm that had increasingly little to do with aesthetics. But the lingering presence of sublime rhetoric in Lyell is striking given that nineteenth-century culture had already begun to sever literature from the study of material things, an increasingly powerful division that later critics projected backward in their construction of the "Romantic movement." The following chapters reconsider that division by realigning the period's literary production with the intersecting scientific culture, and especially with early geology.

Both literature and science devoted serious attention to rugged, mountainous landscapes for the first time in the later eighteenth century, in the wake of the Industrial Revolution. By locating the common origins of geology and Romantic nature poetry in a matrix of cultural practices surrounding landscape aesthetics—including tourism, amateur naturalizing, and landscape design, as well as the consumption of fashionable travel narratives and works of aesthetic theory—I attempt to show that these two, now separate, disciplines diverged in response to a single set of historical conditions. Rocks and mountains marked a frontier that Enlightenment science and neoclassical aesthetics had not crossed. With the new demand for mineral resources, the earth's material presented itself as a foreign substance, provoking both sublime wonder and scientific curiosity. Writers of the period addressed social as well as philosophical dimensions of this newly resonant materiality, making it impossible for us either to discount the exploding demand for fossil fuels and metals or to reduce a complex cultural pattern to this one cause. What is striking is that the description of land*scape* coincides with the study of land*forms:* a range of discourses combine the empirical observation and aesthetic response that later become the separate provinces of earth science and "nature poetry." Both disciplines initially draw on landscape aesthetics for modes of observation, aesthetic categories, and a shared philosophical foundation: the contemplation of the earth's material as foreign to previous systems of human ordering, and hence a new view of nature as the subject of an independent, nonhuman history.

I argue throughout this book that Romanticism and geology spring from a common source, landscape aesthetics; and that rocks become the period's privileged aesthetic objects because its aesthetic discourse negotiates the place of consciousness in a physical environment increasingly understood in geological terms. The first half of the book focuses primarily on the "wonder" that some of the best-known Romantic poets and poems—including Wordsworth's "Resolution and Independence," Shelley's "Mont Blanc," and Blake's *Jerusalem*—bring to geologically significant landscapes. The second half draws on applied geology and poetry by Shelley and Erasmus Darwin to show how notions of economic agency or hegemony are interwoven with this sense of wonder. Many of the chapters, especially chapter 6, also draw on a more eclectic pool of poetry, natural history, and travel narrative to map the complex cultural environment in which economic and aesthetic agency coexisted. The original identity of "landscapes" and "natural resources" complicates our subsequent stronger division between the aesthetic and the practical and should unsettle two consequent assumptions: in literary criticism, the assumption that an autonomous aesthetic sphere really existed, either as a genuine locus of culture or an ideological mystification of the period; and, in environmental discourse, the distinction between two radically incommensurable realities, the purely aesthetic phenomenon of a "preserved," pristine Nature, and the dependence of an industrial society on natural resources and depleted landscapes.

# Acknowledgments

A Long-Term Fellowship at the Huntington Library from the National Endowment for the Humanities was instrumental in helping me to complete this book. The Huntington Library staff were especially helpful and courteous. The University of Missouri–Columbia Research Council provided a timely summer research fellowship at an earlier stage of the project. I also thank the State University of New York Press for allowing me to reprint chapter 2 of *Romantic Science: The Literary Forms of Natural History*, © 2003, which appears in altered form as chapter 4 of this book.

I am grateful to Richard Hamblyn, Howard Hinkel, Haskell Hinnant, Theresa Kelley, Bill Kerwin, Amy Mae King, Michael Macovski, Christine Marshall, Catherine Parke, John Plotz, Brad Prager, Martin Puchner, David Read, Catherine E. Ross, Carsten Strathausen, Patrick Vincent, and Jeffrey J. Williams for helpful comments on parts of the manuscript. Leo Damrosch, James Engell, and Steven Goldsmith acted energetically and graciously as men-midwives at the earliest stage of this project. Bernhard Kendler has been a strong and supportive editor. Louise E. Robbins handled the manuscript with exquisite care and professionalism. Many other friends and colleagues provided support and helped indirectly in innumerable ways.

I am most deeply thankful to my late father, to whose memory this book is dedicated, for his intellectual toughness and zest for the life of letters; to my mother, for her literary sensitivity and the precious gift of the German language; and to my brother and sister, Jacob and Celia, for inspiring me with their independence. My wife, Elizabeth Hornbeck, has been my constant source of light and love since I began this book in earnest, and I thank her above all.

# Abbreviations and Bibliographical Note

BG            Erasmus Darwin, *The Botanic Garden* (cited by part, canto, and
              line number)
Guide         *Wordsworth's Guide to the Lakes*, ed. Ernest de Selincourt
Inquiry       John Whitehurst, *Inquiry into the Original State and Formation*
              *of the Earth*
Lectures      *Humphry Davy on Geology: The 1805 Lectures for the*
              *General Audience*
OMG           Thomas Whately, *Observations on Modern Gardening*
Prose         *The Prose Works of William Wordsworth*, ed. W. J. B. Owen and
              Jane Smyser
PW            *The Poetical Works of William Wordsworth*, ed. Ernest de Selincourt
TE            James Hutton, *Theory of the Earth*

Other works are generally cited by author and short title in the notes; for full citations, see the bibliography. When multiple citations appear in a note, they follow the order of the quotations in the paragraph. Translations of German texts are my own unless otherwise noted.

# ROMANTIC ROCKS, AESTHETIC GEOLOGY

"General Idea of the Appearance and Arrangement of Rocks and Veins," from Humphry Davy, *Elements of Agricultural Chemistry* (1813). Engraving by F. C. Bruce from a painting by Thomas Webster. Courtesy of the Linda Hall Library of Science, Engineering, and Technology, Kansas City, Mo.

# Introduction: Aesthetic Materialism and the Culture of Landscape

## The Romance of the Earth

Rocks are ubiquitous in Romantic poetry: Wordsworth's mysterious boulder in "Resolution and Independence," the pristine summit of Shelley's Mont Blanc, and the "opake hardnesses" of Blake's Cliffs of Albion are some prominent examples. Romantic descriptions of the earth's material differ from much recent environmental discourse by presenting the earth in its otherness, its nonhuman aspect. These metonymic figures register a growing recognition that the earth evolves according to a previously unsuspected internal logic. Rocks stand for the environment as a whole because, paradoxically, the recognition of their alien physicality coincides with the early industrial drive for mineral resources. The aesthetic and economic interests that gravitate toward rock, as the raw material par excellence, equally influence the emerging discipline of geology between 1770 and 1820. Poetry and geology in later eighteenth- and early nineteenth-century Britain are mutually constitutive through the common idiom of landscape aesthetics. Landscape aesthetics incorporates not only aesthetic theory, but also practical accounts of the landscape garden and the scenic tour, popular genres that embrace aesthetic, scientific, and economic questions through their complex materialism.[1] These discourses provide evidence of a literary culture deeply invested in local natural history. The landscapes they anatomize re-

---

1. On landscape aesthetics as a historical rubric, see Elizabeth Bohls, *Women Travel Writers*, 66–67; and Malcolm Andrews, *The Search for the Picturesque*.

main—not unchanged, but still unmistakable after the lapse of 200 years, a geological nanosecond—as concrete historical referents for the poetry.

One such landscape is Middleton Dale in Derbyshire, now bisected by the A623, but still an impressive display of karst topography (limestone pervaded by caverns and underground streams). Thomas Whately, the "supremely influential" theorist of the landscape garden, chose Middleton Dale as the prototype of nature's most "highly romantic" style in 1770. Observing the correlation of strata on opposite sides of the dale, Whately notes that

> the rocks, though differing widely in different places, yet always continue in one style for some way together, and seem to have a relation to each other; both these appearances make it probable, that Middleton dale is a chasm rent by some convulsion of nature beyond the memory of man, or perhaps before the island was peopled: the scene, though it does not prove the fact, yet justifies the supposition; and gives credit to the tales of the country people.[2]

Whately based his aesthetic judgment on stratigraphic correlation, the staple activity of field trips in today's introductory geology courses. While a modern geologist would acknowledge that this practice has a scientific basis, a historian of geology might note that the business of "chasms" and "convulsions" sounds like a poetic cliché of the period. And so it is, but in the form of landscape aesthetics such terms provided a common idiom shared equally by poetry and early earth science (not yet generally known as "geology," though I extend the term deliberately to earlier ideas about the earth). In 1778 John Whitehurst based his respected geology on a volcanic "convulsion of nature" inferred from the same closely observed Derbyshire features. The evocative "convulsions," affiliated with "revolution" in the lexica of poetry and natural history as well as politics, appears in much the same sense as a political image in Erasmus Darwin and many other writers. The familiarity of geological speculation helps to explain why so many observers imagined the French Revolution as a volcanic eruption.[3] Though Whately's lexicon is not political, his classification by style is both geological and liter-

---

2. Richard E. Quaintance Jr. describes Whately as "supremely influential" in his introduction to William Chambers, *Explanatory Discourse*, 6. For the Whately passage, see *OMG*, 94–95.

3. Darwin interweaves accounts of the "Liberty of America; of Ireland; of France" with a theory of "subterraneous fires," according to which volcanoes are "not only . . . innocuous, but useful" (*BG* I.i.152n.). See further Richard Hamblyn, "Private Cabinets and Popular Geology," 188–89; and G. M. Matthews, "A Volcano's Voice in Shelley," 564. Mary Wollstonecraft uses "convulsions" in its broader social/natural sense in her *Historical and Moral View*, 70.

ary, positing the "romantic" as an aesthetic character identified with specific geological features and sharply distinguished from developed landscape.

The character of "romantic rocks" in John Whitehurst's Derbyshire caverns, and even Jane Austen's Lyme, can be traced to Whately's "highly romantic" landscape of Middleton Dale. Whately's description exemplifies an idiom shared by poetry and science and resolves the stock contradiction between science and myth as well. The correlation of strata on opposite sides of the gorge "gives credit" to the "country people['s]" legends about the place, just as it "justifies the supposition" of catastrophic agency, a staple of natural philosophy at that time. These literary and geological senses of "romantic" resonate in rich and complex ways throughout the period. Itemizing the "charms in the immediate environs of Lyme" from Anne Elliot's point of view in *Persuasion* (1818), Jane Austen especially praises

> Pinny, with its green chasms between romantic rocks, where the scattered forest trees and orchards of luxuriant growth declare that many a generation must have passed away since the first partial falling of the cliff prepared the ground for such a state, where a scene so wonderful and so lovely is exhibited as may more than equal any of the resembling scenes of the far-famed Isle of Wight.[4]

By 1818 the natural history of Lyme was a subject of general interest, one reason being the compelling story of the local orphan Mary Anning, who had found a complete Ichthyosaurus skeleton in these cliffs and was now on her way to becoming the "most eminent female fossilist" of her time.[5] By embedding observations on coastal erosion and geological time in a literary description, Austen practices what I shall call "aesthetic geology." Her comparison of Lyme and the Isle of Wight also evokes the competitions staged by travel narratives between the places they describe and other places with a reputation for scenery or geological interest.

Austen's account surprisingly resembles Whitehurst's 1778 description of "romantic rocks." The Derbyshire rocks are somewhat more rugged than those of Lyme, "broken, dislocated, and thrown into every possible direction, and their interior parts are no less rude and romantic; for they universally abound with subterraneous caverns; and, in short, with every possible mark of violence" (*Inquiry*, 63). Whitehurst's geological description, like Austen's, was not recognized as rigorous science by the more modern geology of Austen's day, a science that established itself in part by self-con-

---

4. Austen, *Complete Novels*, 1267–68.
5. The phrase is quoted from a contemporary of Anning's by W. D. Lang, "Mary Anning," 77.

sciously rejecting its literary origins. Whitehurst's aesthetic contemplation of the vestiges of geologic history, again like Austen's, is partly motivated by an interest in "improvement." In Derbyshire the "romantic" dislocations of the strata mark the locations of ore and minerals. Jane Austen, with her persistent critical attention to improvement, observes that the geology of Lyme fosters the economic *and* moral improvement of the district; the bloom of youth restored to Anne Elliot's features at Lyme seems to register a proper balance there (so absent, say, at Sotherton in *Mansfield Park*) between the two kinds of improvement. In Whitehurst's frame of reference, such contemplation belongs to geological inquiry as much as to literary narrative. Whitehurst's theory embodies the contradiction between the empirical and necessarily local practice of fieldwork and the literary practice of composing theories of the earth. His deduction of the processes of earth history from local Derbyshire phenomena adds tremendous aesthetic interest to the description, but is rejected by modern geology on precisely those grounds.[6]

As a physically descriptive adjective, "romantic" in both Austen and Whitehurst refers to the broken or dislocated character of the landforms. At Lyme, coastal erosion breaks up the cliffs, creating Austen's dramatic coastline and orchard soil; in Whitehurst's Derbyshire, the more violent agency of ancient volcanoes has dislocated the strata, creating fissures now filled with ore. In both cases the agency behind these large-scale changes is barely imaginable, sublime, strengthening the landscape's association with the fantastic. Through its fantastic or enchanted appearance, such a landscape belongs to the genre of romance and the literary past. It also belongs to the geological past, a time so remote that its vestiges can be read only as signs of obscure, titanic processes. The time of chivalry provides a literary analogue for the enchanted past of geology, as the Gothic provides an architectural one. The word "romantic" had been applied metaphorically to landscape since the turn of the eighteenth century: "Of places: redolent or suggestive of romance; appealing to the imagination and feelings."[7] The meditation on why such landscapes appeal to the imagination characterizes "Romantic literature"; in both Whitehurst and Austen, by contrast, such appeal constitutes a physical datum and a descriptive category. Friedrich Schlegel established his literary genealogy of Romanticism in 1798, midway between

---

6. As early as 1811, one critic notes that "the Valleys, contrary to Mr. Whitehurst's too confident expressions on the subject, have not the Strata in their bottoms broken up, and deep 'horrid chasms' "; John Farey, *General View of the Agriculture and Minerals of Derbyshire*, 1:490. Cf. John Challinor's dismissal of Whitehurst in *The History of British Geology*, 64.
7. "Romantic," entry 5b, in *Compact Edition of the Oxford English Dictionary*, 2:2572. This definition is traced to Addison (1705). The earliest reference to "romantic literature" cited here (4b) dates from 1841.

Whitehurst's *Inquiry* and Austen's *Persuasion*. The persistence of "romantic rocks" in these and other works helps to explain the renewed currency of the inherited tropes of romance, a renewal cited by Schlegel that remains central to genealogies of Romanticism.[8]

The more familiar literary landscapes of Romanticism also draw on geological categories in order to privilege uncultivated nature, without neglecting the motivations it presents for "improvement." The mysterious erratic boulder in Wordsworth's "Resolution and Independence" evokes the sense of "wonder" common to geological and aesthetic explanations of such phenomena, and uses these associations to address the "more than human weight" of the physicality that haunts the poem. At the same time, the speaker's extraction of a moral profit in the form of the Leech-gatherer's "independence" capitalizes on the association of raw material also activated by the stone. The "unearthly" landscape of Shelley's "Mont Blanc" registers the active undoing of form, revising the old trope of rocks as ruins or mere signs of chaos; but the poem's speaker also mines the landscape for poetic feeling and philosophical imagery, explicitly excluding other human uses. Another description, from Shelley's *Prometheus Unbound*, of "the secrets of the Earth's deep heart, / Infinite mine of adamant and gold," brings out more fully the economic implications of aestheticizing geological forms and processes.

Such images remind us that literature is a product of the earth as well as a product of culture. When Blake exclaims, "They call the Rocks Parents of Men," in a critique of atomist philosophy, he uses rocks to stand for the very notion of an environment causally prior to human existence. The irony of this image turns on the barrenness or inorganic character of rocks. The dominant modern construction of materiality, against which Blake reacts here, relies on the otherness of inorganic matter. What is most essential to materiality in the definitions of early modern philosophy is extension in space, a requisite independent of biological properties. Metaphysics after Descartes turns on the relationships of dependence or independence between matter—thus defined as inorganic—and mind, animate but non-extended. In this respect, "romantic rocks" are part of an ongoing philosophical project to define the place of consciousness in a physical environment. Imaginative literature of the Romantic period is distinguished by its prevailing concern of recreating, constructing, or inventing aesthetic experience. Aesthetic experience (of "nature" especially), rooted in the senses, requires a confrontation with the apparent division between spirit and matter. An object, usually visual, arrests the senses and becomes the occasion for

8. See Schlegel, *Kritische und theoretische Schriften*, 90; cf. 184.

an expression of feeling, a "spontaneous overflow of powerful feelings," in Wordsworth's version. Materiality is essential to aesthetic experience because it makes viable the notion of involuntary aesthetic response.

Rocks become preferred objects of description and explanation, poetic and otherwise, because they instantiate the most basic features of the physical environment. This assumption informs both practical discourses of domestication and the aesthetic idiom of wonder. In fact, aesthetic response to the materiality of rocks and landforms is inseparable from the emerging economic category of natural resources. Humphry Davy, Romantic scientist par excellence, derives his economic definition of geology, as the study of inorganic materials "subservient to different uses," from the established aesthetic value of mountain scenery: "Nature arises subdued by artificial means, not impoverished or deformed, but enriched, and made more beautiful" (*Lectures*, 7). This paradigmatic combination of the sublime and the useful occurs again and again in all the period's discourses on landscape, most often in a context of reflection on the large scale, long duration, and unformed appearance of rocks and landforms. Their primitive materiality is also a hermeneutic temptation, and the idea of rock formations as historical texts features in both aesthetic and practical discourses; the figure of a "rock record" most fully mediates the nonhuman aspect of the environment, assimilating natural history to human history. So, in *Prometheus Unbound*, a subterranean archive of prehistoric fauna and cultures accompanies the "infinite mine of adamant and gold" forming the earth's interior. Against this domestication of the earth, "Mont Blanc," again, stands as a reminder of the absolute otherness and indifference of geologic forms and processes.

The poetry's geological affinities are clearest in relation to earth science produced during the same period, 1770–1820. Two previous studies of Romanticism and geology, Marjorie Nicolson's *Mountain Gloom and Mountain Glory* (1959) and John Wyatt's *Wordsworth and the Geologists* (1995), incorporate these affinities into narratives of influence, situating Romanticism in relation to earlier and later geology, respectively.[9] Nicolson meticulously reconstructs the influence on Romanticism, culturally diffused rather than individually experienced, of the sublime in nature as developed from natural philosophy around 1700. She argues that cosmology from Henry More through Isaac Newton valorizes space as an attribute of God while scientifically establishing its vastness. Inspired by the new cosmology, poetry and aesthetics begin to treat the sublime in nature as a secularized equivalent. By the time of the major Romantics, these associations are customary, occur-

---

9. Dennis R. Dean's pioneering dissertation (1968) examines the period's poetry and earth science side by side.

ring spontaneously in their poetry, and especially around images of mountains and wild landscapes. Wyatt, reversing the trajectory, argues for the influence of Wordsworth's views of nature on early Victorian geology, in a sense confirming Nicolson's view that Romantic poetry embodies the normative secularized wonder motivating scientific inquiry. In the following pages I am less concerned with influence in either direction, or with geology as a "background" of literature, than with a single cultural context in which several discourses converge on a central, historically specific issue. When geology, after 1800, begins to declare its independence from an unscientific past of armchair theories and sensational narratives, it provides a model for discipline formation in the sciences. It also contributes to the redefinition of "literature" that recent critics have seen as central to Romanticism.[10]

## Enlightenment Science, the Sublime, and the Public for Natural History

In 1830, Charles Lyell writes in his *Principles of Geology:*

> We often discover with surprise, on looking back into the chronicles of nations, how the fortune of some battle has influenced the fate of millions of our contemporaries, when it has long been forgotten by the mass of the population. . . . But far more astonishing and unexpected are the connexions brought to light, when we carry back our researches into the history of nature. The form of a coast, the configuration of the interior of a country, the existence and extent of lakes, valleys, and mountains, can often be traced to the former prevalence of earthquakes and volcanoes, in regions which have long been undisturbed. To these remote convulsions the present fertility of some districts, the sterile character of others, the elevation of the land above the sea, the climate, and various particularities, may be distinctly referred.

Surveying the history of science as he embarks on the discussion that will establish the paradigm for modern geology, Lyell notes a fundamental shift in consciousness with ramifications well beyond geology: the growth of interest in nature defined as prehuman and nonrational. Gillian Beer's comment on this development incorporates one of the great commonplaces of the history of ideas: "New organisations of knowledge are particularly vex-

---

10. As Jon Klancher puts it, " 'literature' itself meant one thing in 1780—the whole array of educated genres from natural philosophy and history to poetry and drama—and something very different after 1820 (the restricted category of imaginary [*sic*] genres we know today)" (*Romanticism and Its Publics*, 524). Clifford Siskin also takes up this redefinition in *The Work of Writing*.

ing when they shift man from the centre of meaning or set him in a universe not designed to serve his needs."[11] Beer notes the successive manifestations of this interest in the three canonical scientific revolutions of Copernicus, Darwin, and Freud. In its geological form, this interest pervaded polite culture and even party politics in eighteenth- and nineteenth-century Britain, with government and Tory periodicals featuring articles by Neptunists (who believed the earth was formed by the action of water) and the Whig *Edinburgh Review* among those sponsoring Plutonism (based on heat) over many decades. Lyell famously shifted the terms of this debate to those of "uniformitarianism" and "catastrophism." His debt to Plutonism is clear, however, from the word "convulsions." The persistence of this anatomical metaphor registers a complex cultural legacy: for many decades before Lyell, writers and naturalists had struggled to apply such familiar human terms to landforms and processes increasingly revealed as alien vestiges of a prehuman world. In Lyell's cultural moment, it became possible to deploy the familiar analogy between human and natural history with a full consciousness of its insufficiency. The earlier politics of geological materialism enabled this paradigm shift. Through such tropes as geological/political "revolution," the volatile discourse of the sublime becomes the nexus of two broad contextual rubrics of this study, Enlightenment science and the public sphere.

Rather than a scientific revolution, Lyell's view registers a change in scientific culture created by the persistence of certain literary images. Representations of nature's otherness, such as Whately's "convulsion of nature" and Shelley's "unearthly" landscape, create the idiom for Lyell's prehuman space of "remote convulsions." His foundational *Principles of Geology* owes its object of inquiry not to literature per se, but to a style that pervaded British letters as a whole, imbricating poetry, landscape aesthetics, and premodern geology. A unique matrix of cultural practices produced these converging discourses and their shared style: tourism, mining, naturalizing, exhibition, landscape design, and other forms of "improvement." Enlightenment science marked rocks and mountains as a frontier of knowledge, expressing an alienation that became naturalized in aesthetic form during the Romantic period. Lyell decisively repudiates geology's literary or "fabulous" past and embraces empirical rigor. Yet traditional figures of rocks as foreign objects, and especially the lexicon of the sublime, both enable geology to designate

11. Lyell, *Principles of Geology*, 1:2; Beer, *Darwin's Plots*, 19. In arguing for Lyell's influence on the mid-nineteenth-century American "enthusiasm for geology," Rebecca Bedell credits him with a "complete reorientation of human beings' sense of their place in the world"; Richard Monastersky, "Marriage of Art and Science," A12. Bedell elaborates this point more fully in her book, *The Anatomy of Nature*, 3, 5–7.

its empirical province and give Lyell the materials for his epochal theme, the vastness of geological time. Although this emphasis in Lyell and his predecessor James Hutton has the effect of "shift[ing] man away from the center of meaning," the shift is produced less by newly discovered anomalies in the natural world than by perceptions that had long been latent in images of "rude" and "primitive" rocks.

No one theoretical paradigm adequately explains the shared set of cultural practices shaping Romanticism and geology. It cannot be approached as a "scientific revolution," even in Beer's revised sense, but it also forms a cultural layer too complex to be described as "normal science."[12] The problem of nature—and especially inorganic nature—as a material "other" might be addressed, in Foucault's terms, as an epistemic problem revealing structures common to all discourses in a given period. Certainly, in an analysis of literature and geology circa 1800, "things usually far apart are brought closer" in order to show a historically specific relationship and pattern of knowing.[13] Adorno and Horkheimer's *Dialectic of Enlightenment* is perhaps more germane to the problem of nature's otherness as an analysis of hegemony that begins with the domination of nature, which they present as a demystification of the natural world that remystifies human (and natural) relationships. Both of these models place the problem of scientific revolution within a larger complex of historical change centered on the Enlightenment. Foucault's notion of the Enlightenment as an intermediate stage between the Renaissance and modernity is one version of a common view that there are *two* scientific revolutions, one occurring during the seventeenth century and the other around 1800 with the formation of the modern disciplines. In Adorno and Horkheimer's account, the Enlightenment marks one signal shift heralding the modern demystified view of nature.[14]

Adorno and Horkheimer make continuities visible where Foucault emphasizes the epistemological departure of modern science. Recent discus-

---

12. In Thomas Kuhn's theory, "normal science" presupposes consensus on a certain theoretical paradigm, while scientific revolution is defined as the disruption of that consensus by the discovery of anomalies it cannot explain (*Structure of Scientific Revolutions*, 6). Kuhn points out that in preparadigm science, the consensus "must be externally supplied" (16–17), a helpful consideration for early geology.

13. Michel Foucault, *Order of Things*, x; see also xxii.

14. Compare, for example, Foucault's "post-Classical" Kant (162) with Adorno and Horkheimer's Enlightenment Kant in *Dialektik der Aufklärung*, 6; cf. 32–33. Foucault's insistence on "two great discontinuities in the episteme of Western culture" (xxiii) is echoed in the concept of a "second scientific revolution" summarized by Andrew Cunningham and Nicholas Jardine in *Romanticism and the Sciences*, xix–xx. William Clark, Jan Golinski, and Simon Schaffer's introduction to *The Sciences in Enlightened Europe* is helpful on all these issues, including the influence of Horkheimer and Adorno (10–11) and the overlap between Enlightenment and Romantic science (28).

sions of Romantic representations of nature generally emphasize the hegemonic aspect of the Enlightenment legacy. The sublime, in particular,
seems to have become inseparable from the domination of nature so central
to Enlightenment ideology. Horkheimer and Adorno's critique of Enlightenment, which helped to set the stage for such criticism, describes an aesthetic ideology of nature particularly relevant to the early industrial context
of British Romanticism. The mineral resources vital to this stage are inorganic, so they offer an ideal prototype of the commodification characteristic
of Enlightenment in Horkheimer and Adorno's reading. Lyell's *Principles*
embodies an epistemic shift, completing the emergence of geology as a
modern science, because of this very continuity: it fuses Enlightenment rationalism with a synthetic view of global processes indebted to Romantic
organicism. The experience of nature as landscape and the resulting, all-
pervasive images—of magnitude, formlessness, inscrutable antiquity—are
mainly responsible for the growing interest in geology in the later eighteenth century, for widespread and regularized use of the term "geology"
around 1800, and for its formation as a full-fledged discipline by the early
Victorian period. The same cultural legacy accounts for Lyell's surprising
identification of the "history of nature" with the history of landforms in
particular, as in the passage quoted above.

Close attention to geological and other particularities distinguishes the
"nature writing" of the Romantic period, but this new tendency also reflects
latent counterhegemonies in the materialism of the Enlightenment. Landscape aesthetics in all its ramifications appears retrospectively as a publicity
campaign for geology, the modern scientific discipline that replaces the old
tripartite natural history and redefines nature as the earth's material. This
far-reaching process of discipline-formation seems possible only if "the sublime in nature" was a much more complex and ambivalent thing than is
often allowed, not merely reflecting the domination of nature but also
marking an authentic otherness, as indicated by the vast prehuman
timescale proposed by Lyell. This otherness is the obverse of the apparent
domesticity of inorganic nature. While the sublime as a rhetorical mode
well suits the apologists of industrialization, it performs a wide variety of
other cultural work. As Raymond Williams observes, lived "hegemony . . .
is never either total or exclusive"; the Enlightenment legacy concerning nature is not merely the hegemonic one of an abject otherness, but also preserves the unknowable otherness of the nonhuman and even the counter-
hegemonic analogy of revolution as natural process.[15]

Although "industrial civilization," as Williams argues, "at once produces

---

15. Williams, *Marxism and Literature*, 113.

and limits" Romanticism as counterculture, Romantic nature poetry amplifies the ambivalence of the sublime and its apparently hegemonic legacy.[16]
Wordsworth, Blake, Shelley, and many other poets imagine otherness not
only as subjection but also as indifference, as resistant materiality. Forest
Pyle and Onno Oerlemans are among the recent critics who have noticed
this resistance in their accounts of a Romantic materialism repressed by earlier critical accounts of nature and the imagination.[17] One received narrative
asserts that Romanticism secularizes the sublime features of the cosmos—
the mountains, the ocean, the stars—that appeared as attributes of God in
the wake of Newton and natural theology. But even the most pious poetry
of this earlier period already registers a deep uneasiness about the superhuman scale and alien physicality embodied by the earth's material.[18] The
most pervasive epithet for rocks in the earlier eighteenth century is "rude,"
followed by "savage" and "romantic." These terms reflect a fashionable
Gothic aesthetic, but they also signify a resistance in matter itself to the
transcendent order of Enlightenment cosmology. The epithets persist even
through the secularized cosmology of the Romantic period. Shelley's Mont
Blanc is "rude, bare, and high, / Ghastly, and scarred, and riven." Humphry
Davy, in a geological lecture on the Alps, notes the "sublimity" of the "bare
and confused crags and snowy summits of the primitive mountains."[19]

Such descriptions epitomize a cultural legacy brought to bear on new and
ever-increasing encounters with rocks in mines and museums and on scenic
tours, conveying the lingering sense that rocks stand outside the order of
nature, whether transcendent or autonomous. By marking out a chaotic
space in rocks and landforms, literary description helps to create the
province of geology as an autonomous and imposing field. James Hutton's
*Theory of the Earth* (1795) provides an example from the canon of early geol-

16. Ibid., 114.

17. Oerlemans, in *Romanticism and the Materiality of Nature*, seeks to point out "those moments when the material is not transcended but confronted" (13), adding that this aspect of the
poetry "teaches us that the material resists being read, that it is not a text" (29). Pyle similarly
writes of moments in Keats when the poems' materialism resists their own larger humanist
project; *Ideology of Imagination*, chap. 4.

18. David Mallet's *Excursion* (1728) provides a striking example (I.441–47):

> Up from the centre torn
> The ground yawns horrible a hundred mouths,
> Flashing pale flames—down through the gulfs profound,
> Screaming, whole crowds of every age and rank,
> With hands to heav'n rais'd high imploring aid,
> Prone to th'abyss descend; and o'er their heads
> Earth shuts her ponderous jaws.
>     From Chalmers, *Works of the English Poets*, 14:20.

19. *Shelley's Poetry and Prose*, 99, ll. 70–71; *Lectures*, 75–77.

ogy. In urging the scope of earthquakes and volcanism as evidence for permanent central heat, Hutton asks: "Are those powerful operations of fire . . . which so often have filled us with terror and astonishment, to be considered as having always been?" (*TE*, 1:143). Lacking the means to argue from empirical evidence, Hutton draws on the culturally powerful status of natural disasters as aesthetic objects—established in poems and narratives—to suggest that sublime effects must have a corresponding cause. As the common term of such descriptions, the sublime designates an authentic chaos. A consciousness of real epistemological limits tempers the mystification of human sovereignty now viewed as the subtext of the "natural" sublime: though sometimes aimed, as in Kant, at aggrandizing the subject, the discourse of the sublime persistently privileges objects that are finally unmasterable, such as the Alps, volcanoes, earthquakes, and the earth's interior.

This focus on untamed nature is problematic from the point of view of ecocriticism and other recent environmental writing. Jonathan Bate, for instance, reads Wordsworth as marking off the Lake District as a space outside resource economics, ultimately as a space for preservation.[20] Wordsworth, however, promotes a working—if unobtrusively working—landscape in *A Guide to the Lakes* and much of his poetry; one late poem even expresses great optimism concerning "Steamboats, Viaducts, and Railways" (1833).[21] Bate's reading overstates the case for Wordsworth's ecology, projecting a problematic ideal of wilderness. As Christopher Hitt puts it, "the fundamental problem with the concept of sublime wilderness is that it depends on and reinscribes the notion of nature's otherness, of the separation between the human and nonhuman realms."[22] Hitt goes on to reclaim the sublime from this charge, theorizing otherness without hierarchy through examples from philosophy and contemporary nature writing as well as Romanticism. But the objection to sublime wilderness is worth considering, certainly as a caution against reinscribing an ideological division between the aesthetic and the practical on the natural world. The Romantic-era geological sublime, unlike the idea of wilderness, clearly confronts the coexistence of economic and aesthetic imperatives in a social context. Though early industrial ideas about the earth may have proved environmentally unsound, they also granted nature greater autonomy than we are prone to do. Even the ahistorical sublime of wilderness, as Hitt urges, should be reconsidered before it is dismissed as a symptom of alienation.

20. Bate, *Romantic Ecology*, 19, 21–22.
21. *Guide*, 165. Even when Wordsworth objects to a planned railway in the Lake District, he makes the economic argument that "seclusion and retirement" are "the staple of the district" (*Guide*, 148).
22. "Toward an Ecological Sublime," 603. Hitt goes on to suggest that, contra Kant, "the [sublime] discovery of nature abrogates reason" (617).

Adorno and Horkheimer read alienation strictly as a consequence of the domination of nature. Seen dialectically, "Enlightenment . . . is nature that becomes perceptible in its alienation." Enlightenment thought objectifies nature, and so dominates it, but at the cost of "human alienation from the dominated object," and even the alienation of "human relationships themselves." The Kantian object, in this analysis, exists exclusively as a substrate of hegemonic subjectivity.[23] I would argue that Romantic rocks contest the subjugation of this propertyless "abstract materiality" because they manifest materiality as something that is actually recalcitrant. It is not that nature is given its own history or subjectivity, as Novalis suggested in *Die Lehrlinge zu Sais*, but rather that it proves indifferent if not inimical to subjectivity. Novalis' attempt to reinvent scientific inquiry, however, fruitfully complicates the Kantian legacy of domination. In their reading of Novalis, Andrew Cunningham and Nicholas Jardine point out that *Naturphilosophie* and the myriad protodisciplines of Romanticism also contributed key elements to the intellectual platform of modern science, including the critique of mechanism and emphases on creativity and the historicity of the natural world.[24]

Human proclamations of "the end of nature," by contrast, represent an extreme alienation from nature. This kind of catastrophism conforms remarkably to Adorno and Horkheimer's vision of an unreflective culmination of Enlightenment: "human beings expect that the world, which they cannot escape, will be immolated by a totality—themselves—over which they have no control."[25] Even in an atmosphere of lingering or reviving nuclear tension, arguments positing the exploitation of nature as ineluctable really express the alienation of postmodern culture, a helplessness that is finally an objectification of nature. One symptom of this objectification is the failure to acknowledge environmentalism as an effort to improve conditions for human beings. Gillian Beer points out in a recent essay that nature, from an evolutionary perspective, is essentially deviant and unpredictable. Beer suggests that social constructionism's implicit appeals to nature as a stable ontological ground are thus also a form of objectification or alienation.[26] Romantic recognition of the earth's unpredictability and difference from human interests, on the other hand, permits progressive analogies to human agency. As Hutton and Lyell begin to read evidence of "remote convulsions" in terms of renovative and naturally recurring violence, rather than in terms of divinely ordained catastrophe, poems such as "Mont Blanc" are

---

23. *Dialektik*, 46, 34, 33. Cf. *Dialectic of Enlightenment* (trans. Jephcott), 31, 21, 20.
24. Cunningham and Jardine, *Romanticism and the Sciences*, 7–8. See p. 6 for an impressive list of the period's protoscientific disciplines.
25. *Dialektik*, 35; *Dialectic*, 22.
26. Beer, "Has Nature a Future?" 23. In using "the end of nature," I allude to the title and tenor of Bill McKibben's book (e.g., 58, 63–64).

increasingly able to mobilize the analogy between geological and political revolution.[27]

The episteme and the Enlightenment, though helpful categories for negotiating scientific emergence, are vague with respect to social formations, particularly in comparison to the influential model of the public sphere. Broad histories of print culture, especially Jürgen Habermas' *Structural Transformation of the Public Sphere*, have provided the ground for much subsequent study of eighteenth- and nineteenth-century cultural production. Recent historians of science have turned to the public sphere paradigm as well, and it helps to illuminate the shared practices—literary and nonliterary—informing Romanticism and early geology. In the following chapters, I have in mind the public that read poetry, natural history, travel narrative *and* for which practices like tourism, landscape and garden design, fossil collecting, or mining played some part in lived experience. It is a surprisingly inclusive public, blurring to some extent the gender and class boundaries limiting Habermas' public sphere. This breadth of readership also testifies to the literary and political weight of natural history and other cultural forms excluded from modern definitions of literature. Historians have increasingly emphasized the naturalists' societies, publicly displayed collections, public lectures, and other forums in which the culture of natural history was disseminated in Britain.[28] Popular poetry such as Erasmus Darwin's shows the influence of these institutions, and the widely read genre of topographical poetry served as a clearinghouse for scientific, especially geological, ideas from the early part of the eighteenth century. Here are clues toward answering an important set of questions: How did a public receptive to the exalted mountains of Byron, Wordsworth, and others come into being at all, and how did the complex existing network of discourses concerning rocks and landforms contribute to its orientation?

We can trace the emergence of a public eager for natural knowledge, with Habermas, back to *The Spectator,* for which Addison announces the aim of "bringing philosophy out of closets and libraries, schools and colleges, to dwell in clubs and assemblies, at tea-tables and in coffee-houses."[29] Addison

27. Nigel Leask makes this point effectively in "Mont Blanc's Mysterious Voice," 202. On the expanded semantic field of "revolution," see also Alan Bewell, *Wordsworth and the Enlightenment,* 272. I develop the concept of telluric agency further in "The Style of Natural Catastrophes," which draws on Bruno Latour's and Donna Haraway's arguments for the agency of scientific "objects."

28. Nicholas Jardine, James Secord, and E. C. Spary, eds., *Cultures of Natural History,* extends the emphasis on cultural institutions initiated by David Allen in *The Naturalist in Britain.*

29. *Spectator,* 1:44 (No. 10). Addison studied natural philosophy at Oxford with Thomas Burnet, a fact reflected in his *Essays on the Pleasures of the Imagination* (see esp. *Spectator* No. 420, 413–14). *The Spectator* marks the locus classicus for Habermas' account of the literary culture ("frühe literarische Öffentlichkeit") that becomes the basis for a politically functioning public

does not refer exclusively to natural philosophy, but scientific matters certainly played a role in the coffeehouses—William Whiston, for example, popularized his theory of the earth in lectures at Button's coffeehouse, arranged for him by Addison and Steele in 1714—and continued to do so throughout the century.[30] In this respect, the metropolitan audience for early geology is coextensive with the public sphere designated—if perhaps too summarily—by Habermas.[31] A real increase in the public for natural history seems to have come with natural history's entrance into the secondary school curriculum through the Dissenting academies, after 1760. Natural history (botany, geology, zoology) slowly became less a poor cousin of natural philosophy (the physical sciences) and more an increasingly profitable middle-class avocation. David Allen shows that the establishment of several important natural history societies coincided with this period of growth, giving these disciplines an independent place alongside the more heavily aristocratic Royal Society.[32] Drawing on the many publications and practices surrounding geologically oriented tourism in the second half of the eighteenth century, Richard Hamblyn shows convincingly that "there was an expansion of the franchise on natural knowledge" in this period. A remarkable number of vernacular or "low" mineralogical works, many of them by Dissenters, asserted the priority of local over philosophical knowledge and carried on a "rivalry with officially sanctioned scientific institutions" that made their research "as visible and discrete as that of metropolitan polite culture."[33] The possibilities for women further highlight the wide

sphere (*Strukturwandel der Öffentlichkeit*, 105–7, 89; *Structural Transformation*, 42–43, 30). Though Habermas does not address scientific culture, his public sphere paradigm has become a major reference point for several historical studies, including Jan Golinski's *Science as Public Culture*.

30. James Force, *William Whiston*, 20. On coffeehouse lectures as a forum for geology in particular, see Hugh Torrens, "Geological Communication in the Bath Area," 229–31.

31. Habermas himself comments, in a lengthy new preface written for the 1990 edition of *Strukturwandel*, that his original account does not adequately address exclusions from the public sphere, particularly the exclusion of women (18–20). This limitation has provided the opening for a great deal of revision of Habermas' paradigm. Habermas' apparent assumption of continuity within the British public sphere after 1750 (107; cf. 153–54 and n. 62; *Structural Transformation*, 43, 83) has made his paradigm both very flexible and very debatable for students of that period. Anne Mellor takes up both issues—gender and period—in *Mothers of the Nation*.

32. Allen, *Naturalist in Britain*, 39–43. After 1780 the controversial presidency of Joseph Banks occasioned similar changes in the Royal Society itself. Publications by faculty at the Warrington Academy record the role of natural history in the curriculum. These include J. R. Forster's *Introduction to Mineralogy* (1768) and John Aikin's *Essay on the Application of Natural History to Poetry* (1777). As Anna Letitia Barbauld put it, "Where science smiles, the Muses join the train"; "The Invitation" 109, in *Poems*, 13. Barbauld's poem was later adapted for the Academy's seminal anthology, *The Speaker*.

33. Hamblyn, "Landscape and the Contours of Knowledge," 15, 52, 86. Hamblyn documents the amazing proliferation of nontechnical mineralogical writings at this time and insists

inclusiveness of the geological public. Arguably the most fashionable woman in England, Georgiana, the Duchess of Devonshire, became a life-long devotee of mineralogy while on a continental tour in 1793. Sixteen years later a ten-year-old orphan named Mary Anning began supporting her family by selling Liassic fossils she found in her native Lyme and eventually taught fossil anatomy to "scientific" geologists.[34]

While the public for natural history is too diverse to be constituted as a "counter-public sphere," it does contrast with the model of Habermas with respect to class and especially gender.[35] Habermas memorably describes the "ambivalence" characteristic of the politicized public sphere emerging in the later eighteenth century. His analysis shows that literary culture counts as the qualification for property rights, yet "women and dependents are practically and juridically excluded from the political public sphere, even though female readers . . . often participate more fully in the literary public than the property owners themselves."[36] The geological activity of Georgiana and Mary Anning suggests that in the case of natural history, reader-ship was by no means at odds with participation.[37] In a number of cases, women's involvement in natural history complicates Habermas' distinction between participation in the literary versus the political public sphere. Anna Seward's "Mount Etna," though ostensibly written in praise of Patrick Bry-done's *Tour through Sicily,* ventures independent geological speculations and harnesses the ever-incendiary image of the volcano to an account of the poet's own imaginative agency. Sarah Murray Aust, in her narrative of a solo

---

fruitfully on the importance of the industry-tourism nexus in setting the agenda for earth science.

34. For details on Anning, see the interchapter in this volume. On Georgiana, see Amanda Foreman, *Georgiana, Duchess of Devonshire,* 276. Foreman also describes Georgiana's contact with prominent scientists (280) and the continuation of her mineralogical activity in England (287, 292–93). Cf. Allen, *Naturalist,* 25–26.

35. Habermas' recurring formula for entry into the political public sphere is *Besitz und Bildung* (property and culture). Various kinds of working-class initiatives, however, can be traced in natural history, from botanizing weavers in mid-eighteenth-century Norwich to early nine-teenth-century "workingmen's institutes" in applied science. On the former, see Allen, *Naturalist in Britain,* 39; on the latter, Richard D. Altick, *English Common Reader,* 189. Relevant essays in Jardine, Secord, and Spary, *Cultures of Natural History,* include "Artisan Botany" and "Nature for the People."

36. Habermas, *Strukturwandel,* 121; *Structural Transformation,* 55–56.

37. Women's better-documented involvement in other fields, especially botany, has been widely studied, but there has been little work so far on women and early geology. Mary R. S. Creese and Thomas M. Creese provide a historical sketch in "British Women Who Contributed to Research in the Geological Sciences," and Anne Wallace examines Charlotte Smith's attitude toward geology in "Picturesque Fossils, Sublime Geology?" Marina Benjamin discusses the politics of women's scientific writing in more general terms in the introduction to *Science and Sensibility.*

journey through Scotland, self-consciously enters on the subject of geological theory despite having acknowledged it as a preserve of (male) authority. Logistically doubling this discursive move, Murray Aust advises her (implied female) reader: "Do not suffer the guide to deter you from stepping from stone to stone . . . [or] you will not see the most beautiful part of Dovedale."[38] Amanda Foreman sums up Georgiana's mineralogical activity with an observation that captures a crucial ambiguity surrounding female political agency, natural history, and the public sphere: in the 1790s, she remarks, Georgiana "exchanged political meetings for scientific lectures."[39] Foreman argues persuasively that through her activism, culminating in the 1784 election, Georgiana attained the highest degree of political agency achieved by a woman in that culture. Her turn to natural history, for Foreman, signals a retreat from politics in the face of powerful social and family pressures. This is a deceptively simple substitution.

Women's involvement in natural history—and geology in particular, because of its relationship to materialism—draws attention to a neglected aspect of the public sphere. While the debate about the public sphere certainly helps to illuminate the social role of early geology, the discourse on rocks and landforms also helps to explain what happens to the public sphere in England amidst reaction against the French Revolution. Jon Klancher's widely accepted revision of Habermas describes a less unified, more tenuous public sphere and notes that even this "had largely to be swept away in the 1790s" because of antirevolutionary fervor. Various radical "counter-public spheres" have been posited as emerging in the wake of this fragmentation, but as Kevin Gilmartin cautions, "the revisionist case for diversity in the public sphere needs to remain sensitive to historical variations, or risk becoming as misleading as any insistence on uniformity."[40] Precisely because of their original diversity, the discourse and practices of natural history seem to remain a relatively stable and unified zone within the rifted public sphere of the 1790s. Because it is an ostensibly depoliticized zone, it functions as a virtual public sphere, maintaining the autonomy of the republic of letters against and between the polarized Tory state and radical counter-

38. Sarah Murray Aust, *Companion and Useful Guide*, 1:6–7. This work is remarkable not merely for being a woman's narrative, but for being that of a woman traveling self-consciously, and often pointedly, alone. For Seward's "Mount Etna," see *Poetical Works*, 2:209–14.

39. Foreman, *Georgiana, Duchess of Devonshire*, 292.

40. Klancher, *Making of English Reading Audiences*, 20; Gilmartin, "Popular Radicalism and the Public Sphere," 557. Though the public for natural history remained relatively constant beyond 1790, reactionary suspicion fell on figures of the scientific avant-garde such as Erasmus Darwin and especially Joseph Priestley. Jan Golinski considers the ways in which chemistry became politicized in the careers of Priestley and Humphry Davy; *Science as Public Culture*, chaps. 6–7.

public sphere. Georgiana, for instance, remained in public view because of natural history and later regained political influence. The debate about Hutton's 1795 *Theory of the Earth* became a coded debate about materialism. Not surprisingly, Jean-André de Luc's condemnatory 1796 review essay appeared in the government-sponsored *British Critic* (spanning three issues and more than 100 pages). This sort of reaction, I would argue, is what led radical writers such as Shelley and Erasmus Darwin to deploy images of volcanoes and earthquakes—the secularized and naturalized "convulsions" and "revolutions" of Hutton's theory—precisely for their revolutionary associations. By the 1820s, however, geology had become depoliticized and increasingly excluded women, both prerequisites for the formation of modern science.[41] This genealogy provides one brief hint of the way in which over the previous two generations geological discourse shaped and extended the matrix of the public sphere in England.

The impressive subscription list for John Whitehurst's 1778 treatise on the earth provides a concrete index of early geology's particular public. The interest of this diverse group of more than 500 subscribers to the first edition of Whitehurst's *Inquiry into the Original State and Formation of the Earth* was surely a key factor in Whitehurst's transformation from a little-known provincial researcher to a Fellow of the Royal Society, as his work went through two more editions (1786, 1792). Its publication by subscription, on the one hand, locates the project's public somewhere between aristocratic culture and the bourgeois public sphere; on the other hand, the list is remarkably inclusive, ranging from men in relatively humble trades (for example, a plumber, a "plaisterer," a "sugar-baker") up to the kingdom's highest dignitaries, such as Lord Rockingham and Lady Scarsdale. The bulk of the list conforms to the prescribed mercantile-clerical makeup of the public sphere, but its class diversity and inclusion of women are noteworthy for a work that was specialized for its time.[42] In short, early geology still had the "general intelligibility" absent in a "mature" science.[43] Whitehurst's public

41. Barbara Gates and Ann Shteir argue that "as scientific discoveries continued to unfold in the nineteenth century, science became increasingly masculinized and professionalized," and a feminized natural history became subordinate to the "narrative of science" (*Natural Eloquence*, 9). Gillian Beer surveys some of the feminist scholarship relevant to this point in "Has Nature a Future?" 18–22. Politics is present in other ways in Victorian geology, and its masculinism is complicated, but the contrast with early geology is clear.

42. John Whitehurst, *Inquiry* (1st ed.), unpaginated front matter. Only seven of over five hundred subscribers were women, but this number is significant in comparison to other such lists. The list also represents numerous other trades besides those mentioned above and includes Georgiana's mineralogical tutors (and scientific luminaries) William Hamilton and Joseph Banks.

43. Kuhn, *Structure of Scientific Revolutions*, 20.

overlaps with the early public of Romanticism. The list includes patrons of Wordsworth and Coleridge such as Sir George Beaumont and Josiah Wedgwood, and Erasmus Darwin relied heavily on Whitehurst's theory for geological episodes in his poetry. My contention is that such overlap is typical: mountains are prominent in Romantic poetry largely because Romantic literary culture extends to natural history and other antecedents of geology.

Whitehurst's Derbyshire-based *Inquiry* also offers a suitable occasion for considering the matrix of nonliterary practices in which both natural history and poetry occupied a place. Many travelers bound for the mountains of Scotland and the Lake District caught their first glimpse of sublimity in the steep gorges and gritstone "edges" of Derbyshire's Peak District; for many more, the Peak was wilderness enough.[44] Whitehurst himself calls his theory "subterraneous geography" and explains that one of its primary goals is to stimulate profits in the mining industry. Travelers toured mines as well as caves in the Peak, following a pattern that became known as the picturesque-mineralogical tour. Peak District fluorspar specimens featured in local shops and London collections, both public and private, while lead from the region made its way to international markets and provincial munitions factories. The Peak's rugged landscapes also appeared on the London stage in the 1779 pantomime *The Wonders of Derbyshire*, as did the more exotic basalt formations of Staffa and northern Ireland, which featured in *Nature Will Prevail* and the accompanying pantomime, or "afterpiece," *Harlequin Teague, or The Giant's Causeway* (1782).[45] A sense of the social reality behind these practices—of the public for the politicized sublimity of geology and of the ideologies and institutions of late Enlightenment science—is prerequisite to understanding Romantic poetry as one of several interlocking ways in which the larger culture negotiated the fascination of rocks.

## Social, Philosophical, and Geological Materiality

The first half of this book contextualizes poems of Wordsworth, Blake, and Shelley in relation to what might be broadly termed topographical prose: travel narrative, geology, landscape gardening, and theories of the sublime and the picturesque. These discourses are seldom entirely distinct

44. See, for example, Barthélemy Faujas de Saint-Fond, *Journey through England and Scotland*, 1:chap. 1 and 2:chaps. 17–18; and Murray Aust, *Companion and Useful Guide*, 1:5–12.

45. Both pantomimes were very popular and ran almost continuously for several years (see the entries for each in Charles Beecher Hogan, ed., *The London Stage*). *The Wonders of Derbyshire* was designed by Philippe-Jacques de Loutherbourg, who premiered his famous *eidophysikon* (which featured a number of geological scenes) in 1781.

from one another, in part because they all respond to four essential qualities of the earth's material: it is inorganic, a foreign substance par excellence; it is the oldest and most stable substance; it is massive and immovable; and it is primitive in form, creating an appearance of unformed matter. Whitehurst's "rude and romantic" rocks provide one example of a widely shared description of these qualities. Like the more obviously freighted "convulsion" and "revolution," such commonplaces as "rude and romantic" point toward broader social issues. Austen's account of the Lyme coast in *Persuasion* is one of her relatively few passages of sustained description, but "romantic rocks" (and perhaps "subterraneous convulsions") are also an important subtext in *Pride and Prejudice*. Dramatically reunited at Pemberley, in Derbyshire, Elizabeth and Darcy relieve the "embarrassment" of their renewed intimacy by "talk[ing] of Matlock and Dove Dale with great perseverance."[46] Alluding to the same "Wonders of the Peak" that preoccupied Whitehurst and Thomas Whately, Austen here relies on the social currency of rocky landscapes, on their importance not only for conversation, but also for other genres alluded to in this episode, including travel narrative and picturesque theory. Of the widely fascinating qualities of rocks, their antiquity in particular provides a nexus between the aesthetic and the practical modes of geological discourse. The concept of an ancient "rock record" permits a transition from aesthetic response to economic agency, the main focus of the second half of the book. The rock record, given a modern scientific form by William Smith (1768–1839), remains a basic paradigm of historical geology and a key to the exploitation of mineral resources. Literary readings of the rock record, such as Shelley's in *Prometheus Unbound* (IV.274–318), identify the earth simultaneously as the substance and the text of history, generating a materiality located precisely between the two materialities recently competing for the objects of Romanticism, that of the letter and that of history.

The persistent geographic sense of "romantic" provides a philological lens on three interrelated historical processes, corresponding to the three phases of my narrative in this book: poetry's use of the geological sublime to articulate the otherness of the physical (chapters 1–3); the origin of environmental concepts such as the rock record and "natural resources" (interchapter and chapter 4); and the emergence of modern geology and modern disciplinary boundaries through the interaction of early geology and descriptive poetry (chapters 5–6). The textual materiality of rocks is also historical in that its geological "characters" appear as natural signs of economic realities. Rock formations provide either clues to the presence of mineral resources or, more broadly, instances of naturally occurring "im-

46. Austen, *Complete Novels*, 384.

provement." Wordsworth's critique of applied geology in *The Excursion* III (1814) raises these issues in a way that reflects a shift in the definitions of both geology and literature. In 1770 Thomas Whately produced a reading of the rock record at Middleton Dale without distinguishing between literature and geology. Whately's reading generates an unproblematic narrative of improvement, representing geological process as the source of a "style" of ornament for gentlemen's grounds. But by 1814 and even more by 1830, the discourse and forms of knowledge about the earth registered the tremendous shift in productivity from the landed estate to manufacturing, made possible not only by abundant raw materials but also by a new understanding of those materials as raw, as "natural resources," and by new scientific disciplines. Locodescriptive poetry—going back to the earlier *Excursion* of David Mallet (1728) and before—provides our earliest and most sustained evidence of a systematic interest in rocks and landforms in modern Britain: "not a mountain rears its head unsung."[47] As it developed, such poetry—as in Richard Jago's *Edge-hill* (1767) and parts of Darwin's *Botanic Garden* (1791)—contributed to the increasingly industrial inflection of landscape aesthetics, ultimately precipitating the rupture between a geology of resources and a literature of "pure" aesthetic values.

The term "romantic rocks" designates a predisciplinary construct, a widely understood representation of rocks as the raw material of nature that helped to generate the notion of materiality itself. "Materiality" is a highly abstract term for the most concrete of phenomena, for concreteness itself. "Aesthetic materialism" is another way of expressing this paradox, and also describes an attitude common in Britain during the period 1770–1820. In a different sense, "aesthetic materialism" is a critical method responding to historical and cultural materialism by articulating a historically specific conception of materiality. Though oriented toward the social context of literature, my method differs from other historicist approaches by attempting to locate the sphere identified as material in the period under analysis itself. The material textuality of the rock record locates it between the materiality of the letter, as described by Paul de Man and other deconstructive critics, and the materiality of history, as premised by Alan Liu and other historical Romanticists. The materiality of the earth is not merely textual, however, and my approach also engages broadly with two dominant twentieth-century views of Romantic nature, both of which imagine Romanticism as divorced from the material world. The thinking of an earlier generation, exemplified by M. H. Abrams, understands Romanticism as a reaction against

---

47. Joseph Addison, "A Letter from Italy" (1702), as quoted in William Drummond's *Clontarf: A Poem* (xiii).

mechanism and empiricism. An influential group of more recent critics, represented here by Liu, understands Romanticism as privileging a form of imaginative vision that excludes material and social conditions. More recently, scholarly concern with print culture has added a cultural dimension to the materiality of the text. It is hard to imagine the study of Romanticism now without these categories, but like the earlier, Marxist focus on conditions of production, they pose a limited horizon for understanding materiality. In order to pose a critical horizon beyond print culture, I focus on the deep engagement of the period's literary culture in a set of interlocking inquiries into the physical and the nature of the physical. The rocks and landforms that stimulate these inquiries provide the material touchstone for this analysis and create intertextual, interdisciplinary points of reference. Onno Oerlemans has recently advocated a similar procedure to complicate what he terms the "transcendent" materiality of new historicism, "an idea of nature that begins and ends with language."[48]

My project of linking these social and philosophical dimensions of materiality involves a synthesis of critical and historical accounts ranging from "the materiality of the text" to "material culture." Paul de Man's "Phenomenality and Materiality in Kant" is probably the first essay in the field of Romanticism to use "materiality" as a major category of analysis. De Man argues that Kant's account of the materiality of *nature* (really phenomenality) describes and performs the materiality of the *text*. Materiality here is deliberately abstracted from the natural and social worlds, but this philosophical rigor makes it possible to isolate it as a category, to consider what Husserl called *Materialität überhaupt* (materiality in general) as a literary problem. The flexibility of de Man's concept is aptly illustrated by more recent work on print culture and the materiality of the text, now considered as a social and historical fact.[49] Liu's *Wordsworth*, combining subtle deconstructive reading with historical materialism, draws attention in a new way to social practices adjacent to literary culture, such as picturesque travel and rural administration. One of the essays following in Liu's wake, Stephen Copley's "William Gilpin and the Black-Lead Mine," illuminates the practice of pic-

---

48. Oerlemans, *Romanticism and the Materiality of Nature*, 33. Oerlemans draws on Bruno Latour's notion of the nonhuman to urge that we "restore a balance by including the physical" in our articulations of socially constituted materiality (15). Oerlemans' book (2002) arrived in Missouri as I was preparing my own manuscript for the press; as the first book-length study of materiality and Romanticism, it deserves much fuller engagement.

49. Recent work on print culture in diverse periods fruitfully extends the "materiality of the text" to social practices germane to textuality. A keyword search of the MLA bibliography under "materiality," however, shows that the term has consistently been tied to text or communication over the twenty years it has been used, creating the odd impression that the material world consists only of communicative acts.

turesque travel by showing how the economic and mineral history of a particular locality can assert itself in a narrative professing strictly aesthetic aims. Similarly, I am concerned throughout much of this book to show how the materiality of landscape—of particular landforms in Britain—is registered in literary texts, in terms of both its human uses and its aesthetic particularity. The immensely widespread and diverse practices of natural history, even more than picturesque travel, provide access to material as well as literary culture.

The renewed examination of science as a vital cultural context of literary production, or more properly an integral part of literary culture itself, draws on a relatively long tradition within literary scholarship. M. H. Abrams, for example, represents Wordsworth as "an honest heir" to empiricism, maintaining that his skepticism toward science in the 1802 *Preface* is directed only toward narrowly empirical laboratory practice.[50] The view of Wordsworth as sympathizing with pure as against applied science has been profoundly influential. John Wyatt's *Wordsworth and the Geologists* depicts Wordsworth as exhorting geologists to practice a "noble science" not tied to narrowly empirical concerns, and ecological criticism, too, seems influenced by the view that Romantic poetry grasps the natural world in the form of "concrete experience and integral objects, from which science" merely "abstracts qualities for purposes of classification."[51] On the other hand, Abrams' notion of a compromise with empiricism draws attention to the context in which these negotiations take place: in geology, for instance, the social competition between gentlemanly specialists and field-based professionals was especially active at this time, and Wordsworth very skillfully taps into such defining conflicts. By showing such engagement as typical for the period, Abrams sets a precedent for productive later work on Wordsworth's actual debt to natural history by Alan Bewell, Theresa Kelley, and others.

Scholarship on literature and science has engaged more and more systematically with the history of science. Gillian Beer, in her landmark study, observes that "Darwin's romantic materialism which resulted in a desire to substantiate metaphor, to convert analogy into real affinity, should be understood as part of a profound imaginative longing shared by a great number of his contemporaries." Beer's work is suggestive for a study of Romantic geology for two reasons: because of the interest expressed here in the materiality of practices as well as objects, and because of the historical con-

50. Abrams, *Mirror and the Lamp*, 309. Wordsworth, as "an honest heir" to empiricism, proposes what is for Abrams a representative "solution" to the conflict between empirical science and poetry, which is to say that poetry contains real "data with emotional additions" (314–15).
51. Abrams, *Mirror and the Lamp*, 315. Cf. Wyatt, *Wordsworth and the Geologists*, 179.

tinuity between the stories of geology and biology, mentioned in my discussion of Lyell and aptly expressed in Beer's "plot without man." Beer shows Darwin's work as "embedded in the culture" through its wide readership and through Darwin's own reading, especially in literary texts.[52] Over the past twenty years, other scholars have shown how numerous detailed aspects of natural history were embedded in their particular eighteenth- or nineteenth-century cultural matrix. Much of this work tends to be dominated by the contextualizing imperative put in place by history of science; in *Romanticism and the Sciences* (1990), for example, the institutional structures and political agendas governing scientific practice tend to overshadow the importance of literary culture as the creative context in which much science was carried out.[53] On the other hand, travel narratives have provided an especially fruitful area for interdisciplinary study, and tourism has long been identified as an important factor linking Romanticism and geology. In her study of early scientific travel, Barbara Stafford stresses that "scientific travel added to landscape perception the irreplaceable component of lived experience and engaged contact with material substances," leading to a new sense of "the natural object" as an "independent and powerful presence, possessing a history separate from that of man."[54]

Though the scholarship on literature and science is sometimes too constrained by established history of science, history of science itself, and history of geology in particular, have benefited greatly from the turn toward "practical traditions of research pursued in particular social locations."[55] Morris Berman, in his history of the Royal Institution (where Davy rose to fame), focuses on the conflict between the landed and ascendant mercantile interests; Rachel Laudan illuminates the history of geology through a study of applied chemistry and the bureaucracy in place in Prussia for the administration of applied science; and both James Secord and Martin Rudwick interpret geological controversies through micropolitics within the Geological Society and the social issues underlying them. Secord and Rudwick, like Gillian Beer, study the process by which scientific controversy is resolved and its results become "embedded in the culture." Secord points out that the structure of geological periods, with its inscription of British place-names onto the history of the planet, has become so permanently accepted that it

52. Beer, *Darwin's Plots*, 42, 21, 5–7.
53. See, for example, Christopher Lawrence's remarks on the "political practices" of Davy and Coleridge, in "The Power and the Glory," 223.
54. Stafford, *Voyage into Substance*, 29, xxi.
55. James Secord, *Controversy in Victorian Geology*, 318. Clark, Golinski, and Schaffer's introduction to *The Sciences in Enlightened Europe* reviews such developments in a broad historiographic context.

must be defamiliarized if the "actual practice" of Victorian science is to be reconstructed. Equally stressing the importance of understanding science as a quotidian social practice, Rudwick proposes a synthesis that is especially suggestive for other enterprises of cultural history. The history of the Devonian controversy, he argues, shows that "a consensual product of scientific debate can be regarded as both artifactual *and* natural, as a thoroughly social construction that may nonetheless be a reliable representation of the natural world."[56]

Rudwick's synthesis of "internalist" and "externalist" history of science marks a limit between the social construction of a discourse and the concrete relationship of that discourse to its object. My study also concerns itself with such a limit, but this affinity draws attention to some major differences between the history of geology—however contextual—as a disciplinary construct and my own use of it. It seems unlikely that Rudwick would extend to poetry the privilege of operating as a "reliable representation of the natural world," though this is what much Romantic poetry famously claims to do. A disciplinary construct completed after the Romantic period accounts for the difference between these two models of nature. Rudwick has since applied this difference to an earlier figure as well, arguing that Georges Cuvier's "enduring legacy to geological science" was his empirical rigor and not the "florid prose" that influenced Byron, among others.[57] Such a conclusion reenacts geology's repudiation of its own literariness. In exploring this literariness—which every generation of geologists, up to the Victorian controversialists and beyond, has noticed and rejected in its predecessors—I diverge from the history of geology. I do so not to tip the balance further toward "social construction" but rather to enrich and complicate the sense in which discourses may be understood as "reliable representation[s] of the natural world."

Rudwick's is not the only recent history in which literariness runs afoul of a disciplinary imperative. Questioning Romanticism's importance as a context for the emergence of geology, Michael Shortland protests that "the language and idiom of geology share nothing of the fancies, imagination, and sweet sublimities of the Romantic poets, and everything of the dirt and dangers associated with mining." Shortland is justly suspicious of the yoking of Romanticism and geology as a commonplace of intellectual history, but this commonplace is questionable partly because it has never received really substantive treatment. By attending to working landscapes and class dynamics,

---

56. Secord, *Controversy in Victorian Geology*, xv; Martin Rudwick, *Great Devonian Controversy*, 451.

57. Rudwick, *Georges Cuvier*, 265–66.

as Shortland suggests, we can produce not only a richer history of geology but also a more complex sense of Romanticism, one that shares in the socioeconomic concerns and perhaps even the "predatory sexual combat" with the earth that Shortland attributes to the "geological workmen" arising after 1830.[58]

Not that the Romantics are "geologists." I use "geology" and its derivatives for convenience on the lexical level, but rhetorically there is more at stake in this word choice. I use these terms to refer to geological subject matter rather than method. Until recently, "history of geology" meant the history and prehistory of the modern discipline by that name, not a history of "thinking about the earth." Richard Hamblyn has led the way in the latter direction, showing how geology appropriated and transformed its subject matter: by the early nineteenth century, "geologists . . . convinced themselves of the need to dispense with the gauze of literary and picturesque associations through which the landscape had habitually been viewed." Hamblyn builds his account of "popular geology" on the notion of a new middle-class "connoisseurship of the earth" not colored by "timeworn classical schedules of aristocratic travel"—a connoisseurship practiced by John Whitehurst, among others.[59] The 1778 list of Whitehurst's subscribers, among other documents, suggests that popular geology played a significant role in print culture as well.

The name "geology" did not become established until after the formation of the Geological Society in 1807. From this point, increasingly specialized geological publications encountered a shrinking audience, as illustrated by Hugh Torrens' analysis of the subscription list for a mineralogical work proposed in 1810: "the more geological the contents the more the audience for such books diminished."[60] Only 68 subscribers came forward to support

---

58. Shortland, "Darkness Visible," 39. Though thoroughly documented, Shortland's treatment of Romanticism relies heavily on outmoded clichés, as will be evident from the first quotation (38). Both literary and historical scholarship have shown that Romanticism involves more than "efforts to depoliticize the countryside" (23–24) and that many of its investments in "the secrets of nature" (a phrase Shortland aptly glosses as "a euphemism for the female genitalia," 41) closely parallel those of science and industry. Shortland's polemical purpose becomes clearer when he draws on mid to late nineteenth-century anecdotes to eulogize William Buckland, members of the Geological Survey, and the other "geological workmen" (42; cf. 23, 26), who prove to be more central to his scholarly interest than geologists actually contemporary with Romanticism.

59. Hamblyn, "Private Cabinets and Popular Geology," 200, 188, 195. My phrase "thinking about the earth" alludes to David Oldroyd's *Thinking about the Earth*, a similarly exploratory work; and Stephen Jay Gould's *Time's Arrow, Time's Cycle* also operates to some extent outside the disciplinary paradigm.

60. Torrens, "Arthur Aikin's Mineralogical Survey," 139. After 1807 the Geological Society led the effort to consolidate studies previously pursued under such names as "geognosy," "oryctology," and "mineral geography."

Arthur Aikin's *Proposals for a Mineralogical Survey*, despite a six-year campaign to find the minimum of 300 subscriptions required to finance the ambitious and expensive volume Aikin planned. Torrens points to the contrasting success of an 1839 volume on the same region, arguing that "this was a truly remarkable revolution in the public attitude to geological publication and its financing."[61] I would recast this "revolution" as public acceptance, after a transitional period, of geology's new self-definition as a specialized science. Whitehurst's success, however, indicates the wide circulation of geological inquiry—if not "geology"—as a form of polite knowledge well before 1800. Aikin's story itself shows the intersection of these two paradigms. Ironically, his failed proposal sought to capitalize on the success of his *Journal of a Tour . . . with Observations on Mineralogy and Other Branches of Natural History* (1797), which pursues genuinely geological issues while avoiding self-consciously specialized discourse.

## Aesthetic Objects in Theory and Practice

Paolo Rossi has observed that "so-called scientific progress often consists not so much in a progress within science as it does in taking something that formerly was not science and making it part of science itself."[62] As it evolves, geology absorbs material from several components of its cultural environment, including cosmogony, the "history of nations" (in Rossi's account), and aesthetics. Marjorie Nicolson, arguing that geology and the "natural sublime" interact in this way, dissents from a continuing scholarly tradition by using its geological origins to privilege the "natural" over the "rhetorical" sublime. In doing so, she takes issue with Samuel Holt Monk's *The Sublime* (1935).[63] Monk argues that the eighteenth-century literary fashion for locating the sublime in external natural objects is simply a "debased" version of the rhetorical sublime of Longinus, and an intermediate step toward the Romantic recognition of the sublime as a psychological category, which culminates (for Monk) with Kant. Much of the later work on the sublime—including two very distinguished books, Thomas Weiskel's *Romantic Sublime* and Neil Hertz's *The End of the Line*—has followed Monk's lead in claiming "the sublime" as a psychoanalytic and discursive category. Others, including Frances Ferguson and Jean-François Lyotard, have virtually identified the sublime with Kant as Monk does. Nicolson's argument is useful in reminding us that it is misleading to speak of *the* sublime when we consider its very

---

61. Ibid., 142–43.
62. Rossi, *Dark Abyss of Time*, xiv.
63. Nicolson, *Mountain Gloom and Mountain Glory*, 29–31 and n.

different manifestations in early and late eighteenth-century Britain, not to mention elsewhere. But her book also shows that the use of sublime tropes in connection with natural history, in various cultural locations, provides one meaningful constant: it may be that various writers invoking the sublime were unconsciously following a new psychological bent in culture, but it is certain that in numerous cases, a sublime style was performing cultural work connected with geology. Capitalizing on these insights, my own work approaches the sublime not as a unitary mode of representation, but as a category animated by culturally specific practices, including tourism, mining, naturalizing, exhibition, geology, and poetry. Such practices appear more clearly in relation to literary culture thanks to recent scholarship on Romanticism focused on the "redefinition of literature" that eventually led to the separate articulation of these practices and their "nonfictional" genres.[64]

In 1790 Kant asked impatiently, "who wanted to call deformed mountain masses and the like 'sublime' in the first place?" Two hundred years later the English critic Peter de Bolla argued, in *The Discourse of the Sublime*, that aesthetic theory in eighteenth-century Britain has no object, that it theorizes the general relation between theories and their objects.[65] In the post-Kantian view, "the sublime" does not describe objects in the world or a set of historically specific experiences at all, but rather refers to a kind of language, a set of tropes unfolding a hidden psychological or political narrative. Of course the sublime is these things, but precisely because it *is* tropological, one must look outside theoretical contexts to understand its local and specific meanings. Outside aesthetic theory, the word "sublime" is generally used in eighteenth-century Britain to describe physical objects. By asking what it means to refer to an object as sublime or beautiful, one does not necessarily replicate either Burkean or Kantian assumptions. For me, this has been a historical, to some extent a philological, inquiry with the aim of understanding why certain objects—rocks and landforms in particular—were described as sublime and designated as aesthetic. This recovery of aesthetic objects is informed by an aesthetic materialism often at odds with aesthetic theory.

64. Alan Richardson argues that by the end of the Romantic period, "literature" had become "redefined in terms of certain kinds of fictive or imaginative writing, specifically those which seemed to speak, through 'affections of pleasure and sympathy,' to the 'general and common interest of man'" (*Literature, Education, and Romanticism*, 260). In *The Work of Writing*, Clifford Siskin points out that truth became naturalized as the goal of different forms of writing at the same time that those forms—humanities, social science, natural science—were distinguished as different modes of accessing the truth (19).

65. Kant, *Kritik der Urteilskraft*, §26 (B 95), 343; see *Critique of Judgment*, 113; de Bolla, *Discourse of the Sublime*, 29 and passim. De Bolla argues that the discourse treating the sublime is "constructed . . . upon the recognition that the inquiry has no object" (32).

Aesthetic theory up to Burke is frequently concerned with the physical dimension of the aesthetic, but objects tend to cluster in aesthetic theory, while in applied aesthetics they remain individual, local, contingent. Archibald Alison's abstract and philosophical *Essays on Taste* (1790) provides a striking illustration of this difference, because the essays draw their empirical examples from Thomas Whately's eminently practical *Observations on Modern Gardening*. Poetry and early geology, like such treatises on landscape, are forms of applied aesthetics, in the sense that they draw on aesthetic theory and other systems in order to negotiate local and physical encounters. These discourses do not depend on aesthetic theory, but they do depend on aesthetic categories in order to represent the raw materials that are a shared focus of their projects of explanation and description. Whately, for example, draws on *King Lear* for an image ("one that gathers samphire, dreadful trade") to describe the aesthetic character of a river gorge; the "romantic" character of the gorge, in turn, is an essential premise for geology's account of the ancient catastrophes that produced such landforms.

The verbal and social idiom of such analysis —landscape aesthetics—provides a shared vernacular for geology and Romanticism in their formative stages. Whately's cliff—located on a popular tourist route along the river Wye—also exhibits the representative materiality that rocks and landforms embody in many texts of the period. The cliff becomes an aesthetic object because it instantiates the fundamental properties of matter, most of all in being indifferent to human purposes. The literary allusion, and the aesthetic category that comes with it, seem to provide the explanation the landscape demands. In a sense, the cliff's indifference is propertylessness rather than a property; the perennially fascinating qualities of rocks—their fathomless age, their often immense and irregular forms, their lifelessness—are also, in a way, purely negative qualities or absences that mark the boundaries of human agency. "Nature" and the "aesthetic" are both protean and highly disputed discursive entities; by recovering the geological context of these terms, I hope also to recover some of their power to describe the outside world, a sphere that is merely physical and yet more than human.

# 1

## A Genealogy of the "Huge Stone" in Wordsworth's "Resolution and Independence"

**A** quintessential romantic landscape sets the scene for Wordsworth's famous story of an old man gathering leeches in "Resolution and Independence." Using a stanza form carefully crafted to evoke chivalric romance, Wordsworth produces a geological romance of origins for the old man's condition. The poem draws on the ancient analogy between rock and the human body—one resurfacing in many Romantic discourses—to address the problem of embodiment.[1] It also employs a vocabulary cognate with those of geology, landscape gardening, and picturesque travel, discourses that Wordsworth eventually synthesized in *A Guide to the Lakes*. Midway through the poem, he elaborates the image of a "huge stone," which becomes conflated with the age-worn body of the Leech-gatherer. This "huge stone" is an erratic boulder of the kind perched on numerous "eminences" in the Lake District. Alan Bewell has already suggested the importance of geology for this image in his historically rich reading of the poem.[2] I shall argue here that geology is important in the poem as more than a source of metaphors. Rather than treating geology as a piece of the larger historical context, this chapter contextualizes the poem's geology in particular, showing how it inhabits the period's discourse on rocks. The

---

1. Blake's myth of the Giant Albion revives one strand of this analogy, which can be traced to Michael Drayton's *Polyolbion* (1613–22), material in Raphael Holinshed and other early chronicles, and various antecedents in the ancient world. Romantic discourse on classical sculpture is another important site of analogies between the body and the earth's material. On this point, see my "Stones So Wonderous Cheap."

2. Bewell, *Wordsworth and the Enlightenment*, 265–72.

mysterious erratic boulder evokes the sense of wonder common to literary and geological accounts of such phenomena, and Wordsworth uses these associations to address the "more than human weight" of the problematic physicality haunting the poem. At the same time, he capitalizes on the association of raw material also activated by the stone to extract a moral profit in the form of the Leech-gatherer's "independence" or "moral dignity."

Much twentieth-century scholarship has located the cultural foundations of Romanticism within the aesthetics and politics of the period's voluminous discourse on landscape. The well-traveled terrain of the picturesque and the landscape garden is worth revisiting, however, as a forum for ideas about the earth's material. The topos of the erratic boulder illustrates a practical dimension of the discourse on landscape. In "Resolution and Independence" as well as in geological and other topographical prose, the erratic boulder becomes an occasion for intervention in landscape, suggesting a grouping of all these discourses under the category "applied aesthetics." Wordsworth, Humphry Davy, William Gilpin, and others harness aesthetic discourse to the explanation of geological phenomena so consistently as to illustrate a broad consensus concerning the earth's material. Many of the specific attitudes shared or contested among writers on landscape also inform the poetry of Shelley and Blake, the other canonical reference points of this and the next two chapters. My focus on representations of the earth's material permits a thematic and historical specificity often lost in readings of the aesthetics and/or politics of the Romantic landscape. The influence of these discourses cannot be reduced to causal constructions of intellectual history or philology. Wordsworth, for example, derived his erratic boulder, the "huge stone . . . / Couched on the bald top of an eminence" (lines 57–58), from many decades' worth of geological theory accumulated in bits and pieces of reading and conversation. The point here is not so much that the image of the rock has its sources in the history of geology, but that geology is sufficiently mixed and culturally embedded with aesthetic categories to make it possible to produce such images spontaneously. The early industrial emphasis on raw materials and the widespread confidence in exploiting them generates responses in the poetry that are more specific, more tactical, but also often more unconscious, than the poets' more general and ideological reactions to industrialization itself. It is above all in the construction of materiality, in the physical contingencies of particular objects in the poetry, that its response to current ideas about the components of landscape becomes visible.

The geological components of landscape, in particular, acquire the status of representative objects and a representative materiality. The most widely shared tropes concerning rock involve substance or solidity, as when Dr.

Johnson "refutes" Berkeley's radical idealism by kicking a curbstone.[3] Rock's hardness, in turn, embodies the challenge to cognitive ordering represented by ideas of rock as unformed matter or "primordial substance." Marjorie Nicolson has shown that rocks and mountains become conventional signs of the vastness and stability of external nature in the course of the eighteenth century.[4] This stability appears as obduracy in works of applied aesthetics such as Thomas Whately's *Observations on Modern Gardening* (1770), where rocks confront the "landscapist" with his most severe physical limitations: "They are themselves too vast and too stubborn to submit to our controul" (*OMG*, 99). Since the antiquity of rocks poses the greatest challenge for the emerging discipline of geology, their stability across time presents a similar obduracy. Thus Humphry Davy declares that geological strata provide evidence of "a former order of things," and that "the connection between their causes and effects is mysterious, but apparently within reach of our faculties, and it is displayed in characters which can be deciphered only with difficulty, but which express sublime truths" (*Lectures*, 77–78). Wordsworth draws extensively on both geology and landscape gardening for his *Guide to the Lakes* (first published 1810). William Gilpin, especially, among Wordsworth's acknowledged predecessors, regulates intractable features of landscape such as ungainly mountains by means of rigid aesthetic protocols. Wordsworth does not extend this "tyranny" of the eye (1805 *Prelude* XI.179) quite as far in his *Guide*, but his writing is by no means free of it, and the discourse of the picturesque, along with geology and landscape architecture, clearly shapes his attitude toward rocks. Yet Wordsworth also emphasizes more than Gilpin the physical rigor of traveling in the mountains, a rigor analogous to the resistance experienced by the geologist and landscape designer.

### Things and Sense

Before recontextualizing the "huge stone" of 1802 in relation to geology, the landscape garden, and the picturesque, I will make an initial pass through "Resolution and Independence" in order to link the middle stanzas to the structure of the whole and introduce the more general issues of physicality pervading the poem. No sooner does the Leech-gatherer appear—at

---

3. James Boswell, *Life of Johnson*, 333; see also 412.
4. In *Mountain Gloom and Mountain Glory*, Nicolson charts the history of this trope and provides voluminous catalogues of canonical passages to illustrate its flowering in the Romantic period.

the critical moment when it seems that a reliance on nature is at odds with self-sufficiency—than he is transformed into a "huge stone," a "thing endued with sense" (line 61).[5] Soon afterward he becomes a cloud (75), and later his voice "seems like a stream" (107). The speaker, still caught up in his confusion about things and sense (a mature "abyss of idealism," perhaps; *PW*, 4:463), is forced to render the Leech-gatherer a feature of the landscape. The "huge stone" is a source of "wonder to those who do the same espy / By what means it could thither come, and whence" (59–60; see the perched boulder in the upper right of the frontispiece to this book). Wordsworth harnesses this wonder to the old man, who is "incorporated with the beautiful and permanent forms of nature."[6] The lesson that finally emerges for the speaker is the lesson of matter: the impulse to transcend "fleshly ills" predicates the mind's independence of the malaise of materiality. This malaise takes the form of a muteness of the object world to which the speaker tries to assimilate the Leech-gatherer, but it also takes the form here of the living body, of the damaged body that emerges so clearly from Dorothy Wordsworth's account of this incident in her journal.

The incident undergoes a profound metamorphosis from the journal entry to the first manuscript of the poem, which continues through the published versions. Dorothy Wordsworth explicitly describes the material conditions of the Leech-gatherer's existence, introducing him as "an old man almost double": "He had been hurt in driving a cart, his leg broke, his body driven over, his skull fractured."[7] The body registers these marks of the physical in a way that is perhaps too legible for it to qualify as a poetic image; the old man's bent stance and his "feeble chest" (92) survive in the poem, though both construed purely as signs of age. Alan Bewell remarks that this stance "would have evoked the revolutionary image of the laboring poor," as well as figuring the archetypal position involved in "gathering a bare subsistence from nature." Bewell's account indicates that this is not

---

5. *PW,* 2:235–40. All subsequent references are to this edition unless otherwise specified. A series of substantial revisions may be traced from the now-mutilated original manuscript of 1802 onward. I discuss below the first set of changes, between this ms. and the first publication in 1807, but there are also major changes from the 1807 version to that of 1820, the one cited in my text. One entire stanza is deleted from the earlier version; its most interesting phrase for my purposes is "naked wilderness." See Wordsworth, *Poems, in Two Volumes,* 125.

6. Wordsworth, *Lyrical Ballads,* 245. Cf. Wordsworth's usage of "forms" in *Guide to the Lakes,* in the Owen and Smyser *Prose,* 2:229, l. 2395; 201, ll. 1516–17; and 217, l. 2036. Dorothy Wordsworth mentions in the journal entry relating their encounter with the Leech-gatherer that William is just then working hard on the *Preface* to *Lyrical Ballads,* which strengthens the link I am suggesting here between the preface and the poem; *Journals,* 1:63.

7. D. Wordsworth, *Journals,* 1:63.

merely an elision of social conditions, which he sees as present in the form of an elliptical response to "postrevolutionary trauma."[8] If the specter of Revolution remains necessarily elusive, it seems clear that the "huge stone," as the locus of materiality, gradually absorbs the old man's bodily and social conditions. This absorption creates the space for the moral allegory: in his geological form, the man's "moral dignity" can be extracted, as becomes clear from Wordsworth's letter on the poem to Sara Hutchinson. On the one hand, the final poem registers that leeches have grown scarce (125), that the Leech-gatherer is "old and poor" and his employment "hazardous and wearisome"; but it insists that "in this way he gained an honest mainte-nance" (stanza XV). Dorothy Wordsworth, however, records that "he lived by begging," a conclusion supported by the astronomical increase in the price of leeches: "Leeches were formerly 2s.6d. [per] 100. They are now 30s."[9] The partly mutilated initial manuscript of the poem may have con-tained more of these hard facts. The legible portion contains the germs of the later version's morally corrective self-sufficiency ("I yet can gain my bread"), but also notes that "he seem'd like one . . . / For chimney-nook, or bed, or coffin meet."[10] Alongside their concern with the occupational haz-ards of poetry, the published versions retain only a handful of metaphoric approximations of this physicality, the most condensed of which is perhaps the allusion to the "pain, or rage / Of Sickness," which "a more than human weight upon his frame had cast" (70). This weight comes to rest in the "huge stone," which figures the weight of the material world from which the Leech-gatherer has wrested his deeply problematic "independence."

The stone illustrates most clearly the "subreption" (to adapt Kant's term) by which the old man is incorporated with the forms of nature. The formal sign of the subreption is the metaphor's abrupt beginning, most pronounced in the standard text of 1820 that I am quoting. (The 1807 text includes an

---

8. Bewell, *Wordsworth and the Enlightenment,* 267, 269, 261. The old man's posture also represents the evolutionary position of the ancient amphibian, reconstructed from fossil evi-dence, that Bewell sees the "sea-beast" occupying in the evolutionary narrative informing the extended metaphor of stanzas IX–X (268). Robert Essick's explanation ("Wordsworth and Leech-Lore") of the traditional practice of gathering leeches by allowing them to become at-tached to one's legs suggests another possible occupational influence on the old man's body.

9. D. Wordsworth, *Journals,* 1:63.

10. *Poems, in Two Volumes,* 323, l. 132; 319, ll. 57–58. Three other manuscript variants are interesting: extended passages on the old man's peddling and the scarcity of leeches (pp. 129, 323) and a canceled stanza of description. This stanza begins:

> He wore a Cloak the same as women wear
> As one whose blood did needful comfort lack;
> His face look'd pale as if it had grown fair,

and goes on to describe a "bulky Pack" that sounds like a peddler's pack (p. 126 n.).

additional stanza, later deleted, between the two in question here.) Stanza VIII concludes: "the oldest man he seemed that ever wore grey hairs." The next stanza begins: "As a huge stone is sometimes seen to lie, / Couched on the bald top of an eminence." The metaphor—almost an epic simile—is extended as well as detached, so that the poem's return in the next stanza to the old man enhances the credibility of the myth of origin that the intervening stanza can now be seen as offering. The rhetoric of this explanation of the stone that becomes an explanation of the man is strikingly opposed to what Wordsworth aspired to in the original draft of the poem. As he explains to Sara Hutchinson, citing lines later removed from the poem: "What is brought forward? 'A lonely place, a Pond' 'by which an old man *was*, far from all house or home'—not stood, not sat, but '*was*'—the figure presented in the most naked simplicity possible."[11] The final version of the poem diverges most clearly from this ideal in using "seemed" or "seems" rather than "was" three times between lines 56 and 64.

Wordsworth continued to revise the poem because the image of the old man, in its "naked simplicity," was not as profoundly "impressive" for his readers as it was, intrinsically, for him. Hence his need for an answer to the unanswered question of the draft: "how came he here, thought I?" The strong parallel between lines 50–51 ("now whether it were by peculiar grace") and 60 ("By what means it could thither come") amplifies the question and links the provisional answer about the stone ("like a sea-beast crawled forth") most directly to the question about the man. The "oldest" of line 56, though it evokes the Wandering Jew, also evokes geological time, and a theory of spontaneous generation much older than geology, according to which the sea-beast would be generated by the action of the sun on the slime of the moor below. Completing the subreption, or rhetorical substitution, the old man becomes a link in the chain of an indefinite regress of images, from the stone to the sea-beast to the man to the demiurgic cloud, back to the man again, and so forth (stanzas X–XI). These images are strange to sense, "more than human," "not all alive nor dead," contorted even in the familiar locus of the body. Taken together, they defamiliarize the natural world in order to register its otherwise inarticulable materiality. Wordsworth's explanation of the metaphor in his 1815 *Preface* confirms the complexity of the mythmaking he has elected the poem, in the course of revision, to perform: the "intermediate image" of the stony sea-beast provides the "near[est] resemblance to the figure . . . of the aged Man" (*Prose*, 3:33) because the sea-beast, at the morphological stage of crawling forth, provides the paradigm case of a "thing endued with sense." The Leech-gatherer

---

11. Dorothy Wordsworth and William Wordsworth, *Letters*, 366.

somewhat resembles another of Wordsworth's "things" (Lucy in "A Slumber") in that he seems not to "feel the touch of earthly years."

There are formal and psychological reasons, too, why the old man is such a thing, the reverse of a soul encased in a body. Such a thing is necessary not only to provide the ideal ("incorporated with the . . . forms of nature") that Wordsworth had set forth in the *Preface* of 1800, but also to address the crisis posed by the speaker's problems with perception and futurity—his alienation from nature. A crisis lyric pattern that produces an animated "thing" resembles an act of conjuring, an activity also associated with the stanza form. The instability of the poem's initial landscape not only leads to a displacement of the ego onto it (as Thomas Weiskel has shown most effectively), but also sets the tone for the complex function of sense impressions throughout the poem. The landscape cannot provide a reliable material and affective base for the poem's ideas, nor can the speaker's intermittent connection to it ("I heard the woods and distant waters roar, / Or heard them not," 17–18) generate a coherent narrative. His initial reaction to the Leech-gatherer's speech is fully consistent with this initial pattern: "his voice to me was like a stream / Scarce heard; nor could I word from word divide" (107–8). But the old man has begun to satisfy both cognitive and narrative exigencies by showing, as a "thing endued with sense," conformity with morphological principles in his series of transformations. The reliable return of his narrative relieves the problematic authority of the speaker, evident in the latter's inept attempts at conversation (84–89) as well as the almost Humean atomism of the sense impressions (17–18) and consequent lack of long-term stability (34–35).[12]

The poem's archaizing also addresses the problem of authority by conveying the sense of a refined allegory. It follows the rhyme scheme of the Troilus stanza or rhyme royal, which amplifies the moral dimensions of the allegory. Still more significant is the rich and complex Spenserian ambience. The poem's Spenserian diction comes the closest to restoring the Leech-gatherer to a kind of individual human agency: like a Spenserian conjurer, he "con[s]" the waters like a book (80), his eyes contain mysterious depths (91), and he exhibits powers of dream-projection and hypnosis (stanza XVI).[13] Wordsworth concentrates his most deliberately archaic diction

---

12. Lewis Carroll's "The White Knight's Song," as suggested by one of its alternate titles ("Haddock's Eyes"), plays wickedly on Wordsworth's speaker's failure to grasp the significance of the Leech-gatherer's answer to his repeated question, "What is it you do?" (119). See *The Annotated Alice*, 307 ff. In this discussion I am indebted to Thomas Weiskel's reading of the poem, *Romantic Sublime*, 60–62; see also 137–39.

13. Samuel Schulman traces the thematic influence of Spenser and especially Thomson, identifying the conjurer element of the Leech-gatherer (for whose border dialect, says Schul-

around this effect, though his extension of a conjuring power to the speaker (who "bespeaks" the man) shows that this power is fluid, and may suggest a relationship to the poetic act of bestowing meaning. Steven Knapp explains the Leech-gatherer as a "materialized deposit" of the stanza form, nicely combining formal and geological questions.[14] The concluding hexameter line borrowed from Spenser does more, though, than motivate the old man's "pointless repetition." Its narrative pooling effect, which requires each stanza to begin anew, also provides a cyclical principle more legible than nature's cycles of generation and decay.

The "firmness" of the old man's mind is remarkable only in relation to his "decrepit" body. Wordsworth invokes nature's permanence, through the "huge stone," to explain that body's survival. The hypothetical causes adduced to explain the physiological condition of the old man's body can be translated readily into a geological lexicon of cosmic upheaval in the distant past (66–70):

> His body was bent double . . .
> As if some dire constraint of pain, or rage
> Of sickness felt by him in times long past,
> A more than human weight upon his frame had cast.

I have already quoted these lines to suggest a link between the "huge stone" and the figure of tremendous weight. The proximity of this passage to the epic simile of the stone (the previous stanza) equally supports a transfer of the quality of duration, which is not coincidentally one of the three requisite qualities for a sublime experience of mountains prescribed in "The Sublime and the Beautiful" (*Prose*, 2:351). A number of other images in the remaining stanzas also suggest such a transfer: while weaving in additional comparisons, descriptions of the Leech-gatherer's immobility, his features, and his "lofty" speech all bear some relation to the initial figure of the boulder. Again, the old man dramatizes the qualities of this rock in the service of a geological explanation of materiality, rather than being explained himself as the product of "revolution" seen as a geological metaphor for political upheaval. The huge stone also figures perfectly the solitude pervading the poem, a solitude the speaker finds both inspiring and disturbing (127, 131). Another poem of 1802, "There is an Eminence," offers suggestive parallels in this respect.

---

man, Wordsworth "substitutes a Spenserian dignity") with the wizard in *The Castle of Indolence* ("Spenserian Enchantments," 40).

14. Knapp, "The Sublime," 1019. One of the first critics to notice the poem's many images of flooding and residual deposits was Geoffrey Hartman, *Unmediated Vision*, 33–34.

## "Wonder to Those Who Do the Same Espy"

At bottom, this "huge stone" (57) is a figure for problematic physicality, both as it affects the material conditions of the old man's existence and more generally as it affects the aesthetic experience of nature. The man brings out the salient qualities of the stone, including its antiquity, its mass, and its inexplicability, that qualify it to stand for materiality in general. These qualities inform a general attitude toward rocks in the period, especially the idea of rock as the most primitive form of matter. Wordsworth's huge stone has a historical reality closer to the landscapes analyzed by geologists and aestheticians than (for example) to the "imagined pinnacles and steeps" invoked by Keats.[15] To show these connections, I shall make a second "pass" through the poem with more frequent detours, staging a dialogue among the voices of early geology (Humphry Davy, John Whitehurst) and landscape aesthetics (William Chambers and especially Thomas Whately) that will become increasingly familiar to readers of this book. Poetry, geology, landscape architecture, and picturesque theory all display distinctly aesthetic and practical modes in the conversation about erratic boulders. Yet the aesthetic and the practical become clearly interdependent in these accounts of the physical environment, shedding light on the complex ideology of this distinction.

The Wordsworth poem most obviously instantiates the aesthetics of poetry, responding purely to the otherness of the nonsentient. Paul Fry's broad, Heideggerian theory of poetry touches on similar themes: poetry, he says, records a "cognition-neutral state" of "astonishment at the blank opacity with which the world discloses its being."[16] I am concerned to show how such an aesthetics is historically specific and involved with geological objects. In verbal "compositions" of alien, impenetrable rocks and mountains, materiality is constructed as an absolute otherness that guarantees the authenticity of aesthetic response, whether this takes the form of creative ecstasy or of revolt against materiality itself, as in Blake. "Resolution and Independence" draws first of all on the wonder that attaches to geologic phenomena (abstracting, for the moment, from the practical or ethical claims of the allegory). As elaborated in the letter to Sara Hutchinson and the 1815 *Preface*, the poem shifts its focus in the course of revision from an

---

15. "On Seeing the Elgin Marbles" 3, in *Complete Poems*, 58.
16. Fry, *Defense of Poetry*, 7. Fry designates Wordsworth the "laureate" of this "opacity" (7), later defined as "the glimpse of the nonhuman, specific to literature" (31).

image that proves insufficiently compelling "in its naked simplicity," the old man by the pond, to a more compelling geological image of naked simplicity itself.

The poem returns to description in the account of the "huge stone" after being diverted from its initial course by the speaker's anxiety about his reliance on nature. This return signals the beginning of the quasi-theodical justification of nature by which the appearance of the old man redeems the landscape and restores the poem's connection to it. On the one hand, description returns in the form of a metaphor or myth of origin for the mysterious appearance of the old man. On the other hand, the old man himself embodies the metaphysical problem of the rock and of rock in general. Wordsworth's parallel expressions of indeterminacy concerning both objects (50–51 and 60) support this reversal of the expected metaphoric relationship. The old man figures the otherness of the rock through this mystification of his origins, through his absolute antiquity (56), his primitive condition, and his "naked simplicity." The "whole body of the man" preoccupies the speaker, while his voice "seems like a stream scarce heard" (stanza XVI); the phantasmagoric vision of this battered and decrepit body translates into human terms the endurance of the glacial erratic in its exposed and precarious position. As the limit case of bodily endurance, the Leech-gatherer's survival hints at the unimaginably larger scope of geologic time, while also borrowing the permanence of the rock and other "forms of nature" for its idea of independence.

The sense of wonder in Wordsworth's poem, characteristically for its time, responds to a conventional set of qualities in rocks—inexplicability, colossal duration, immense size, and primitive form. The same qualities also suggest the resistance of rocks, and the physical universe they stand for, to all kinds of inquiry. This opacity generates a practical mode within literary discourse about rocks in which both sovereignty over nature and the idea of minerals as resources are at issue. The interest involved here and the question of power, even in so abstract a case as the "Power" of Shelley's "Mont Blanc," provide counterparts to the wonder and aesthetic attitude of "Resolution and Independence." The poem itself allegorizes this tension by capitalizing on its rock of a man, mining him for moral profit. Early geology is concerned primarily with the age and stability of rocks and landforms, as opposed to their simple otherness. Of the period's four major categories for rocks—otherness, antiquity, magnitude, formlessness—the second is literally a practical concern as well as an aesthetic one. Geology's most obvious kinship with poetry and aesthetics resides in its epistemological attitude toward these qualities, in its attention to the knowability of rocks and their status as resources. In this practical respect, the precise observation and hints of positive knowledge in "Resolution and Independence" exemplify important parallels between the literary discourse and geology proper.

Some of the poem's observations suggest both a geological method and a background in geological theory. Alan Bewell stresses the importance of geological "revolutions," in particular the Deluge, as heuristic analogues for the French Revolution, whose influence Bewell sees this poem (among others) as exorcising in the form of "postdiluvian/postrevolutionary" trauma. The Deluge had long been put forward as the only agency capable both of depositing large masses of nonnative stone (now regarded as glacial erratics) and of transporting marine fauna into various landlocked regions where their fossil impressions were found. With this background in place, the method of observation can be seen as geological: "Wordsworth reads the pain and violence of the man's past through his silent, stony features . . . deciphering the revolutions that have given rise to 'a huge stone.' " At the same time, "Wordsworth emphasizes the fictionalizing . . . procedures" of scientific inquiry, alluding to the "history of mind as it moves from superstitious to factual fictions." One of these "factual fictions," the sea-beast, merits closer attention. Referring to the commentary on this image in Wordsworth's 1815 *Preface*, Bewell remarks: "the closest thing to this intermediate image of a 'sea-beast' assimilated to a 'stone' is a fossil." To illustrate, he cites examples of fossilized "sea-saurians" and other large unclassified marine fossils found at inland sites throughout the eighteenth century.[17] Under this general category, an especially topical link between human beings and marine fossils occurs in Humphry Davy's lectures on geology.

In one of his popular Royal Institution lectures, Davy debunked an early eighteenth-century explanation of large marine fossil fauna proposed by Johann Jakob Scheuchzer. Scheuchzer's essay, subtitled "an account of the remains of a man who had witnessed the deluge," analyzes the fossil remains of a creature that Scheuchzer supposes to have been an antediluvian man, depicted in an engraving accompanying the publication (Fig. 1.1). Davy doubts whether the "bones were human" and adds dryly, "at least it must be allowed that the antediluvians, if they had such skeletons, must have been very different from the present inhabitants of the globe" (*Lectures*, 80). Davy grounds his enlightened skepticism primarily on an analysis of the sandstone in which this fossil was found, but it must also rest, in part, on the morphology and incipient evolutionary biology informing Wordsworth's

17. Bewell, *Wordsworth and the Enlightenment*, 272, 268, 265, 266.

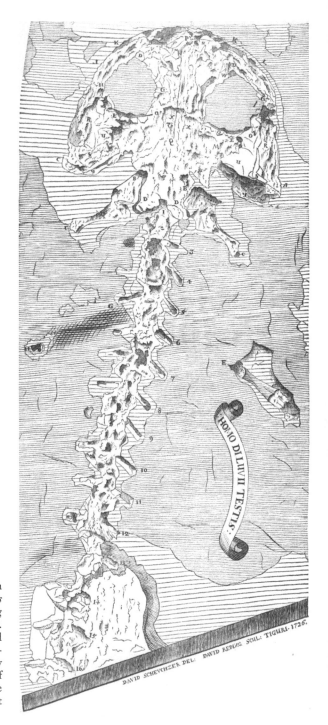

1.1 "Homo diluvii testis," in Johann Jakob Scheuchzer, *Homo Diluvii Testis* (1726). Engraving by David Reding from a drawing by David Scheuchzer. Courtesy of the Rare Book and Special Collections Library, University of Illinois at Urbana-Champaign. Humphry Davy comments: "The antediluvians, if they had such skeletons, must have been very different from the present inhabitants of the globe."

image.[18] Whatever Wordsworth's investment in these "factual fictions," "superstitious" ones like Scheuchzer's also play an important role in his geology. Scheuchzer's narrative conflates a "sea-beast" with a man, exemplifying the sort of theory that would have been equally present with more recent ideas in the "geological unconscious" informing the production of Wordsworth's image. The "sea-beast," as a "thing endued with sense," is supposed to explain how the erratic boulder arrived at its improbable location. "Sense" in this respect is motility, and Wordsworth's explanation seems to capture an evolutionary moment (the *Ur*-"crawling forth") by grafting the old science onto the new. His emphasis in describing the old man shifts from immobility (75) to mobility (103, 129–31), further elaborating these parallels. The "aged Man . . . is divested of so much of the indications of life and motion as to bring him to the point where the two objects [stone and sea-beast] unite and coalesce in just comparison" (*Prose*, 3:33). Ultimately this poetic juggling act motivates an explanation that might seem rather fanciful by the standards of geology itself.

I deliberately emphasize the affinities between Wordsworth's explanation and geological explanation in a strict sense. Though the scope of the former is widened by self-consciously poetic imagination, the two have common objects and common premises. The common object here is the fossil record, and the common premise is that of a legible history inscribed in it. The idea of a rock record, as Bewell's argument helps to illustrate, is an important analogue for the human history that can be seen as inscribed in rocks as a final instance of textuality. In Wordsworth's poem, the rock itself remains illegible but eventually becomes legible through its corporeal double, the Leech-gatherer. Wordsworth's catastrophism ("pain, or rage . . . long past") makes the method of reading geological, as does the flash of recognition at the end of the poem, in which the "firm mind" crystallizes out of the "decrepit Man" (138) in the same way that sacred or philosophical narrative crystallizes out of the strata in early geology. This legibility will be clearer in other examples to follow, but even in this poem it is evidently the counterpart of the resistance and inscrutability privileged as aesthetic qualities. The sense of wonder gives way to the imperative of instrumental reason. But conversely, the practical mode within geology itself—the urgent striving to make minerals more accessible as resources—also gives way to the sense of wonder. The substance of history equally functions as the mute substance of nature in "aesthetic geology," referring broadly to the vividly imaginative aspects of early earth science.

---

18. On Davy's correspondence with Coleridge and Wordsworth, and especially his influence on the 1802 *Preface* to *Lyrical Ballads*, see Roger Sharrock, "Poet and the Chemist."

John Whitehurst's *Inquiry into the Original State and Formation of the Earth* (1778) provides a topical example of aesthetic geology, related to the other geological problem of "Resolution and Independence." Whitehurst, a precursor of geological Plutonism, would offer the following account of Wordsworth's "huge stone" (and "by what means it could thither come, and whence"): such erratics, he writes, were "ejected from their native beds by subterraneous blasts," sometimes traveling "fifteen or twenty miles."[19] This is still catastrophism, though the catastrophe here is fiery rather than watery. Wordsworth certainly knew of James Hutton's *Theory of the Earth* (1795), the *organon* of Plutonism, and he probably would have shared his contemporaries' disapproval of Hutton's dismissal of sacred history—though it seems implausible to think of Wordsworth as a partisan of one school or another.[20] Whitehurst provides an early instance of the positivism of Hutton and Davy, coupled with a prominent strain of economic interest. In his preface he proposes a new science or subspecies of geology, to be called "subterraneous geography" (*Inquiry*, ii). This geography was to provide a blueprint for the location and discovery of subterranean mineral wealth. His announcement of a rigorous scientific method also shows a connection to mining: he wants to limit the scope of geology to "facts and laws of nature," minimizing inference in favor of empirical data, some of it "collected from experienced miners." But along with this new emphasis, Whitehurst offers a fantastic mise en scène for the "ejection" of erratics worthy of Thomas Burnet and the "learned fabulists" of the science's infancy: in the vicinity of these erratic boulders, the strata "are broken, dislocated, and thrown into every possible direction, and their interior parts are no less rude and romantic; for they universally abound with subterraneous caverns;

---

19. *Inquiry*, 63–65. Historians of geology have pointed out that Whitehurst's science is limited by the local scope of his fieldwork. He did not travel much outside Derbyshire. This local character gives the geology a particular aesthetic and material interest, but it prevented Whitehurst from seeing that erratic boulders in fact travel much farther than twenty miles.

20. Bewell suggests that Wordsworth is invested in a "proof" of the Deluge qua Revolution (266). Hutton's Plutonism—the antecedent of the currently accepted theory of a molten core—had left diluvialism behind as early as 1788, building its uniformitarian explanation partly on Whitehurst's suggestion that the Deluge itself was caused by volcanoes. While the Neptunists held to the notion of catastrophic floods, by the early nineteenth century the belief in a single Deluge, corresponding literally to Genesis, had given way to more complex accounts. Goethe was a vocal partisan of Neptunism, but there is no such evidence in Wordsworth's case. Wordsworth admired the style of Thomas Burnet's much earlier *Sacred Theory of the Earth* (1684) and corresponded occasionally with geologists, but without taking a clear position. See further John Wyatt, *Wordsworth and the Geologists;* and Nicolaas Rupke, "Caves, Fossils, and the History of the Earth." Wordsworth had at least indirect knowledge of both Whitehurst and Hutton from Erasmus Darwin's *Botanic Garden*, in which both are extensively cited.

and, in short, with every possible mark of violence" (63). Cited previously for its use of "romantic," this passage also demonstrates, through the topical example of erratics, how the aesthetic mode is inscribed in the very style of geological writing. This mode persists thirty years later in Davy—witness his notion of rocks as "characters" that "express sublime truths"—as does the aesthetic interest in alien, primeval landscape.

Wordsworth may have had his own image of the "huge stone" in mind when attempting a description of erratic boulders for the earliest version of his *Guide to the Lakes* (1810). The Leech-gatherer's pool, though technically too low in elevation to qualify as a "tarn," provides one thematic link to this description, which occurs in the passage on tarns: "The water, where the sun is not shining upon it, appears black and sullen; and, round the margin, huge stones and masses of rock are scattered; some defying conjecture as to the means by which they came thither; and others obviously fallen from on high—the contribution of ages!" (*Guide*, 41). This is a comparable instance of aesthetic geology, a formalist reading of landscape that cultivates a sense of wonder. Wordsworth's self-conscious, almost apologetic recommendation of the tarns contrasts forcibly with his enthusiasm about lakes, mountains, and views elsewhere in the *Guide*. Other elements of the scene amplify the sense that this is a primitive, alien landscape, a haunt of "unknown powers" (in Davy's phrase). "A not unpleasing sadness," the same passage continues, "is induced by this perplexity, and these images of decay; while the prospect of a body of pure water unattended with groves and other cheerful rural images by which fresh water is usually accompanied . . . excites a sense of some repulsive power strongly put forth."

This description shares the poem's powerful sense of solitude and brings out the barrenness and "desolation" implicit in its contrast between this and the fertility of the initial stanzas. Here this barrenness takes the rhetorical form of negation ("not unpleasing," "unattended"), which mimics the otherness of the objects in question: these landscapes are instances of a mere physicality, a "repulsive power" that may be described only as "other than" or as "naked wilderness," to use the term from a canceled stanza of the poem. The dwindling of the leeches brings out the association of a "death of nature" possibly present in both landscapes.

This barren character of the rocks, and of the composition they form in conjunction with the water (but without vegetation), corresponds very nearly to the aesthetic protocol established by treatises on the landscape garden. Thomas Whately observes that "mere rocks . . . may surprise, but can hardly please; they are too far removed from common life, too barren, and unhospitable, rather desolate than solitary, and more horrid than terrible; so austere a character cannot be long engaging, if its rigour is not softened by circumstances" (*OMG*, 93). The style here is reminiscent of Dr. Johnson's criticism, and there are similar passages in Whately's essays on Shakespeare. The attitude toward rocks is also reminiscent of the early Augustan suspicion described by Marjorie Nicolson as "mountain gloom"— but these echoes of earlier taste overlie a much deeper aesthetic affinity: the early industrial relation to the earth's material draws on a repertoire that extends from these earlier tropes to the celebration of nature's grandeur typically associated with Wordsworth. Even as mountains become aesthetic objects, the Enlightenment emphasis on their problematic physicality remains necessary to register their inorganic nature (as in the passage above) and the problem of their origin (as in the *Guide* passage). This is not to deny historical difference or the development of ideas, but to point out continuities in the language used to describe landforms from the early eighteenth through the early nineteenth centuries.[21] These continuities reflect the interaction between large-scale economic change, as mineral-driven industry approached the importance of agriculture, and the cultural forms for processing this change, which coalesced into modern geology over roughly the same period.

Though similar to Wordsworth's, Whately's formalist reading of landscape gestures more visibly toward a social context. His notion of rigor captures something of the metaphysical severity of rock, while the observations that illustrate it resemble those of the geologists in their aesthetic mode. His examples are typically "rent in the mountain by some convulsion of nature" (*OMG*, 95–96), producing "wonder" and "curiosity" excited especially by such phenomena as erratic boulders : "some . . . are wonderfully upheld by fragments apparently unequal to the weight they sustain" (113). Whately appeals strongly to the "vast and rude" formula (101) that has been recast in more modern terms by Wordsworth's time, but his aesthetic response shows a deep affinity with Wordsworth, both in its notion of slightly uncomfort-

---

21. A telling difference between Whately and Wordsworth is the latter's well-known exclusion of an ironworks (among other things) in his description of a scene along the Wye (in "Tintern Abbey"). Relating how aptly "machinery" may be "accommodated to the extravagancies of nature," Whately argues that "a scene at the New Weir on the Wye, which in itself is truly great and awful, so far from being disturbed, becomes more interesting and important, by the business to which it is destined" (*OMG*, 108–9). The "natural dusky hue" of the woods and cliffs is dramatically deepened by the machinery devoted to this "business," an "iron forge, covered with a black cloud of smoak, and surrounded with half-burned ore, with coal, and with cinders." (The New Weir is between Ross and Monmouth, a short distance north of Tintern Abbey.) While the "awful" sublimity of the cliffs along the Wye provides a constant between the two descriptions, the spread of such "dark Satanic mills" between 1770 and 1798 is surely one reason why they might no longer seem "compatible with the wildest romantic situations" (*OMG*, 110) by the time of the poem.

able rigor and in the leading concept of "character," expressed in Wordsworth's "not unpleasing sadness." Whately states, in these passages, the alterity of rock in its purest form, building his protocol out of qualities that could serve as a gloss on Wordsworth's "repulsive power": it is "unhospitable" and "far removed from common life." Such terms as "wonderfully" designate the aesthetic mode of this formalist reading. But the practical context of a social environment is present in the forms themselves: characters such as Wordsworth's melancholy provide the organizing principle for the garden and landscape compositions Whately recommends. These compositions extract such characters from "unimproved" nature, amplifying and ordering them as prescribed by the ideology of improvement: "to supply [nature's] defects, to correct its faults, and to improve its beauties" (1). In his practical mode, Whately reads "unimproved" nature for models of improvements to be carried out within the confines of improved landscape. In his account of Dovedale, for example, he draws on the geological process of stratification as a paradigm for the "arrangement" of rocks (113).

This practical mode of landscape gardening is at its most radical (and impractical) in William Chambers, whose work on gardens closely follows Whately.[22] The constantly expanding scope of the improvements Whately suggests in "Of Rocks" seems to indicate that he has forgotten, or means to qualify, the initial caveat that rocks are "too vast and too stubborn to submit to our controul." Chambers, however, omits any such caution altogether. Chambers' sketch of a massive, isolated boulder in the *Dissertation on Oriental Gardening* (1772) shows very clearly the transformation of a protocol for aesthetic response into a prescription for composition, the transformation of wonder into technical mastery. In describing Chinese improvements of lakes, Chambers gives an account of massive boulders rising from the water: "when they are large, they make in them caves and grottos, with openings, through which you discover distant prospects."[23] Here, however, there is no "wonder . . . / By what means they could thither come, and whence": the boulders themselves are "large artificial rocks, built of a particular fine colored stone, found on the sea-coasts of China, and designed with much

22. Apart from this largely polemical and derivative foray into landscape architecture, Chambers was the most important civic architect of his day, as illustrated by the comprehensive 1996 exhibition at the Courtauld Institute, housed in one of his major surviving public works, Somerset House. See the accompanying publication, John Harris and Michael Snodin, eds., *Sir William Chambers.*

23. William Chambers, "Designs for Chinese Buildings" (1757), reprinted in John Dixon Hunt and Peter Willis, eds., *The Genius of the Place*, 286. Whether or not this is authentic description, it became reality when Franz von Anhalt-Dessau completed *Der Stein* in his garden at Wörlitz in 1796, following Chambers' recipe and even adding details from other parts of the work, such as an erupting volcano.

taste."[24] Chambers' description deemphasizes the artificial origin of these "erratics" by discussing them together with natural features of lakes, while leaving no doubt about their massive size: "[They] have in them caverns for the reception of crocodiles, enormous water serpents, and other monsters . . . and grottos, with many shining apartments. . . . They plant upon them all kinds of grass, creepers and shrubs which thrive on rocks . . . with some trees rooted into the crevices."[25] Here is a full complement of the "softening" circumstances suggested by Whately—as well as a precedent for the "organic" linkage of sea-beasts and huge stones. Art triumphs over the primitive substance of nature, retaining the "vast" but leaving very little of the "rude."

The later discourse of the picturesque inherits ideas about vast, rude rocks from both landscape gardening and picturesque travel. Picturesque theory, in its concern with the shapes of geological features, is another structural element in the development of a geological discourse, as well as an influence on Wordsworth. Uvedale Price's *Essay on the Picturesque* (1794) draws on geological features for a uniquely systematic account of ugliness and deformity. Both these categories derive from an aesthetic mode of picturesque theory that confronts in yet another way the alien physicality of rock. "The ugliest forms of hills," Price declares, "are those which are lumpish, and, as it were, unformed. . . . When the summits of any of these are notched into paltry divisions, or have such insignificant risings upon them as appear like knobs or bumps . . . they are then both ugly and deformed." Other instances of ugliness (not strictly derogatory in Price) with "some mixture of deformity" include the "slime" deposited on riverbanks and "huge shapeless heaps of stones" like those surrounding Wordsworth's tarn—and like the Elgin Marbles, in the judgment of those who opposed their purchase by Parliament. Price may well have the Lake District in mind when he observes that frequently "persons who come from a tame cultivated country . . . mistake barrenness, desolation, and deformity, for grandeur and picturesqueness."[26] Price at his most extreme reacts *against* the aspects of geologic phenomena that generate a sense of wonder in Wordsworth and

---

24. William Chambers, *Dissertation on Oriental Gardening*, 66. The earlier, slightly more authentic description details this process ("Designs for Chinese Buildings," 285). Professor Lydia Liu informs me that there was in fact such a practice, though some of Chambers' details are probably exaggerated. There was a superb exhibition of the similar but smaller-scale "scholar's rocks" at the Arthur M. Sackler Museum (Harvard University) in 1997 while the first draft of this chapter was being written. There are many Western precedents for Chambers' interest in Chinese gardens; the contemporaneous fads of *conchyliomanie* and *chinoiserie* had left their mark on French gardens earlier in the century.

25. Chambers, *Dissertation on Oriental Gardening*, 67.

26. *Sir Uvedale Price on the Picturesque*, 149–50.

other contemporaries, though this reaction leads to increasingly subtle re-flections on the compatibility of ugliness—if not deformity—with the sub-lime and the picturesque. But his discussion as a whole presents another prototype for the desolate landscape of "Resolution and Independence."

Price in this aesthetic mode acknowledges the influence of William Gilpin, whose *Observations* on the landforms of the Lake District (1787) are replete with instances of "irregular," "disgusting" forms and excessive rude-ness. Though affiliated with Gilpin's travel writing, Price's *Essay on the Pic-turesque* is equally a treatise on landscape gardening in the tradition of Whately (and, like Chambers' *Dissertation*, a polemic against the school of Capability Brown). Alan Liu and others have shown the rich political im-plications of Price's and Knight's picturesque theory.[27] It also holds, in its practical mode, particular economic implications: ugliness and deformity disrupt the sublime but encourage "improvement," so that the aesthetic protocol and the strategy for improvement become identical in practice. In the often-cited case of visual command over a landscape, by contrast, the practical agenda of the aesthetic remains more abstract, a theoretical equiv-alence. Discussing the topographical scars caused by flooding and by min-ing activity, Price notes that "when the rawness of such a gash in the ground is softened, and in part concealed and ornamented by the effects of time and the progress of vegetation, deformity, by this usual process, is converted into picturesqueness." The passage continues in the practical mode: "This connection between picturesqueness and deformity cannot be too much studied by improvers, and, among other reasons, from motives of economy." The natural process can be taken as the model for a landscaping technique by which "deep hollows and broken ground," rather than being filled, are simply left open: "to dress and adorn them costs little trouble or money."[28]

Despite his vehement opposition to the laissez-faire policy that will later characterize Price's government of nature, Chambers' essay offers a striking analogue for this "economical" suggestion of Price. Chambers suggests the following application of "Chinese" method to the English landscape: "hills might, without much difficulty, be transformed into stupendous rocks, by partial incrustations of stone, judiciously mixed with turf, fern, wild shrubs and forest trees."[29] This is a naturalized version of the extravagant ar-tificial boulder. The "stupendous rock" emphasizes the concern of land-scape architecture, as a practical discipline, with the massive size of rocks;

27. See chap. 3 of Alan Liu's *Wordsworth*, "The Politics of the Picturesque"; and the 1994 collection, edited by Stephen Copley and Peter Garside, bearing the same title.
28. *Price on the Picturesque*, 151.
29. Chambers, *Explanatory Discourse*, 132.

Wordsworth's "huge stone" draws on this preoccupation, as well as on geology's concern with their age and on picturesque theory's visual emphasis on their formlessness, in exploring the boundaries of materiality. The wonder inscribed in Chambers' phrase, however, becomes an effect of technique and a testament of the sovereignty over nature that is the substance of his fantasy here. His success in making it a plausible fantasy complicates the relationship between wonder and sovereignty that is a structural feature unifying all the discourse on landforms. The aesthetic response in "Resolution and Independence," too, generates a kind of technical mastery by prompting a geological reading of the old man-stone as a redemptive figure of stability, a "resolution" of the poem's crisis.

In Chambers' case, the wonder and the sovereignty of the "stupendous rock" are both avowedly factitious. Writers on the landscape garden and the picturesque necessarily realize that, in practice, human intervention produces "terror" in rocks and similar aesthetic effects. In such a realization, however, literary and philosophical aesthetics merge with the ideology of improvement. Rock remains the most intractable substance of nature considered as a human environment, one in which both subjugated and wild domains can appear as resources. The value of wilderness as an aesthetic resource explains the continued striving after "terror" or some version of it by enlightened improvers, poets, and theorists alike over a period of many decades. The loss of rational control or subjectivity entailed by this terror is always to some degree a ritual enactment of the real impossibility of ultimate control over nature.

## Wonder and Sublimity

I privilege this sublime moment as a conscious gesture of dissent against the Kantian tradition of stressing its subsumption by a narrative of self-recovery and self-assertion. Frances Ferguson gives an especially forceful account of what is politically at stake in this subsumption. Ferguson targets poststructuralist critiques of Kant, which, she argues, privilege an "empirical infinite" that limits individual agency.[30] Such a critique, however, is also possible with reference to the materiality of nature rather than the materiality of language. If the abdication of imagination in favor of practical reason represents a purely formal recognition of coherence, preserving a politically viable agency, it is also true that this Enlightenment model of agency helped to precipitate the ongoing environmental crisis. If it is overly sanguine to

30. Ferguson, *Solitude and the Sublime*, 21 and passim.

speak, with Jonathan Bate and others, of a "Romantic ecology," it is nevertheless true that English Romanticism remains quite independent from
Kantian instrumentality and preserves a space for the abdication of agency
over natural processes. The geological sublime holds wonder in tension
with the impulse toward mastery. At this stage, environmental circumstances do not yet require the discourse on landforms to acknowledge a contradiction between awe and agency, or even between the admiration and the
domination of nature.

I emphasize "wonder" in this and the next two chapters as a more flexible
alternative to "the sublime." Although Steven Knapp and others have found
plausible connections to the sublime in "Resolution and Independence,"
Wordsworth himself avoids direct allusion to it in favor of the "wonder" attributed to those who "espy" geological phenomena. Keats' sonnet "On
Seeing the Elgin Marbles" further illustrates the distinct connotations of
"wonder"—"these wonders" is its only description of the Marbles—in a
context seemingly far removed from the one at issue here. I choose this term
partly to avoid being too narrowly specific: the cultural consensus concerning rocks, whether sculpted by Phidias or by glaciers, is merely a lowest
common denominator, and the deviation from such assumptions is frequently more interesting—as when Blake exclaims, in *Jerusalem*, "they call
the Rocks Parents of Men" (67), invoking the fascination of rocks for his critique of materialism. Romantic poetry anatomizes, contests, or celebrates
the primitive materiality of rock, connected through its indifference to biological organization with the influential Lockean definition of materiality as
mere extension in space. The materiality of the deeply weathered
Parthenon sculptures, for example, provokes a range of responses, from
Keats' "wonder" to Richard Payne Knight's "disgust." The consensus here
concerns not the aesthetic value of the carved human figures, but the decisive importance of their material nature, their rockiness. The underlying
connection between this sonnet and "Resolution and Independence," then,
resides in the way Keats assimilates the Parthenon sculptures to a Romantic
landscape of "pinnacle[s] and steep[s]" resembling both the starkly grand
Lakeland scenes of Wordsworth and the spectacular Mediterranean scenes
of Byron or Shelley.[31]

While encompassing "the sublime" and its derivatives under a more traditional rubric, "wonder" in a postmodern context also recaptures elements of
the sublime obscured by the investments of recent criticism. Hartmut
Boehme draws attention to one such element by redefining the sublime as an
experience of geological otherness, thus privileging the sublime of self-loss

---

31. This is the argument of my "Stones So Wonderous Cheap."

or "ex-centric being" marginalized by Ferguson and others.[32] Boehme's paradigm (based largely on Novalis) is more effective for Keats than for Wordsworth; the form of wonder in Keats is typically a sublime moment in which the unity of apperception ceases to operate. I make use of the Kantian term "unity of apperception" to indicate the distance between my construction of "wonder" and the Kantian sublime. Kant's "aesthetic judgment" is of limited application in literary contexts because of the mediating role—bridging the gap between pure and practical reason—that he assigns to it in the front matter of *The Critique of Judgment*.[33] The failure of imagination is not only the source of pleasure in the sublime, but is also required, in terms of *The Critique of Pure Reason*, for the synthesis of the manifold not to be disrupted, for the subject not to overflow and dissolve in representations. Since, for Kant, the "I think" must accompany all representations (because representations must be conceivable in their subject), experience is unified a priori, and this "unity of apperception" enables consciousness to recognize representations in the form of concepts. In arguing that empirical consciousness has no relation to the identity of the subject, Kant dismisses a hypothetical "multicolored" self that surprisingly resembles Keats' "chameleon poet."[34] If the geological objects of Romantic aesthetics are not, like the sublime objects of the Third Critique, replaceable and inessential, it may indeed be useful to think of the experience of wonder or sublimity as not consisting of representations, or not taking place in a subject. This seems momentarily true when "the light of sense goes out" for the narrator of Wordsworth's *Prelude* (book VI) as he crosses the Simplon Pass, though here "Imagination" famously restores unity of apperception. Keats more readily embraces the dissolution of subjectivity. E. T. A. Hoffmann, describing the allure of the netherworld of the mines and its effect on a romantic young man, is probably the most explicit: "His ego dissolved among the splendid rocks."[35]

The sense of wonder throughout the period is generally attributed to a unified, conventional set of qualities in rocks; but the range of response, which belies the metaphor of a "consensus," comes with the diversity of forms—aesthetic, practical, ideological, scientific, or some combination of these—through which it is expressed. In this chapter, my aim has been to illustrate this range of response via the concrete example of the erratic boul-

32. Boehme, "Das Steinerne," 135–41.

33. *Kritik der Urteilskraft*, 1790 introduction, 224–25; cf. *Critique of Judgment*, 435.

34. *Kritik der reinen Vernunft*, 136 (B 132), 137 (B 134); *Critique of Pure Reason*, 153–54. The "B" refers to the original pagination of the second edition.

35. "Sein Ich zerfloss in dem glänzenden Gestein"; Hoffmann, "Die Bergwerke zu Falun," in *Ausgewählte Werke*, 2:189. One of many relevant passages in Keats is a moment in *Endymion* in which, far from being required for a coherent aesthetic experience, self-consciousness appears as a void Endymion must flee. See *Endymion* II.274–86 in *Complete Poems*, 95.

der in "Resolution and Independence." The Elgin Marbles are "erratics" in a very different sense, but they are also "huge stones" in the parlance of the day, indicating how wide a range of geological objects was subject to the same set of responses. By invoking the trope of "rude rocks" for a collection of architectural sculpture, Keats merely reverses the long-established trajectory of architectural metaphors for natural rock formations. In "Ode: The Pass of Kirkstone," written the same year as Keats' sonnet, Wordsworth refers to a pile of unhewn stones at Kirkstone Pass (a name that itself represents this paradigmatic conflation of nature and artifice) as a "monument," borrowing his sister's description, "a monument of ancient grandeur" (*Prose*, 2:369). All three poems draw on the commonplaces adhering to "rude rocks," but they differ in their negotiation of the relationship between the "rude" and the "polite." In "Resolution and Independence," the rough but enduring stone, in its metonymic relation to the old man's battered body, authenticates the inspirational narrative. In Keats' sonnet, a fragmentary syntax registers the "rude" marble, while the delicate symmetry of the sonnet form registers the sculpture, contained in the "Elgin Marbles." Many more poems from the period stage this tension between geologic and poetic form in a striking variety of ways.

A parallel tension between economic value and aesthetic value resides in a wide range of geological objects, articulated by naturalists and "improvers," as well as the poets, in disparate—if generally cognate—ways. Just as the resistance of Wordsworth's "huge stone" is needed to set off the Leech-gatherer's hard-won independence, the resistance of "stupendous rocks" is required to mark and to stimulate the progress of landscaping, of positive geological knowledge, and the technological mastery of nature. Because it represents the ultimate frontier of industrial development, the mineral world comes to stand for the province of materiality in general, the material "worked on," always with varying degrees of control, in all these areas of culture. Pristine mountain scenery is a crucial resource for sensibilities ranging from Wordsworth's to Byron's, a resource whose aesthetic viability depends on successful competition with potentially intrusive forces such as industry or tourism. The aesthetic interest of geological phenomena is thus defined economically, as the exclusion of economic interest. The metaphysical resistance of mountain scenes, as the last bastions of the purely aesthetic, is constructed from the physical resistance of rock. Yet the purity of the aesthetic can also depend on the perfect convertibility of aesthetic plenitude in the landscape into economic wealth, a tradition in English poetry extending back at least as far as Spenser's "land of gold." My reading of "Resolution and Independence" opens out onto this larger context, not merely because geology is part of the historical background, but because geological description is the forum for the period's environmental discourse. The economic and aesthetic factors at work in this discourse operate visibly in Wordsworth's poetry, which, in turn, contributes to the emergence of a concept of environment. The poem's brief geological moment—its account of a "huge stone"—both helps to explain how it generates the material for its moral object of stability and provides an avenue for reconstructing the period's broad definition of geology itself.

# 2

## Geological Otherness; or, Rude Rocks and the Aesthetics of Formlessness

**R**ock provides an index of the substantial reality required of the world if it is to sustain human bodies and provide building materials for human culture. The various accounts of "huge stones" described in chapter 1 register this primitive reality in their aesthetic response to the otherness of rock. "Astonishment" underscores this connection, as does the etymology connecting "aesthetic" and sensation.[1] To express the *fundamental* character of rock—as literal and metaphorical ground—it will be useful to borrow Hartmut Boehme's epithet "das Menschenfremdeste" ("that which is most foreign to the human").[2] As the basic inorganic solid substance, rock embodies the nonhuman and thus the type of the external object within a dualistic epistemology. When addressing the primitive reality of rocks and landforms, poems and narratives also rely on a strong distinction between the aesthetic and the practical. Geological features provide a literal ground for this distinction, however unstable and ideological it often proves to be. Raymond Williams' influential argument, that "natural beauty" was "invented" along with the aesthetic to separate consumption from production (the practical), remains compelling.[3] But there is still

---

1. See the entry "astone," in *Compact Edition of the Oxford English Dictionary*, 1:131. Cf. "astonied," "astonished." The latter is a variant not occurring before 1500. The conjectural derivation of "astoned" from the Latin "thunderstruck" involves a conflation of "petrified" with the French *étonner*, dating back at least to the fifteenth century. For "aesthetic" and "material" (derived from "timber") see *Compact Edition*, 1:37 and 1742, respectively.
2. Boehme, "Das Steinerne," 120.
3. Williams, *Country and the City*, 120–21.

heuristic value in considering the aesthetic qua aesthetic within a historical context. Aesthetic discourse on rocks articulates the sense of a real, physical world, a planetary depth of experience that can either enable or confound economic agency. Some of the most famous episodes of Romanticism invest the description of landforms with the power to negotiate the real, as in Wordsworth's Simplon Pass (*The Prelude* VI) and Shelley's "Mont Blanc." Similarly, the erratic boulder enters the landscape of "Resolution and Independence" partly to anchor the initial, hallucinatory appearance of the old man to something real, and partly to figure the reality of the body as against the ideality of "so firm a mind."

My reliance on the category of "the other" for describing aesthetic response might seem to problematize the heuristic isolation of aesthetic from economic interest, or of wonder from hegemony, to use a term almost inseparable from "the other" in political discourses. In its original, Hegelian sense, "the other" is by definition a slave, albeit a slave who "works through" his or her subjection.[4] While the myriad psychoanalytic and postcolonial readings of otherness are often less sanguine about the other's capacity for agency, they all deal with human subjects and power relationships. If "otherness" applies to inanimate, inorganic objects, it would seem to follow that some form of subjugation is at issue, such as the exploitation of natural resources. My emphasis, however, is on the other-than-human rather than the other-than-self. The Hegelian dialectic of self and other, or master and slave, turns on the "truth of self-certainty," just as the Kantian sublime turns on the integrity of the self. Psychoanalytic and postcolonial theory also emphasize questions of identity-formation, accounting in part for the great influence of "the other" as a theoretical category.[5] By contrast, the dwindling or dissolution of subjectivity is at issue in period accounts of geological otherness, as in literary accounts of the sublime. On the one hand, such claims—that one is overwhelmed by the Alps, for example—are a fashionable by-product of better roads and increasing tourism. More significantly, though, a great many such accounts associate the materiality of rock with limitations on human agency. The ultimate object of wonder is the rocks' resistance to aesthetic categories themselves: they do not simply become humanized or sacralized others, but stand outside this structure and give the lie to technological progress, artistic endeavor, and other forms of human

4. G. W. F. Hegel, *Phänomenologie des Geistes*, 134, ll. 37–39 (§195). The earlier translation, "lord and bondsman," seems closer to *Herr* und *Knecht* than does "master and slave," since Hegel does not use *Sklave*. The etymological link between *Knecht* and "knight" reinforces the association with vassalage.

5. See, for example, Leela Gandhi, *Postcolonial Theory*, 39–40; and Slavoj Žižek, *Sublime Object of Ideology*, 118–24, 195–99.

ordering. This chapter draws on poems by Wordsworth and Shelley, as well as prose by Dorothy Wordsworth, Kant, and Thomas Burnet, that demonstrate this anxious fascination with the limits of form.

These texts are all versions of travel narrative, and their use of geological formalism to articulate an absolute otherness cannot be wholly separated from the ways in which aesthetic discourse also makes a subjugated other out of nature or uses geological otherness as a figure for categories of race and gender. These connections derive from the "excursion poem," a tradition informing such works as Wordsworth's 1814 *Excursion*, his "Ode: The Pass of Kirkstone" (1817), and Shelley's "Mont Blanc" (1816). These poems in many ways resist and revise their eighteenth-century predecessors, such as David Mallet's *Excursion* (1728), in which earthquakes gestate in the earth's "infernal womb," and James Thomson's *Summer* (1727), which situates volcanic activity in Africa in order to figure the "horrid" excesses of this "world of slaves."[6] The Wordsworth and Shelley lyrics are embedded in travel narratives, but only the *History of a Six Weeks' Tour* containing "Mont Blanc" actively addresses the cultural otherness absent from the poem itself and from Burnet's Alpine narrative as well as the Wordsworth materials. *History of a Six Weeks' Tour*, largely the work of Mary Shelley, thus merits special attention as an index of the social and cultural unconscious of geological otherness. "Mont Blanc," however, also shares with Wordsworth's Lakeland excursions the impulse to demystify the exotic spectacles created by eighteenth-century tourism, to renew a sense of topography as prior to and outside culture. While this desire is subject to ideology critique as a form of imperialist deterritorialization, it also leads to a culturally specific perception of the geological as a separate sphere, without which both Romanticism and modern earth science are inconceivable.

## Alien Physicality

"Ode: The Pass of Kirkstone," appended by Wordsworth to the last edition of *A Guide to the Lakes* (1835), offers the most concise of the *Guide*'s "close encounters" with alien exteriority. Its first stanza sketches a scene cognate with the "huge stone" stanza of "Resolution and Independence" and the descriptions I have linked with it. The stanza gathers many of the attributes contributing to a sense of the radical alterity of rocks: barrenness, rudeness, vastness, inexplicability, and immense antiquity. The speaker's

---

6. Mallet, *The Excursion* I.520; Thomson, *Summer* 1096–1100. Such geographic inscriptions of race and gender are largely sublimated in the poems by Wordsworth and Shelley.

aesthetic response seems ritualized, repeated "Oft as I pass along the fork /
of these fraternal hills,"

> Where, save the rugged road, we find          5
> No appanage of human kind;
> Nor hint of man, if stone or rock
> Seem not his handy-work to mock
> By something cognizably shaped;
> Mockery—or model roughly hewn,          10
> And left as if by earthquake strewn,
> Or from the Flood escaped:
> Altars for Druid service fit;
> (But where no fire was ever lit,
> Unless the glow-worm to the skies          15
> Thence offer nightly sacrifice;)
> Wrinkled Egyptian monument;
> Green moss-grown tower; or hoary tent;
> Tents of a camp that never shall be razed;
> On which four thousand years have gazed!          20
>       (*Prose*, 2:251)

Lines 6–7 in particular illustrate the sense of Boehme's term "das Menschenfremdeste."

A tension ensues as the speaker endeavors to attach some limited human
qualities to the recalcitrant rock—somewhat like the "rocks that muttered"
in *The Prelude* (1850) (VI.630–40). In the ode, despite the humanizing figures and the human history rehearsed in stanza three, the rock seems to resist humanization and remains "savage" (line 48). The rocks here are not
transformed into legible "characters," like those in *The Prelude*. Instead they
become the subject of numerous disjunctive propositions and competing explanations. The apparent spontaneity and disorder of the ensuing
metaphors work against the regularity of the stanza form. The unusual term
"appanage" (5) evokes both barrenness and strangeness: the rock cannot
provide sustenance, nor is it a "natural or necessary adjunct" of "human
kind."[7] It stands apart from the category of economic resources, but this indifference makes it an aesthetic resource that "thrills" the bosom (2). The
basic alterity of the rock, stated as baldly as possible in lines 6–7, gives way
first to the modifying "if" clause (7–9).

The remainder of the stanza offers a series of speculative "hints of man,"
a chain of association assimilating human to natural history. The whole suggests an attempted humanization that fails, leading to the rejection in the

---

7. See the entry "appanage" in *Compact Edition of the Oxford English Dictionary*, 1:95.

following stanzas of the scene at the pass itself in favor of the panoramic and cultivated landscapes below. The nearest thing to a "hint of man" is the resemblance of some of the stones to human "handy-work," such as the resemblance of the "kirk stone" to a chapel, which gives the pass its name (47–48). The stanza concludes with four possible figures for the rocks, but first introduces the catastrophism merely hinted at in "Resolution and Independence": the fashioning of "cognizably shaped" objects seems to have been interrupted by the more elemental and violent agency of earthquake or deluge (9–12). The "mockery" of "stone or rock" participates in the agency of catastrophic subterranean forces. But the "or" disjoining "model" and "mockery," which like the following "or" seems to pair two competing alternatives, also carries the suggestion that any possible kinship with the rock is negated by its link to these forces.

The boulders at the pass appear as models for a succession of human structures—but each time, the poem immediately retracts these suggestions or renders them ambiguous. As a result, the resemblance seems unattainable and the questions raised by the "ors" unanswerable. The first suggestion, that of the Druid altar (13), is qualified by the rock's indifference to human purposes, then qualified again by a second nonhuman agency (the glow-worm). Wordsworth similarly qualifies the image of the "hoary tent" by removing it from human context and human time (18–19). The tower (18) seems in the process of being reclaimed by nature, a ruin being perhaps the human production most "cognizable," and certainly most topically linked, with the rocks. For the "wrinkled Egyptian monument" (17), he draws on an ancient association between the face of the earth and human physiognomy, attempting to find a shape cognizable not only with human "handy-work," but with human form itself.

And yet the Horatian regularity of the stanzas reminds us that all this strangeness or otherness is contained by the poetic structure. This structure expresses the compensating and redemptive order the speaker perceives in the landscape below, particularly in the final stanza: "Thy lot, O man, is good, thy portion fair!" (80). The chthonic irregularity thematized in the first stanza represents an incongruity within this overarching structure, and thus performs the irreducible materiality of rock, its ultimate resistance to form. The breathless succession of increasingly inadequate architectural metaphors registers this resistance, which leads the poem to reject the rocky scene in favor of a moralized landscape that *can* accommodate the pattern of disillusionment and recovery. This is a vision of nature as coherent narrative: "the greenness tells, man must be there" (66). Wordsworth stations a "Genius" at the pass to obscure "care" and "guilt" with his mists (35–38) while showing the human landscape bathed in the radiance of its fruitfulness. This allegory absorbs the initial chaos of the pass, which becomes a "hill / of duty" (55–56) and a station from which to view the "cultured Plain" with its denizens "Joy" and "Faith" (72–77). The "Genius" of the pass is ultimately a figure for the "repulsive power" of alien exteriority, which forces positive attention (like the Kantian sublime) to the moral sphere, and can then be dismissed: "Farewell, thou desolate Domain!" (71).

The uneasy coexistence of human and natural history in the first stanza stands as an attempt to render the otherness of rock, which prompts the turn to the moral sphere that is the real occasion of the poem. Wordsworth's "wrinkled Egyptian monument" expresses this coexistence with admirable compression, since it stands for the oldest known human production and this antiquity is reinforced by wrinkles, the most pronounced sign of age on the human body.[8] This hybrid image is another remote approximation, another metaphor bouncing off the rocks, but its incongruity aptly images the incommensurability of human with geological time. The period invoked in the stanza's closing line—"four thousand years"—is a figure prescribed by sacred history, and so in some sense the closest human approximation of geological time (although strict biblical dating of geological events was widely questioned by this time).[9] The progress of the poem shows these approximations to be inadequate, to point the way to a socially grounded aesthetic. The *phenomenological* failure of the rock to show a human face occasions a sort of transforming theodicy of the rugged landscape on the basis of its *moral* function, but the link between human and natural history persists in the allusion to Agricola's campaign in Cumberland (41–47) and especially in the inescapable suggestion that by contrast to the rocks at the pass, *any* "appanage of man" must be a "portion fair." This cynicism seems worthy of the Solitary, Wordsworth's misanthrope, whose description of rocks in *The Excursion* (III.124–52) bears a close resemblance to the ode's initial description.

Wordsworth added this ode to the *Guide* in 1835, placing it after the account of two excursions taken from Dorothy Wordsworth's journal, and remarking: "The following verses . . . after what has just been read may be acceptable to the reader, by way of conclusion to this little volume" (*Prose*, 2:251). "What has just been read" is the second excursion, which leads to the banks of Ullswater via the Kirkstone Pass and includes an account of the

8. One analogue for this striking image is Keats' addition of wrinkles to a Greek monument, the statue of Saturn whose brow "wrinkled as he fell" in *Hyperion*, composed not quite two years after Wordsworth's ode. Another is the "statue" of Hermione in *The Winter's Tale*, whose wrinkles turn out to be real.

9. See further John Wyatt, *Wordsworth and the Geologists*, 36–37.

pass sharing some details with the ode. In a discussion of the transformations of this narrative (originally dated 1805) and its relation to the ode, Eric Walker formulates a useful distinction between Dorothy's "feminine" sublime and the "masculine" sublime introduced in William's revision.[10] In this context, Dorothy's prose presents a competing aesthetic response to the rocks at Kirkstone Pass: "At such a time, and in such a place every scattered stone the size of one's head becomes a companion. There is a fragment of an old wall at the top of Kirkstone, which, magnified yet obscured as it was by the mist, was scarcely less interesting to us when we cast our eyes upon it than the view of a monument of ancient grandeur has been.—Yet this same pile of stones we had never before even observed" (*Prose*, 2:369). This description locates the human and cultural connotations of the rocks that the ode vainly seeks by means of its more extravagant metaphors. Dorothy Wordsworth singles out the fragments of a wall, an unambiguous sign of human "handy-work" that climactically dispels her initial anxiety about bad weather at the pass.

Dorothy's emphasis on the mist accounts for the most important differences between this passage and William's description of the same scene in the ode. Because of the mist, the details of the scene—such as the fragment of the wall and perhaps even its human affiliation—come into sharp relief. The mist is largely absent from the ode, however, and good visibility is of great importance throughout the poem, both at the pass and to permit the panoramic vision of the following stanzas. A slightly revised version of Dorothy's narrative precedes the ode in the *Guide* text (*Prose*, 2:244–45). As Walker observes, William's revisions tend to excise "human community" and "human sympathy." But the final text retains its emphasis on the mist and the old wall, in contrast to the ode, where the social appears only in stanza three with Agricola's legions. The ode, then, deliberately constructs a scene devoid of "companionship." As Theresa Kelley puts it, the ode attempts "to people the landscape in spite of itself," but is finally forced instead to favor the beautiful landscape below. She argues that this attempted sublimity characterizes the *Guide* as a whole: "The sublime is implicit in the *Guide* account of a place so chaotic, archaic, and primitive that it has never accommodated human life."[11]

---

10. Eric Walker, "The Pass of Kirkstone." The journal entry was revised a number of times by both Dorothy and William before it was published in the *Guide*, as indicated by *Journals* MS. 13 (see the textual note in *Prose*, 2:361–63).

11. Walker, "The Pass of Kirkstone," 117; Kelley, *Wordsworth's Revisionary Aesthetics*, 39–40. I do not agree with Walker that William systematically "reshape[s] Dorothy's prose toward the expression of imaginative power" (120), or the masculine sublime. The heightened otherness of the rocks in the poem is a function of the ode's exalted philosophical idiom. Though William's

In his account of the relationship between the sublime and geological phenomena, Hartmut Boehme argues that the Romantic sublime can be divided into two alternative reactions to the inhospitable qualities listed by Kelley. Boehme sketches the Kantian sublime in broad outlines as a "proto-industrial" attitude, linking it to the history of science and technology and the colonial project of "dominion over peoples and over nature."[12] Kant reacts to those natural forces not yet subdued by technology, epitomized by the Lisbon earthquake of 1755, by educating the imagination to an interiorized sovereignty over the genuine terror they provoke. But Romanticism, Boehme argues, develops an alternative reaction that counters the demystification performed by Enlightenment science and philosophy with a remystification of rock. "Dialectical mutual recognition" of an independent other in nature supplants the "devaluation of nature as subjugated other." Novalis and other Romantics, Boehme argues, restore a voice and a face to the earth's material ("das Steinerne"), which becomes a "thou" ("Du").[13] This duality of "thou" and "it" corresponds suggestively, if not precisely, with the "feminine" and "masculine" sublimes emerging from the two accounts of the Kirkstone Pass.

Boehme posits a form of poetic recognition that does not compromise the otherness of rock as *das Menschenfremdeste*, the archetype of the nonhuman; the humanization he ascribes to the "dialectical" Romantic sublime is not a projection of human qualities onto inanimate nature. This alternative sublime takes the form of *Selbstbegrenzung* (self-limitation), establishing connection to the world through loss of identity.[14] Boehme cites the precedent here of metempsychosis, and one might add the divine inspiration so prominent in Longinus' account of the rhetorical sublime and also important in eighteenth-century accounts such as Robert Lowth's. The poet, then, gains access to the voice of nature without the "technological violence" of projection, but at the risk of a kind of Orphic madness. Boehme's paradigm case is Novalis, who recognizes that "poetry is ex-centric being; a knowledge of the other [or the strange] presumes self-estrangement."[15] Romanticism inherits this *Fremdsein*, or otherness, from the kinds of tropes familiar in English literature since the Alpine journals of John Dennis and Thomas Gray. But the

---

1823 revisions of the prose tend slightly in this direction, he does not revise further when adding the poem in 1835, and thus distances it from the journal's more prosaic work.

12. "Beherrschung der Völker und Beherrschung der Natur"; Boehme, "Das Steinerne," 124.

13. Ibid., 137 ("dialektische gegenseitige Anerkennung") and 123 ("die Absetzung vom erniedrigten Anderen der Natur").

14. Ibid., 128, 135.

15. Ibid., 135: "Poesie ist ex-zentrisches Dasein; Erkennen des Fremden setzt eigenes Fremdsein voraus."

inherited idea of reading this geology as physiognomy is reversed in a kind of internalized mimesis: "The sublime has a petrifying [astonishing] effect," writes Novalis in *Die Lehrlinge zu Sais* (in a passage that provides one of the epigraphs for this book). The subject of the geologic sublime envisions a bark of stone ("Steinrinde") growing over his own body, registering "the identity-dissolving power of the other."[16] Novalis thus makes the identity between the sublime and the petrifying a literal one: aesthetic response becomes a perfect mimesis, uniting the cause and effect of the sublime.

I dwell on this argument because it throws into sharp relief the difference between two fundamental aspects of aesthetic response to the mineral world. These two attitudes become polarized in Boehme's account because he attends exclusively to the sublime mode of discourse about rocks. Nature in Boehme's Romantic sublime provides "the indispensable answer," an answer to the question of presence and the confusions of self-consciousness.[17] But these forms of aesthetic and epistemic interest coexist and interact with economic and ideological interest in the broader range of discourses about rocks and landforms. This complexity is especially apparent in the relevant cultural practices: for the geologist, the discovery of "beautiful systems" and "subterranean powers" (in Humphry Davy's words) can lead to the discovery of mineral wealth; for the "gardenist" (to use Hugh Walpole's term), wild and grand scenes in unimproved nature become models for improvement, for physical intervention in landscape; and for the picturesque traveler, these scenes give way to the sovereignty of the frame and the pencil. Conversely, the quest for such sovereignty in all three discourses can lead to unexpected experiences of wonder, such as Gilpin's "deliquium of the soul."[18] A brief encounter with the Kantian sublime will help in mapping the complex relationship among the sublime, materiality, and geological objects. Kant's insistence that aesthetic judgment is "disinterested" has been systematically exploded by recent critics, most dramatically by Lyotard in "The Interest of the Sublime." Where Boehme insists on the relevance to Kant of imperial and industrial economic interests, Lyotard posits a symbolic economy in which the Kantian faculties themselves are a form of capital. But a limited degree of "aesthetic interest," of the wonder Boehme claims for poetry, does coexist in Kant with practical interest. While Kant's version of "disinterested" or "purely" aesthetic experience is obviously an ideological abstraction of the period, reducing the aesthetic to pure interest of any sort is a naive inversion of the same dualism.

---

16. Novalis, quoted in ibid.: "Das Erhabene wirkt versteinernd," 136; "die identität-sauflösende Macht des Anderen," 139.
17. Boehme, "Das Steinerne," 138.
18. William Gilpin, *Three Essays*, 49.

Kant interconnects at least two forms of interest when he translates the technological domination of nature into the moral register, as the subjugation of "nature within." His 1790 account of the sublime concludes by restaging, in a more concrete form, the scenario previously enacted by the faculties under the abstract rubrics of number and power: "The wonder, which borders on fright, the horror and the sacred shudder, which . . . seize the spectator at the sight of heaven-ascending mountain masses . . . are . . . only an attempt, to make a venture with imagination in order to feel its capacity to link the agitation it has caused with psychic equilibrium . . . and thus to triumph over . . . the nature within us, and therefore also the nature without."[19] Kant emphasizes the imagination's capacity to bring agitation or affective response into equilibrium or a "state of rest," just as the security of the observer in nature countervails any real terror. The capacity of practical reason for sovereignty over both natures thus emerges as the real object of wonder, generating a new vision of nature that replaces the received protocol of terror induced by mountain-masses with rational equilibrium. If mountains ascending toward heaven, combined with the sacred shudder of their votaries, suggest the theological aspect of the sensuous imagination, both the geography and the theology are corrected in Kant's revised phenomenology of the sublime. This revision is the project of the "General Comment" concluding "The Analytic of the Sublime."

The revised phenomenology displaces cultural encodings, including economic interest, and advances a departicularized vision of primitive reality that appeals explicitly to poetry. This vision represents a conjunction of aesthetic and epistemic interest by attempting to depict the world as it would appear to aesthetic judgment alone. The sublime heavens identified by a pure aesthetic judgment can be represented only ("bloß unter dieser Vorstellung . . . [gesetzt]") in the form of an all-encompassing vault, in their barest appearance. Likewise, the ocean must be represented as it would appear to the pure (ostensibly *dis*interested) perception of a poet, "nach dem, was der Augenschein zeigt." The mineral world, hitherto the most prominent source of images for Kant's analysis of the sublime, is eliminated from this revised canon. Kant's choice of poets to stand for immediate intuition operating independently of empirical concepts ironically resembles Boehme's notion of a poetry that abjures the traditional practice of projection and personification. Kant in fact makes a point of *depopulating* these

---

19. "Die Verwunderung, die an Schreck grenzt, das Grausen und der heilige Schauer, welcher den Zuschauer beim Anblick himmelansteigender Gebirgsmassen . . . ergreift, ist . . . nur ein Versuch, uns mit der Einbildungskraft darauf einzulassen, um die Macht eben desselben Vermögens zu fühlen, . . . und so der Natur in uns selbst, mithin auch der außer uns . . . überlegen zu sein"; Kant, *Kritik der Urteilskraft*, 359 (B 117); *Critique of Judgment*, 129. I differ from Pluhar on certain points.

new scenarios, opposing them to landscapes peopled by mythological figures or scientific concepts. Rocks are excluded, it would seem, because they are aesthetically overdetermined, "corrupted" by aesthetic tropes. And yet it is precisely rock that is already barren, lacking the populations ("vernünftige Wesen" and "Wassergeschöpfe") and practices (communication and trade, in the case of the ocean) that aesthetic judgment must eliminate because they suggest a teleology in nature.[20]

Kant finally concedes that aesthetic pleasure in the sublime does require an alloy of sensory perception ("Wohlgefallen" predicated on "Sinneninteresse").[21] His reorganization of perception in this passage and the resulting transformations arise from this necessity of "Sinneninteresse," which Paul de Man terms "a quasi-theological necessity." Despite the emphasis on pure aesthetic judgment in the revised aesthetic protocol, judgment is inevitably corrupted ("verunreinigt") to some degree by the interest of sense. Kant's revision, accordingly, serves to minimize that corruption: he calls for a "purified" sensation that registers only what is etched upon it by pure experience, by mere ocular vision ("was der Augenschein zeigt"). Hence the twofold (geographic and theological) correction here of the previous passage: rocks or "mountain-masses," in their rude immensity, too readily satiate the "interest" of sense. Because they already resist empirical concepts ("allerlei Kenntnisse," the projections of science and myth), they cannot be stripped down and depopulated as the heavens and the ocean can by the Kantian program of reform, the renunciation of excessive sensory interest.

The theological correction similarly imposes a kind of asceticism on perception, pushing it in the direction of intellectual intuition (itself unattainable). The shudder of terror has disappeared from the revised scenario, and with it the sacral ("der heilige Schauer"). The need for an alloy of sense may be a theological one arising from the fallen condition of perception, but the sensuous imagination must limit itself strictly to the most primitive determinations of phenomena, such as distinguishing between rest and motion. Kant subsequently cites the prohibition of graven images as the most sublime passage of Moses, confirming these restrictions on the sensuous imagination.[22] Here he concludes that the "style" of nature in its sublime produc-

---

20. Kant, *Kritik der Urteilskraft*, 360 (B 118–19); *Critique of Judgment*, 130.

21. Kant, *Kritik der Urteilskraft*, 361 (B 120); *Critique of Judgment*, 131. Cf. the "Transcendental Aesthetic" of the First Critique. Kant's notion of the poet as agent of a pure, even dispassionate perception may seem less strange, and more "romantic," by comparison to the poet in Coleridge's "The Nightingale: A Conversation Poem," "to the influxes, / of shapes and sounds and shifting elements / surrendering his whole spirit" (27–29). To perceive the nightingale stripped of its literary connotations, such as "melancholy" and the Philomela myth, requires a defamiliarization like Kant's treatment of the ocean.

22. Kant, *Kritik der Urteilskraft*, 360 (B 119), 365 (B 124); *Critique of Judgment*, 130, 135.

tions is naive simplicity—"Einfalt (kunstlose Zweckmaessigkeit)"—as it is likewise in the "second nature" of the ethical sphere.[23] The revised scenario of heavens and ocean appears in this light as a type of stylistic analysis. In a formulation that stands out from the grammar of the passage, Kant ascribes agency to perception itself: "was der Augenschein zeigt." Rocks lend themselves too readily to the creation of graven images, interfering with the work of pure perception. The "style" of heavens and ocean, interpreted as nearly as possible through the lens of pure sensation, is that of simple exteriority. If the tactile and visual strangeness of rock prohibits Kant from *making* it strange as he does heavens and ocean, this helps to confirm what is for me the primary sense of the passage: that phenomena at the level of materiality are strange to sense.

The use of the terms "phenomena" and "materiality" in this context inevitably alludes to Paul de Man's influential reading of the "General Comment." De Man's linguistic materiality seems far removed from the earth's material. But his reading offers some suggestive parallels for a reading differently invested in materiality. He gives the "quasi-theological necessity" for sensory pleasure as one of several examples of the disruptive role of phenomenal extension in the Kantian sublime. For de Man, Kant's endeavor to account for the phenomenality of the sublime produces a "formal materialism," a system that constitutes by means of linguistic performance the only possible spatial manifestation of the occult power of reason apparently called for by the conflicting ideas of a dynamic sublime on the one hand and extension on the other. He sees Kant's "nonteleological vision of nature" as the core of this performance.[24] Like the depopulated ocean, Kant's example of abstracted body parts illustrates such a vision. For de Man, this dismemberment signals a "disarticulation" of the critical philosophy through a "fragmentation of sentences and propositions into discrete words . . . or finally letters."[25] Kant's contemporaries, however, looked to rocks for the dwindling of legible meaning to its irreducible core. By moving from mountains through the heavens and ocean to body parts, his conclusion on the sublime points toward a corporeal ground of illegibility. De Man draws on Wordsworth to illustrate what kind of "poets" Kant means in his explanation of "dismemberment": we must regard individual limbs "the way the poets look at the oceans severed from their geographical place on earth."[26] On the surface, Wordsworth's fragment "The Sublime and the Beautiful"

23. Kant, *Kritik der Urteilskraft*, 366 (B 126); cf. 387 (B 154); *Critique of Judgment*, 136; cf. 158.
24. De Man, "Phenomenality and Materiality in Kant," 137–38, 136, 143.
25. Kant, *Kritik der Urteilskraft*, 361 (B 119); de Man, "Phenomenality and Materiality in Kant," 144.
26. De Man, "Phenomenality and Materiality in Kant," 135–36, 142.

shows him to be a very different kind of poet, a geographical one; yet de Man's stylistic analysis of Kant provides a lens on subtler affinities between Kant's nature and Wordsworth's.

Wordsworth's analysis of "the sublime as it exists in a mountainous landscape" appeals to a primitive reality attained, like Kant's, by perception purified through a critique of aesthetic conventions. But the experience in Wordsworth derives its alien qualities from the properties of mountains themselves: stability and permanence, physical magnitude, enigmatic and irreducible materiality. In some respects, Wordsworth's positioning of the observer in his "mountainous region" resembles the protocols of the picturesque. Since he must compete with the picturesque writers for territory, both literary and geographic, he anxiously differentiates his own method, which he says will work only "if our minds be not perverted by false theories."[27] Wordsworth stations his observer "so near that the mountain is almost the sole object before our eyes, yet not so near but that the whole of it is visible." (Kant cites René Savary's account of the pyramids to the same effect.)[28] The point of this stationing is that the rocks in question must fill the frame and not appear, as in the picturesque arrangement, "so distant that . . . they are only thought of as the crown of a comprehensive Landscape." Only the proper foregrounding provides the exact recipe for the sublime, as distinct from the simply "grand." The object fills the eye, just as the "consummation of the sublime" will fill the mind with an "image of intense unity" (*Prose*, 2:353–54). The dimensions of Wordsworth's chief example, the Langdale Pikes, acquire a peculiar importance: by excluding the rest of the landscape, the mountain makes it strange, disrupts its composition. While such an effect must depend largely on perspective, what is essential about "the Pikes of Langdale and the black precipice contiguous to them" is that they are large enough to accomplish this without narrowing the range of vision, and that they are composed of naked rock, devoid of "softening circumstances," in Whately's phrase (*OMG*, 93).

By entailing this unfamiliar character on the scene, Wordsworth assimilates the observer here to the "Person unfamiliar with the appearances of a Mountainous Country," for whom the sublime is "more conspicuous" than for a seasoned observer (349–50). The operation of the "law of sublimity" as

---

27. *Prose*, 2:351. Wordsworth specifies at the end of the fragment that these "false theories" are "caprices of vanity and presumption derived from false teachers in the philosophy of fine arts & of taste." In an imagined commentary on the same scene on the shores of Windermere by one of these false teachers, he marks their identity by italicizing the word "picturesque" (360).

28. Kant, *Kritik der Urteilskraft*, 338 (B 87–88); *Critique of Judgment*, 108. The example is ironic in light of Kant's dismissal of aesthetic conventions in the "General Comment."

it is "collected from . . . objects" requires that the forms of rocks be *arresting* to the observer, and a lack of familiarity is the first prerequisite for the primitive sublime here, as opposed to the "awe" and "admiration" that come with familiarity (353). Wordsworth's emphasis on form or outline is more concrete, more empirical, than Gilpin's or Hogarth's: rocky forms can be determined in a fundamental way that precedes the application of concepts, as in the Kantian paradigm of heavens and ocean, and so generate a sense of basic exteriority or primitive reality. Wordsworth's two types of form, sinuous and angular (352), correspond suggestively to Kant's restrictions on the ocean: to the poet's dispassionate vision, it is simply either calm or agitated. Wordsworth, like Kant, offers a reproducible technique of defamiliarization, but differs in his approach to the unrepresentable. For Kant, who wants to minimize the role of sensation, mountains themselves are too saturated with convention to represent the placid sovereignty of reason triggered by the sublime. For Wordsworth, it is merely a matter of banishing the *wrong* conventions (that is, the picturesque) in order to apprehend unrepresentable but fundamentally exterior phenomena. "Form" in this sense most immediately causes the "sensation of sublimity," which is "found to resolve itself into three component parts: a sense of individual form or forms; a sense of duration; and a sense of power" (351).

Wordsworth's poetry offers many additional instances of the agency of mountains and other geological objects that seem to derive their power from materiality itself, like the stones at Kirkstone Pass and the cliffs in the 1799 lyric on the "Influence of Natural Objects." The peculiar agency of the hills, stars, and cliffs in the more famous version of this poem (1805 *Prelude* I.428–89) is perhaps the least conventional aspect of its description of winter pastimes. The "distant hills" send forth the "alien sound / Of melancholy" (470–71), and a sense of the earth's rotation comes through in "the solitary cliffs / Wheeled by me" (484–85), lines that translate the gravitational force of planetary mass into the bodily idiom of dizziness. A narrative of boyhood from *The Excursion*, more nearly contemporaneous with the "Sublime and Beautiful" fragment (ca. 1812), complicates the relationship between feelings and the objects of an "unproductive . . . rugged" landscape (I.110):

> In such communion, not from terror free,
> While yet a child, and long before his time,
> Had he perceived the presence and the power
> Of greatness; and deep feelings had impressed
> So vividly great objects that they lay
> Upon his mind like substances, whose presence
> Perplexed the bodily sense.
>
> (133–39)

The departicularized description here accentuates the power these "objects" have through their bare existence: it is enough that they are large and barren. At the same time, the reciprocal "defacement" of setting and (auto)biographical narrative call to mind de Man's analysis of prosopopoeia. This hint of reciprocal estrangement unites the Ode's "wrinkled Egyptian monument" with the Medusalike power of Novalis' petrifying sublime and the arresting forms of the Langdale Pikes with the "sinister" connotations of the epitaphic injunction, "Pause, Traveler!"[29]

The mere physicality of this and other Wordsworthian landscapes captures the interest pertaining to raw materials. The centrality of these concerns leads him, in a letter of 1805, to make the uncharacteristic argument that landscaping carries greater accountability than the other arts: "if this be so when we are merely putting together words or colours, how much more ought the feeling to prevail when we are in the midst of the realities of things" (*Guide*, 144). In identifying these realities as hills, streams, and trees, Wordsworth unconsciously echoes Thomas Whately's vindication of landscape architecture as being "as superior to landskip painting, as a reality to a representation" (*OMG*, 1). This connection underscores the durability of Whately's paradigm, of the seminal cultural moment in which the substance of landscape became identified with reality. This identification is reinforced in the ensuing decades as raw materials increase in social importance. The progress of the "improvement" that Whately endorses in 1770, extending to agriculture as well as to manufacturing, also reinforces the identification of these materials, especially rocks, as "rude" or unformed, a quality prerequisite to all the scenarios of primitive reality considered so far.

## Chaos and Rude Forms in Rocks

The otherness of rock finds expression in the stubborn, enigmatic materiality, the alien exteriority, developed in literary figures of the earth's material. By virtue of this otherness, rock stands as the dominant instance of the primitive reality of things, of the illegible substance of nature. The aesthetic is the sphere designated for the creative approximation of preconscious, prelinguistic experience, of the physical content of sensations, such as the tactile solidity and resistance of a stone. It is primarily within this sphere,

---

29. Paul de Man, *Rhetoric of Romanticism*, 78. I am indebted to Brad Prager for these associations, which trouble the Romantic insistence on a distinction between unhewn and hewn (or graven) stones.

then, that an illegible substance of nature can be posited. During the heyday of natural history, scientifically oriented discourse competed directly with literary explanations of natural phenomena, and there are also instances of legible rock in poetry and aesthetics. Before turning to the pervasive imagery of the rock record, let us consider what may be the most famous poetic meditation on geological otherness or unreadability. Percy Shelley's "Mont Blanc," in its complex context of Alpine literature and philosophical discourse, augments the absolute otherness attributed to rocks with a dynamic agency. Shelley's explanation of formlessness—that it is produced by active deformation—also competes with geological explanations, but preserves the "extatic wonder, not unallied to madness" attached to such chaotic scenes as the Alps.[30]

"Mont Blanc" (1816) displays the three component parts of Wordsworth's mountain sublime of 1812—power, duration, and individual form. The ambiguity surrounding "Power" in Shelley's poem readily indicates his differing treatment of these materials. In the sonnet "To Wordsworth" (1814), Shelley laments that the "poet of Nature" no longer stands like a "rock-built refuge" for "truth and liberty," and Mary Shelley comments, responding to *The Excursion*, "he is a slave."[31] Indeed, the "power" of the mountain's "voice . . . to repeal / Large codes of fraud and woe" (lines 80–81) seems a projection of the liberating voice that Wordsworth, according to the sonnet, has lost. Despite many obvious differences between the poets, there is a substantive generic connection between "Mont Blanc" and the Kirkstone Pass ode. Both are embedded in coauthored travel narratives and engaged in dialogue with the prose of both the poet himself and his coauthor (though Dorothy's role remains unacknowledged in the *Guide*). "Mont Blanc" concludes *History of a Six Weeks' Tour,* just as the ode concludes *A Guide to the Lakes,* by dramatizing the aesthetic experience of the completed tour through the privileged example of "rude," formless rocks. Many readers of Shelley's poem have addressed its relationship to Wordsworth, emphasized particularly by Robert Brinkley in his discussion of the poem's three major cruxes: the relationship between mind, Power, and the "universe of things" at the beginning, the meaning of "but for such faith" in the middle of the poem, and the thrust of its concluding question.[32] The recuperative move-

30. Shelley, *Complete Works*, 6:137.

31. *Shelley's Poetry and Prose*, ed. Fraistat and Reiman, 92. The editors quote Mary Shelley's journal entry of September 14, 1814, in their note to this sonnet.

32. Brinkley, analyzing the layers of revision preserved in manuscript, argues that Shelley's revisions "echo Wordsworth, both Tintern Abbey and the Intimations Ode, so that epistemology becomes an expression that is pervaded by allusion" ("Spaces between Words," 247). Phrases such as "flood of ruin," however, also allude to the important generic context of Alpine

ments concluding sections III and IV—the beneficent "majestic River" (123) and especially the promising if problematic "faith so mild" (77)—have received more critical attention than the description of destructive geological agencies in which they are embedded.[33]

Shelley's account of form and deformation in these descriptions marks a salient difference from Wordsworth. This difference is partly a product of differences between the Alps and the Lake District. It is also due, however, to Shelley's adaptation in "Mont Blanc" of the old trope of mountains as ruins, as evidence of violent upheaval in the geologic past. The resulting sense of primeval chaos in the poem helps to distinguish it from other Romantic descriptions of such scenery. Shelley's notion of destruction as a dynamic process operating on the rocks stands opposed to the suspended animation suggested by the "stationary blasts" of Wordsworth's account of the Simplon Pass (*Prelude* VI) and other such images. What appears in "The Sublime and the Beautiful" simply as "abrupt lines" appears in Shelley as a rude and ravaged formlessness that becomes the chief sign of geologic otherness in "Mont Blanc." The poem builds its emphasis on deformation in a way that recalls two other Wordsworthian moments. In the manuscript, Shelley twice uses and then cancels an image shared with the Kirkstone Pass ode, the idea that landforms "mock" "human handy-work." In one canceled line, "Frost . . . makes mockery of human forms / Columns & pinnacle & pyramid"; elsewhere it "mocks human art."[34] Since Shelley minimizes this trope in his final text, giving ice and rock a more autonomous destructive agency, it is tempting to think of Wordsworth's ode (1817) as a rebuttal. Shelley's responses to Wordsworth, however, are better documented, and a related scene from *The Excursion* may have influenced Shelley's draft. In this scene, the Solitary identifies a "mass of rock" that seems "rudely to mock the works of toiling Man," and he points to individual examples resembling pyramids and columns—the same architectural metaphors that Shelley strikes from the lines quoted above.[35] In his printed text, these metaphors are deferred to section IV, reserved for the glacier's direct interaction with

---

literature going back to Rousseau and beyond. Some critics see Shelley as rejecting Wordsworth along with the bulk of this tradition and its conventional piety. See, for example, Jerrold Hogle, *Shelley's Process*, 78.

33. See, for example, Nigel Leask, "Mont Blanc's Mysterious Voice," 196; and Earl Wasserman, *Shelley*, 235, 238.

34. See lines 63–71, Bodleian Ms. Shelley adds. e. 16, as reproduced in Brinkley, "Spaces between Words," 258.

35. Wordsworth, *The Excursion* III.127; see also 149, 131. The architectural imagery is first introduced and canceled in line 60 of the draft (Brinkley, "Spaces between Words," 253), then in the canceled line quoted in my text, and finally appears in line 104, where it stands. "Column" is not restored to this version, but intriguingly appears in place of "law" in line 141 of the fair copy in the Scrope Davies Notebook (*Poetry and Prose*, ed. Fraistat and Reiman, 101n.).

human "work and dwelling." This deferral considerably heightens the poem's sense of dynamic geological process and nonhuman agency.

Mont Blanc in Shelley's scene is not just resistant to form, but a vigorous agent of its destruction. Its "subject mountains their unearthly forms / Pile around it, ice and rock," forming a "desart peopled by the storms alone" (62–67):

> —how hideously
> Its shapes are heaped around! rude, bare, and high,
> Ghastly, and scarred, and riven.
>
> (69–71)

The scene is depopulated, like the Kantian scenario, and, but for the suggestion of Power, devoid of teleology. Otherness here finds expression in the paradox of "unearthly" geologic form. In the journal-letter to Thomas Love Peacock from which Shelley draws many of the poem's images, he writes that the ice forms "pierce the clouds like things not belonging to this earth," echoing an image of Mary's in her narrative of their 1814 tour printed in the same volume: "it requires an effort of the understanding to believe that [the Alps] indeed form a part of the earth."[36] The poem's Alps assume the character of an alien landscape above all through their "shapes," whose violent disorder intrudes into the verse: following the strong caesura of line 70 ("around! | | rude"), the iambic basis of the meter is suspended. A spondaic foot follows the caesura, and the line is artificially end-stopped by the trochaic foot ("ghastly") that begins the next line. The imitative, "inorganic" form of these lines explicitly links poetic and geological form, creating a way for the poem to transvalue the conventional catalogue of sublime Alpine features. The journal-letter's description of Shelley's first view of Mont Blanc provides a clue to his ambition here: "Nature was the poet, whose harmony held out spirits more breathless than that of the divinest."[37] The poem trumps other such claims by the sheer energy of its deformation, using strange and disorderly materials to challenge other representations and disrupt the personification of nature.

"Mont Blanc" also departs forcefully from poetic conventions governing the depiction of violent environmental upheaval, drawing instead on natural history. Shelley's apocalyptic vision in section IV seems to compete consciously with stale eighteenth-century conventions—such as the "frozen horror" of the tourist on the precipice—substantially raising the stakes of such an encounter. The tide of destruction derives much of its force from

---

36. Mary and Percy Shelley, *History of a Six Weeks' Tour* (1817), in Shelley, *Complete Works*, 6:141, 102.

37. Shelley, *Complete Works*, 6:137.

Shelley's naturalistic observation that glaciers do in fact move, or "creep / Like snakes that watch their prey, from their far fountains, / Slow rolling on" (100–102). At the glacier's source stands a "city of death," or rather, Shelley corrects himself, "a flood of ruin":

> Vast pines are strewing
> Its destined path, or in the mangled soil
> Branchless and shattered stand.
>                                                     (105–11)[38]

The violence escalates and extends to the animal and human kingdoms (114–200):

> The dwelling-place
> Of insects, beasts, and birds, becomes its spoil;
> Their food and their retreat for ever gone,
> So much of life and joy is lost. The race
> Of man, flies far in dread; his work and dwelling
> Vanish, like smoke before the tempest's stream,
> And their place is not known.

Three motivations for this apocalyptic dénouement can be quickly established. The structural parallel between the remoteness of Power from "all things that move and breathe" (94) and the glacier's indifference to the organic sphere suggests that the progress of the glacier is mapped out as another illustration of Power.[39] But the natural history of the glacier itself, augmented by observation (a second motivation), also provides inspiration for the very idea of an indifferent Power—or so it would seem from Shelley's Wordsworthian headnote to the poem. Finally, the temper of the times is apocalyptic. Wordsworth locates a more Hebraic "great Apocalypse" in a nearby landscape; Mary Shelley's *The Last Man* and Byron's "Darkness" provide more germane analogues, and geology's apocalyptic tendencies in the second and third decades of the nineteenth century are perhaps most germane of all. The revival of catastrophist, diluvialist science in the wake of Georges Cuvier's influential paleontology provides the geological context

---

38. Cf. the letter to Peacock (ibid., 139) and the draft's "A ~~city mountain~~ city of death" (l. 105).

39. This would be the familiar logic of philosophically oriented readings of the poem. But Shelley's Power is as much local and telluric as it is a quasi-platonic "Principle of Permanence," in Charles H. Vivian's phrase (see *Shelley's Poetry and Prose*, ed. Reiman and Powers, 571). From a geographic perspective, the epistemological apparatus introduced by Earl Wasserman, to prove that "reality is . . . not something given *ab extra*" in the poem, seems an unnecessarily labored attempt to navigate the mountain's exteriority (*Shelley*, 227).

here. Cuvier's English disciples included Byron, William Buckland, and James Parkinson, whose works Shelley owned. Shelley also shares his most dramatic biblical allusion—"their place is not known" (120)—with Thomas Burnet, one of the earliest catastrophists.[40]

Shelley himself invokes this long-standing connection between apocalyptic scenarios and empirical natural history in the journal-letter to Peacock, drafted, like the poem, during the week of July 21, 1816. This stylistically self-conscious letter takes naturalistic description as its object. Having been "warned" by "the raptures of the travellers," Shelley declares, he will "simply detail to [Peacock] all that [he] can relate," rather than "exhaust the epithets which express astonishment." This complex rhetorical stance leads Shelley into the precincts of natural philosophy as well as natural history (and occasionally back into the pitfalls of rapture). A more frequently noted protestation from the letter locates Shelley's science in relation to Buffon's *Histoire naturelle:* "I will not pursue Buffon's sublime but gloomy theory— that this globe which we inhabit will at some future period be changed to a mass of frost" by the encroachment of polar ice and Alpine glaciers.[41] He does seem to pursue the theory in the poem, where it accompanies images from the letter and informs the climactic destruction of human habitats (lines 118–19). The poem's geology, as amplified by the letter, can be read at least two ways. On the one hand, Shelley rejects the currently accepted theory of H. B. de Saussure in favor of one closer to Buffon's by-now-archaic armchair theorizing. This lack of scientific ambition seems confirmed by Shelley's preface to *History of a Six Weeks' Tour* (which includes the letter as well as the poem): "few facts relating to [the Alps] can be expected to have escaped the many more experienced and exact observers, who have sent their journals to the press."[42] This stability of facts provides a pretext for preferring the earlier theory, which lends itself better to an apocalyptic vision. On the other hand, original description is an equally pressing problem: hence the letter's project of observation, and hence, I will argue, a second way of reading the poem's geology. "Mont Blanc" turns in spite of itself to an unobtrusively original natural history of the mountain and especially the glacier. In both the letter and the poem, this natural history provides a sustained focus, harnessing both direct observation and a more current knowledge of geomorphology to the formal project of rejuvenating spectacular description.

40. Burnet alludes to the same passage of Psalm 103 in *The Sacred Theory of the Earth*, 306. Buckland was the first professor of geology at Oxford, whose ideological distance from Byron should indicate that catastrophism was politically adaptable.

41. Shelley, *Complete Works*, 6:134, 140.

42. Ibid., 6:139–140, 87.

Buffon's theory—though the letter proposes *not* to pursue it—enables not only certain images in the poem but also the letter's more empirical observations about the glacier. Shelley has both literary and scientific reasons to insist on the glacier's relentless advance, as against Saussure's theory—later proved correct—of alternating advance and retreat: an inexorable glacier embodies the power of deformation so central to the poem's vision of the earth's material, and it confirms Shelley's own careful observations, leading him to connect Alpine and continental glaciation as no geologist had yet done. Saussure and others had noticed the power of Alpine glaciers to transport material (stressed by Shelley in the letter and lines 109–20 of the poem), while Buffon had theorized a global ice age without basing it on glacial transport or other observed Alpine phenomena. Shelley, unworried by Buffon's lack of empirical rigor, bases a global phenomenon (especially in lines 112, 120) on his own local observations, anticipating a development in geology that began only with Louis Agassiz's work in the late 1830s.[43] The point here is not that Shelley was "trying" to do science—he wasn't—but that the work of the poem, as it was then conceived, encompassed what we might consider a specifically scientific contribution.

The letter to Peacock makes it clearer exactly what Shelley observed, while the manuscript revisions help to clarify how these observations shape the poem. Adhering for the moment to his self-appointed task of neutral description, Shelley tells Peacock of his visit to the Glacier de Boisson (or Bossons) on the afternoon of July 23. "These glaciers flow perpetually into the valley," he observes, "ravaging in their slow but irresistible progress the pastures and the forests which surround them, performing a work of desolation in ages, which a river of lava might accomplish in an hour, but far more irretrievably; for where the ice has once descended, the hardiest plant refuses to grow." Shelley supports these inferences with an account of the rate and mechanics of the glacier's movement and another observation of its power to transport "enormous rocks [up to forty cubic feet], and immense accumulations of sand and stones." His more figurative account of the "work of desolation" also offers the suggestive comparison between glacial ice and lava, related to the poem's question whether an earthquake or volcano created the landscape around Mont Blanc (71–74). Shelley leaves the glacier convinced that "in these regions, everything changes, and is in mo-

43. Agassiz developed the first systematic and influential theory of continental glaciation, based unlike Shelley's or Buffon's on the deduction from field evidence of ancient episodes of glaciation. See Gordon Davies, *Earth in Decay*, 264–65. Saussure's theory, adopted by Playfair and others, does not extrapolate from Alpine to continental glaciation. Patrick Vincent has pointed out that in 1816 the Chamonix glacier was approaching a 200-year peak in its growth ("What the Mountain Should Have Said").

tion."[44] While his description also alludes to the trope of frozen torrents, the naturalist's perception of underlying dynamism contradicts and ultimately supplants the static tropes inherited from "the raptures of the travellers." Published narratives by two of Shelley's contemporaries, who followed exactly the same itinerary a year later, differ on the extent and destructiveness of the glacier's agency. Thomas Raffles, though full of the pious raptures Shelley despised, is curiously close to him, especially to the poem, in his accounts of such "tremendous agents of destruction." Louis Simond, whose intellectually progressive persona more closely resembles Shelley's, endorses Saussure against Buffon and recommends lighting wood fires around the glaciers to stop their merely "accidental encroachments": "This seems so obvious, and the experiment so easily made, that it is a matter of surprise it has not occurred to these poor people looking on in stupid despair."[45]

Shelley works his emphasis on dynamic geological process into the poem in a series of careful revisions. Under the influence of a long tradition in topographical writing, he had originally incorporated conventional architectural metaphors into five lines of the first sustained description of Mont Blanc and its "subject mountains" (60–71 in printed text). In the manuscript, two of these lines are struck out, then repeated, then struck out again: the sequence "Columns & pinnacle & pyramid" and the suggestion that "Frost here makes mockery of human forms."[46] Shelley systematically deleted all these references to finished, intentional forms in order to emphasize the process of deformation, and perhaps to resist the humanizing tendency of these standard tropes, especially as reactivated by Wordsworth's *Excursion.* The printed text reintroduces these images in a later section of the poem for the distinct purpose of describing the mountain's relation to "mortal power" (103–4)—a relation that is precisely one of negation rather than aestheticizing homology.

Shelley's emphasis on the power of geological deformation in shaping the landscape of Mont Blanc opens onto numerous contexts, including aesthetic, political, and theological engagements along with the formal and sci-

44. Shelley, *Complete Works,* 6:139, 141.

45. Raffles, *Letters during a Tour,* 197; Simond, *Switzerland,* 1:295–96. Sounding more Huttonian than Shelley, Simond insists that "it does not follow . . . that the encroachment will be permanent, for . . . the principle of dissolution will ever be found commensurate," and points to compelling evidence in the form of wooded terminal moraines far beyond the glacier's current edge (294). Raffles, however, notes that the "principal village" in this valley had been completely buried by an avalanche 200 years before (*Letters during a Tour,* 197). Concerning the Glacier de Bossons, he simply observes, "There is too much of death in it ever to awaken pleasing sensations" (190). I am indebted to Patrick Vincent for recommending these sources.

46. I rely again on Brinkley's transcription of the manuscript in "Spaces between Words."

entific issues discussed so far. The tone of the letter to Peacock shifts as Shelley moves from science to myth in his discussion of Buffon, and then changes sharply as he falls at the end into a polemic against the tourist industry. Nigel Leask has shown how Shelley's resistance to this "commercialised sublimity" depends on his rejection of pious, conservative Alpine lyrics—especially Coleridge's "Hymn before Sunrise"—and his interest in Huttonian earth science.[47] For Leask, "the poem's difficult task is to win back the interpretation of the natural sublime from myth-making poets, pious tourists, and geological catastrophists." Relying on John Playfair's popularizing interpretation of Hutton's *Theory of the Earth*, Leask suggests that Shelley adopts both the "self-regulating . . . economy" of nature and the enlightened critique of superstition characterizing this new geology and late Enlightenment rationalism more generally.[48] While Leask points up some evidence of such an interest, particularly in the image of the river as "breath and blood of distant lands" ("Mont Blanc," 123–26), his argument is most compelling in its political and theological aspect. He points out that by the time of the Restoration in France, Mont Blanc's "cultural significance was predominately fideistic and legitimist" in the manner anticipated by Coleridge's "Hymn." To challenge the scientific and aesthetic doctrine of divinely ordained catastrophe is therefore to propose "an anti-catastrophist politics of reconstruction, challenging the prevalent post-Napoleonic castigation of revolution and revolutionary doctrines of perfectibility." Shelley's insistence on the mountain's otherness, his naturalistic production of an inhuman spectacle, prohibits an unqualified acceptance of Leask's suggestion that the poem "finds a positive meaning in the inexorable and destructive agencies of ice and water." But his strong emphasis on Huttonian geology has the advantage of illuminating the large political stakes in Shelley's representation of the mountain *qua* mountain. In the journal-letter, this complex cultural politics—in an account of the mountain's *social* landscape, absent from the poem—also informs the discussion of Buffon, whose "wild fictions" Shelley does not simply reject with Playfair.[49]

47. Leask, "Mont Blanc's Mysterious Voice," 184, 182 ff. Simond estimates that as of 1817, between 4,000 and 5,000 tourists are visiting Chamonix each summer; *Switzerland*, 1:308.

48. Leask, "Mont Blanc's Mysterious Voice," 196, 188.

49. Ibid., 185, 202, 196. Leask overstates Shelley's endorsement of Playfair (195) and, in general, of the "view from the top" supplied by modern science since H. B. de Saussure's pioneering ascent of the mountain in 1786 (186). Shelley no more "endorses" Playfair than Buffon, nor is the use of Buffon entirely "flippant," since as Leask himself notes (195), Shelley uses Buffon indirectly to contradict Saussure (*Complete Works*, 6:139–40). Cf. Erasmus Darwin, *BG* I.ii.13–20, as well as Simond's sympathetic argument that the appearance of stratification in Chamonix is a "fatal blow" to Hutton's theory (*Switzerland*, 1:309).

In this letter Shelley famously characterizes the inhabitants of the Alpine landscape as "half deformed or idiotic." This identification surprises, not least because local opinion about glaciers seems to confirm Buffon's "sublime but gloomy theory," and Shelley endorses that opinion against Saussure's. Nonetheless, the repugnant fascination of goiter-afflicted mountaineers forms a fine but discernible thread running through *History of a Six Weeks' Tour,* as well as a topos informing Simond's and Raffles' ambivalence toward the natives of Chamonix. Shelley half-jokingly presents the landscape to Peacock as evidence for the "supremacy of Ahriman . . . above all these deadly glaciers, at once the proofs and symbols of his reign;—add to this, the degradation of the human species—who in these regions are half deformed and idiotic." Conventionally, local hardships such as the glaciers were represented as instilling republican virtue, and critics have read even the Ahriman scenario as compatible with republican sentiment.[50] But other passages in *History of a Six Weeks' Tour* make it clear that much of the Shelleys' travel in the Alps was conceived as a literary pilgrimage, not at all consistent in its pursuit of the ideals—democracy, philanthropy, atheism— recorded by Shelley in the guest book of the Hôtel de Londres. Looking back on the 1814 journey in his preface, Shelley self-consciously places Mary's narrative in the shadow of *Childe Harold,* canto III.[51] Another passage details Shelley's and Byron's June 1816 tour of the sites featured in Rousseau's *Julie,* elevating the novel itself over any intrinsic local interest. In this passage, Shelley limits his sociological observations to the area's children, who are "in an extraordinary way deformed and diseased. Most of them [are] crooked, with enlarged throats." While he attributes the particular wretchedness of Evian's inhabitants, still subjects of the King of Sardinia, to the evils of despotism, generally Shelley's politics are more ambivalent than his homage to Rousseau and the early Wordsworth might suggest. He fails to mention that Chamonix itself—the geographical focus of the poem and of much of the prose—never was Swiss territory, a distinction politicized by numerous English writers in their contrasts between the free and virtuous Swiss and the "slovenly" Catholic Savoyards. The poem's suggestion that "the race / Of man flies far in dread" before the glacier (117–18) reiterates the journal-letter's connection between "deadly glaciers" and afflicted inhabitants, and together these two passages articulate a resistance not only to the idealized landscape of Shelley's predecessors, but also to

50. Shelley, *Complete Works,* 6:139–40. Hogle reads this passage as "dethron[ing] all . . . supreme beings" (*Shelley's Process,* 85). Leask points out that "the mountain had formerly represented a potent symbol for the revolutionary followers of Rousseau (including the young William Wordsworth)" ("Mont Blanc's Mysterious Voice," 185).

51. Shelley, *Complete Works,* 6:87.

their idealized Swiss mountaineers, supposedly imbued with a heroic instinct for liberty.[52]

Shelley's reaction to the people of the Alps, in fact, seems more of a piece with Thomas Gray's reaction to "the Alpine monsters" in 1739: "The creatures that inhabit them are, in all respects, below humanity."[53] Shelley's letter reacts to the otherness of Mont Blanc's social landscape through a rhetoric of domestication that persists as a subtext of the poem and resonates in other passages of *History of a Six Weeks' Tour* as well as in *Frankenstein* (the story of another type of "Alpine monster"). The poem draws only on the letter's geological images of deformation and deformity, subsuming cultural otherness, while the letter domesticates it explicitly. During their first visit in 1814, Mary Shelley reports, the "brutal rudeness" of some "exceedingly disgusting" Swiss fellow-passengers "provoked S—— to knock one of the foremost down." Percy's later assertion that the natives are "deprived of anything that can excite interest or admiration," while less demonstrative, seems a by-product of tension between the lexicon of wonder and the mandate of description. Shelley in this letter insists simultaneously that the landscape is alien, "not belonging to this earth," and domestic, having tourist institutions "just as at Keswick, Matlock, and Clifton" and, even in geological terms, "differ[ing] from that of Matlock in little else than in the immensity of its proportions, and in its untameable, inaccessible solitude."[54] His suspicion of tourism leads Shelley, surprisingly, to minimize the geological distinctions so fervently celebrated earlier in the same letter, and above all in the poem. Matlock can also be specifically linked through its lead mining to Servoz, where Shelley observed lead mines on the way to Mont Blanc. Mary

---

52. Ibid., 6:131, 125, 126. Early English examples include James Thomson's *Winter* 414–23 and Milton's sonnet "On the Late Massacre in Piemont." The Rousseauvian accounts include George Keate, *The Alps*, 11–16; the Alpine episodes in Ann Radcliffe's *The Romance of the Forest* (1791); and *Childe Harold's Pilgrimage* III.62–76. The contemporary accounts of Raffles and Simond are instructive here as well. Unlike Shelley, both record their crossing of the border between Switzerland and Savoy (formerly Sardinian, now French) in a variety of symbolic ways. Simond—not English, but certainly Protestant—is the more hostile of the two, comparing the locals unfavorably with the Swiss on many accounts: they are more intimidating to tourists, their poverty is more squalid (*Switzerland*, 1:289–90), their chalets are "vastly inferior" (299), and their cheesemaking is more "slovenly" (304), to name a few. At the same time, this valley produced the "heroes" who guided Saussure on the first instrumented ascent of Mont Blanc (299), also an occasion for much praise in Raffles' more forgiving account (*Letters during a Tour*, 196; cf. 214–16). Saussure himself, according to Kant, reported that "the good and otherwise sensible Savoyard peasant did not hesitate to call anyone a fool who fancies glaciered mountains" (*Critique of Judgment*, 124). Thanks again to Professor Vincent for guidance on these issues.

53. *Correspondence of Gray*, 1:260–61.

54. Shelley, *Complete Works*, 6:105, 140, 141, 142, 135.

Shelley notes that he actually toured one of these mines, and this fact, together with his analysis of the "quackery" of local curiosity merchants, *aubergistes*, and guides, suggests a significant engagement with socioeconomic organization that is largely absent from the poem. Mary herself stresses the locals' resourceful adaptation to the rigors of the Alpine winter, in a telling contrast to the imminent threat to "[man's] work and dwelling" posed by the landscape in the poem (118). Their tactics range from collecting government bounties on wolves to working seasonally as hotel porters in Paris, while those who stay at least enjoy the benefit of "cheap firing."[55] Why is it, then, that Percy Shelley translates such resourceful and independent, if not idyllic, citizens simultaneously into monsters, people rather like the inhabitants of Matlock, Derbyshire, and (in the poem) mere ciphers of the glacier's destructive power?

Both Swissness and Englishness are, from a certain point of view, monstrous literary and ideological constructions that can be purged by the power of geological agency. Their republican elements, Shelley's writing suggests, are negatively counterbalanced by their conventionality. Shelley's "othering" of the Swiss (and Savoyards) is of a piece with his domestication of the landscape via mining and tourism, since Servoz and Matlock resemble each other in negative respects, such as their corrupt tourist industries. These cultural questions about the natives—to judge from the poem's selective reliance on the letter—seem to interfere with Shelley's vision of the landscape. This interference, however, represents a fundamental literary problem: the burden of the past here takes the form of idealized representations of the Swiss, representations that are suspect because of their association with an English "liberty" that Shelley had always questioned. The monstrosity of the Savoyards—as opposed to the Swiss, a distinction Shelley seems deliberately to collapse—reflects the distortion inherent in the ideology of English liberty, which in turn is one of the "codes of fraud" the mountain's voice has the power to "repeal." Yet it repeals them in a violent, destructive way, enveloping the local population and demolishing form and ideology altogether. In the poem, geological deformation supplies the final instance of otherness and gives Shelley the language to carve out new space in the well-traveled terrain of Alpine writing. The consequence, though, is a highly problematic idealism. Leask, insisting that Shelley endorses Hutton-

---

55. Ibid., 6:136, 142; Mary Shelley, *Journals*, 1:119–20. Mary also learns from the guide, Ducree, that "the women do almost all the work such as reaping making hay &c," while the men serve as guides when possible, often accumulating sufficient fees to save for winter. Characteristically, Simond shares Percy Shelley's skepticism toward the tourist industry (*Switzerland*, 1:306) while Raffles is more credulous (*Letters during a Tour*, 217).

ian geology in order to find "positive meaning" in "destructive agencies," goes on to read *Frankenstein* as a critique of this positivism.[56] Such a critique is certainly present in the novel, but Percy Shelley's dalliance with positivism and Mary's evident objections to it are only part of the story.

Shelley drafted "Mont Blanc" on the same excursion, probably largely on the same day, July 24, on which Mary first mentions working on her "story."[57] The novel shares the poem's interest in Alpine mountains as a privileged locus of materiality. In one crucial passage, these rocks become touchstones for the otherness of Frankenstein's creature. When Victor Frankenstein returns to the Alps, the source in a loosely Wordsworthian sense of his ideals and sublime ambition, the Creature confronts him there with the fact that he is bound—to his creation if not (Prometheuslike) to the mountain itself. Remarkably, the Creature is *at home* in the Alps, proof in itself of his monstrosity: Victor's first hint of a "superhuman" presence is the fact that "he bounded over the crevices in the ice, among which I had walked with caution."[58] Victor has crossed the Mer de Glace to look back at Montanvert and Mont Blanc towering above it, a vantage point to which the Shelleys themselves attempted to climb on July 24, 1816. Mary Shelley's brief account presents the evening spent writing as an enforced consequence of the rain that drove them back when they had ascended halfway. In describing the next day's successful ascent, Percy draws on their shared image of "unearthly forms," a paradox transferred to the Creature when Victor notices his "unearthly ugliness."[59]

The parallel to the poem's ambivalent Alpine revisionism, then, lies in the fact that Frankenstein is inspired by the idealized sublimity of the Alps to produce a terrifying other who is truly at home there—on "another earth," as he observes shortly before the metaphor becomes a reality, "the habitation of another race of beings." Natural and human otherness reinforce each other. Considered as pure materiality, the mountain's aestheticized otherness becomes threatening. Similarly, Frankenstein responds to his creature here in terms of unformed (though animated) matter, as "a filthy mass that moved and talked." Raffles' religious inflection of the trope of "unearthly" landscape sheds light on Frankenstein's disappointment as well as Percy's narrative: "there was nothing . . . at all terrestrial" about the scene, Raffles reports, "and imagination peopled it with beings pure as its spotless soil"—

56. Leask, "Mont Blanc's Mysterious Voice," 199.
57. M. Shelley, *Journals*, 1:118 and n.
58. Mary Shelley, *Frankenstein*, 127; see also 125 on the Alps and youthful ambition. Thanks to Brad Prager for the suggestion concerning Prometheus ancient and modern.
59. M. Shelley, *Journals*, 1:118; Shelley, *Complete Works*, 6:141; M. Shelley, *Frankenstein*, 127.

beings rather unlike the "disgusting" spectacle of local people with goiters.[60] For the Creature, sublime mountains and glaciers are mere objects of utility, his "refuge," demonstrating the superhuman powers that place him over against the abject inhabitants of the poem's glacial landscape.[61] Mary Shelley, like Percy, restores the mountain's otherness to displace the connotations of orthodoxy attached to it. But as Fred Randel suggests, the Creature's appeal for a mate in the hut on Montanvert also gives Frankenstein a choice between sterile otherness and generative "sympathy."[62]

The association between rocks and materiality as such is common to the novel and the poem, and powerful enough to be noticed in readings not particularly attuned to the geological context. Christoph Bode notes the poem's evocation of "unalloyed materiality," and Frances Ferguson calls Shelley's mountain "the ultimate example of materiality."[63] Jerrold Hogle suggestively assimilates the mountain to a human other in reading it as a figure for the "primal conditions" of birth and death.[64] For Ferguson, however, "Mont Blanc" is "a poem about the impossibility of seeing the mountain as alien," in the same sense in which we can see the poem's "letters . . . as an example of the materiality of language but cannot see them as language without seeing them as implying something more than matter." Thus the attempt to think the mountain without a signified, "without metaphysical attributes, fails," and the poem "merely inverts" the "myth of natural religion that is attached to Mont Blanc." The "deformity and idiocy" of the Alpine population "merely provide, in human form, a repetition of the mountain's role of pure materiality." In Ferguson's view, Shelley treats this materiality as the antithesis of the sublime and thus "endorse[s]" the Kantian sublime in order to "domesticate the material world for the purposes of aesthetics." The poem, then, deliberately "mistak[es] the activity of the material world for agency, by taking it to be as intentional as any human activity might be." I question what seem to be the crucial presuppositions of this argument: that language is the privileged form of physicality; that "nature needs us in order to be perceived as destructive" or sublime; and above all that agency is by definition human.[65] Shelley elaborates a geological agency of deformation in order to question and rejuvenate the sublime, shifting the environmental upheaval associated with the Alps from the cosmological past into the pres-

60. M. Shelley, *Frankenstein*, 123, 174; Raffles, *Letters during a Tour*, 185, 226.
61. M. Shelley, *Frankenstein*, 128, 174.
62. "The effect is to displace a religious experience with a sympathetic one"; Fred Randel, "*Frankenstein*, Feminism, and the Intertextuality of Mountains," 527–28.
63. Bode, "Kantian Sublime," 343; Ferguson, "Shelley's *Mont Blanc*," 202.
64. Hogle, *Shelley's Process*, 77.
65. Ferguson, "Shelley's *Mont Blanc*," 204, 211, 204, 203, 209, 213, 210.

ent. Shelley's glaciers are nonhuman agents of violent deformation, rather than legible (or illegible) monuments of it. The poem approaches Mont Blanc as "matter" manifestly *not* "designed by its perceiver"; doubt is required as much as faith for "man" to "be 'reconciled,'" as Hogle puts it, "with the 'nature' that is actually perceived (ll. 78–79)."[66] The famous concluding question to the mountain—"And what were thou, and earth, and stars, and sea, / If to the human mind's imaginings / Silence and solitude were vacancy"—cannot be, from this point of view, purely rhetorical. The question about this question—whether it is rhetorical or not—remains a major crux for readers of the poem because, like the philosophical chestnut about the tree falling in the forest, it presents an antinomy.[67]

Shelley embraces both elements of earlier natural history and geology's more recent classification of the Alps as "primitive" or original mountains in his response to "the naked countenance of earth / . . . even these primaeval mountains" (98–99), but rejects the religious content of all these representations.[68] Differences arise on the question of what such landforms—putatively unaltered since the creation, glimpses of the earth's "naked" materiality—"teach the adverting mind" (98–100). In teaching the mind that "Power dwells apart," Shelley's description articulates a nonhuman agency, a modern vision of geologic process as dynamic and continual. The Power that asserts itself symbolically through this process "dwells apart" not only from Thomson's "Eternal Mind" (*Winter* 577), but equally from the "awful Power" and "Eternity" invested in such scenes by Wordsworth and Coleridge. Coleridge's "Hymn before Sunrise, in the Vale of Chamouni" (1802)—which Shelley seems to have been reading at the time he wrote "Mont Blanc"—makes an especially telling contrast. Coleridge's poem is organized around the opposing trope of temporal stasis, the common element in his catechism of all the features of the landscape:

> Who gave you your . . .
> . . . unceasing thunder and eternal foam?
> And who commanded (and the silence came),
> Here let the billows stiffen, and have rest?"[69]

66. Hogle, *Shelley's Process*, 80.

67. I mean this in the Kantian sense of dilemmas of pure reason (*Critique of Pure Reason*, A 422–23). Bode ("A Kantian Sublime," 357) and Ferguson, because "nature needs us" ("Shelley's *Mont Blanc*," 210), take the question as rhetorical, while Hogle (*Shelley's Process*, 86) and Brinkley are both more inclined to heed the question's "suspension of belief" ("Spaces between Words," 263). Wasserman responds by quoting Shelley's own *Essay on Christianity*: "where indefiniteness ends idolatry and anthropomorphism begin" (*Shelley*, 233).

68. On "primitive" mountains, see Humphry Davy's overview of this argument (*Lectures*, 53–55), citing the Alpine studies of Jean-André de Luc, H. B. de Saussure (65), and others.

69. Coleridge, *Poems*, 376–80, here ll. 44–48. Raffles, perhaps unconsciously quoting Coleridge, exults that he can "commune with Him who bade these billows stiffen, and these sum-

Other parallels include the "jagged rocks," "for ever shattered and the same for ever" (42–43), and the "Ice-falls," "motionless torrents . . . / stopped at once amidst their maddest plunge!" (51–53). The dynamism here is "evermore about to be," the deformation monumental. Unlike Shelley, "Coleridge never was at Chamouni, or near it, in his life!" as Wordsworth was to complain in 1844.[70] Coleridge tacitly acknowledges this lack in his third-person headnote, which contrasts markedly with Shelley's claim to have written "under the immediate impression" of the scene.[71] Like Shelley in the context of *History of a Six Weeks' Tour*, however, and like Wordsworth in the Kirkstone Pass ode (and elsewhere), Coleridge draws literary raw material from an unacknowledged female contributor, Friederike Brun. This common element complicates the relationship among the poems. Because Coleridge's poem has the most strongly mediated relation to travel narrative, it brings out a potential allegory in the other two: concern with the materiality of rock in all three poems can also be read as a concern with the materiality of the poem and its philosophical idiom, its composition out of the observations of others.[72]

One of the earliest English Alpine writers, Thomas Burnet, maintains that an "actual view" of the Alps is crucial for stimulating rigorous thinking and writing about them, as Wordsworth implies much later in his criticism of Coleridge. Marjorie Nicolson, in her history of the literary representation of mountains, focuses heavily on Burnet because the germ of his theory of the earth lay in a face-to-face encounter with the Alps on his Grand Tour of 1671.[73] Burnet's mixture of aesthetics and geology makes him a pivotal figure: his vision of "mountain gloom" already anticipates "mountain glory." Throughout her book, Nicolson argues that "mountain gloom"—the set of seventeenth-century poetic conventions embodied in an image from Marvell of "unjust," "hook-shoulder'd" mountains—gradually gives way to

---

mits rise—and say, as the eye embraced the mighty whole, 'My father made them all'" (*Letters during a Tour*, 208). An indignant footnote to this passage reports: "Yet, amid these scenes . . . a wretch has had the hardihood to avow and record his atheism," referring to a Greek entry, presumably by Shelley, in the album of the hospice on Montanvert. Many readers of "Mont Blanc" note its implicit dialogue with Coleridge's poem. Nigel Leask reviews the evidence that Shelley was traveling with a copy of *The Friend* in which this poem was reprinted ("Mont Blanc's Mysterious Voice," 185), and Keith Thomas surveys the increasing number of English Alpine lyrics at this time ("Coleridge, Wordsworth," 101–2).

70. *Wordsworth's Literary Criticism*, 250.

71. Shelley, *Complete Works*, 6:88.

72. Critical debate continues over Coleridge's plagiarism of Brun's "Chamouny beym Sonnenaufgange" (1795), itself an avowed imitation of Klopstock. Keith Thomas argues persuasively that Coleridge essentially "translates" his 1802 ascent of Sca Fell into an account of Mont Blanc ("Coleridge, Wordsworth," 93–98).

73. Nicolson, *Mountain Gloom and Mountain Glory*, 207. Nicolson adopts these terms from Ruskin's *Modern Painters*.

"mountain glory," the entirely different convention of being "ravished" by mountains under the sign of "the vast, the grand, the majestic." Large-scale shifts in philosophy and theology, playing themselves out in influential writers beginning with Burnet, cause this change, completed by about 1740: " 'Mountain Gloom' was gone. . . . Mountains had ceased to be monstrosities."[74] Burnet shows a new aesthetic sensibility, but his morbid fascination with deformity is more representative and, I would argue, the stronger part of his legacy. This fascination is active in "Mont Blanc," and the tension between Shelley's alien Alpine landscape and Coleridge's theological one is more than a matter of intellectual history. Shelley's use of dynamism shows how intimately geological innovation was related to revisionary poetics. But his account of primitive, alien forms or even formlessness also reflects the durability of such images. There are clear, sometimes uncanny, parallels between Burnet's descriptions and Shelley's, despite the 135-year gap between their publications. I would argue, *contra* Nicolson, that the images she associates with "mountain gloom" have an exceptionally long cultural incubation period. Shelley's empiricist headnote seems to confirm Burnet's assertion that "there is nothing doth more awaken our thoughts or excite our minds to enquire into the causes of such things, than the actual view of them."[75] The deep structure of the Alpine encounter, persisting through its historical variations, seems to dictate that it be understood as a face-to-face encounter with the other.

Burnet's "actual view," in his own historical context, unsettles the presupposition of a smooth and uniform antediluvian globe and the theological construct of monumental divine agency. Reacting to the shock of his Alpine experience, Burnet asserts that "there is nothing in Nature more shapeless and ill-figured than an old Rock or a Mountain, and all that variety that is among them, is but the many various modes of irregularity; so as you can not make a better character of them, in short, than to say they are of all forms and figures except regular." To emphasize the unearthly character of this landscape, Burnet offers hypothetical visions of how it might appear to one of two observers. In one instance, he speculates on the reactions of "a man . . . carri'd asleep out of a Plain Country, amongst the *Alps*, and left there upon the top of one of the highest Mountains." If persuaded that the sight "on every hand of him" of "a multitude of vast bodies . . . in confusion" were not a dream, this observer might suppose himself to be on a different planet; or "he would be convinc'd, at least," that the "Rocks standing naked round about him" were features of a part of the world more

74. Nicolson, *Mountain Gloom and Mountain Glory*, 212–15, 34, 215, 345.
75. Burnet, *Sacred Theory of the Earth*, 110.

"strangely rude, and ruine-like" than anything he had previously imagined. In the second instance, Burnet imagines the appearance of the sea floor drained of all its water, personifying the earth's destructive force, unlike Shelley, in an obviously gendered way. When he pictures this "vast and prodigious Cavity," he exclaims, "naked and gaping at the Sun, stretching its jaws from one end of the Earth to another, it appears to me the most ghastly thing in Nature." Burnet's "all forms and features except regular" makes the definition of form purely negative; here this negativity is realized in the cavity or "open Hell" of an encompassing chaos. Burnet's aesthetic geology theorizes rocks as raw, unformed matter. Like Shelley in "the naked countenance of earth," he draws on the analogy between the body and the earth: his islands, anticipating Kant as well, are "scatter'd like Limbs torn from the rest of the Body"; his mountains are "wild, vast and indigested heaps of Stone and Earth"; and almost all landforms, it seems—at least when devoid of vegetation—are in "every way deform'd and monstrous."[76]

Burnet sees in such landforms, especially in the prototype of the Alps, a limit case of chaos and deformity. As the "greatest Examples of Confusion," rocks embody formlessness in their composition; in their "nakedness," they dramatize the recalcitrance of raw matter. Theologically, these features function as a text or monument of God's wrath.[77] But in relation to cosmology and aesthetics, their signification is more equivocal. The vision of a dry ocean, for example, personifies (in the form of gaping jaws) the destructive force of the Deluge in a way that suggests a continuing process. Naked and virulent landforms become agents of ruin as well as signs of it: " 'Tis another chaos." In "Mont Blanc," too, chaotic landscape is naturalized as an agent of violent upheaval rather than standing as a monument of it. Such landscapes both perform and register the displacement of the organic by the inorganic, like a second, slow, but all-consuming Deluge. In Shelley's words, "the rocks . . . / . . . have overthrown / The limits of the dead and living world" (111–13). This apparent chaos of the mineral world remains the sign of its otherness until long after geology has begun to establish a complex kind of order in it, because disorder expresses the inorganic nature of rock. The Huttonian or Plutonist geology that influences Shelley is in this respect a

76. Ibid., 112–13, 111, 102, 91, 84, 104.
77. By emphasizing continuities between Burnet and the Romantics, I have had to deemphasize an important difference that might be equally revealing, the obvious role of gender in the earlier construction of geological otherness. If mountains are monuments of God's wrath, then one cause of this would be the association between femininity and sin. The more strongly anatomical character of geological otherness in seventeenth-century images (see Nicolson, *Mountain Gloom and Mountain Glory*, for examples) seems to encourage more specific analogies between the earth and the *female* body. These analogies, as we shall see, are more prominent in Blake.

logical successor to Burnet's diluvialism, because of the gradual recognition that like other inorganic bodies, rocks evolve under inconceivably high temperatures. In fact Burnet himself calls for molten rock as his sacred history concludes with a general liquefaction: "the exterior region of the Earth is melted into a fluor, like molten Glass, or running metal. . . . This huge mass of Stone is soften'd and dissolv'd . . . and swallowed up in a red Sea of fire."[78] In "Mont Blanc," Shelley envisions a plutonic *origin:* "Did a sea / Of fire, envelope once this silent snow?" (73–74).

The comparison suggests that the notion of an alien destructive agency, which seems so novel in Shelley, is a traditional element of the reaction to geological chaos and deformation. This agency, though construed differently in most respects by Shelley and Burnet, is in both neither human nor properly divine (as Burnet's many theological critics might testify). Burnet's theory remains relevant for Romanticism because it offers a construction of materiality that is opposed to the aesthetic and provides the tension or resistance to aesthetic categories necessary for such "gloomy" episodes as the desolation of "Mont Blanc" or the disappointment of the Simplon Pass in *The Prelude.* Nicolson remarks that the "fierce pleasure these poets felt in the irregularity of Nature" violates the norms of their Augustan predecessors and distinguishes the new vision of a "solemn Nature whose majesty is enhanced rather than marred" by the presence of "rough, jagged, monstrous stones."[79] My view is that developments in empirical science and technology, and not just theology or philosophy, helped to create the pleasure and the majesty of these landscapes, but also generated a deeply alien other. Stones remain (or again become) "rough, jagged, and monstrous" in some Romantic poetry because of their power to mar a constructed transcendence. In the Kirkstone Pass ode and "Hymn before Sunrise," the transcendence subsumes geological chaos and otherness, but the rude rocks, standing for materiality in general, provide the resistance necessary to mark that transcendence. In "Mont Blanc," they pose a more formidable limit.

Kant's *Critique of Judgment* provides parallel instances of an ambivalent fascination with the sublime *in* natural phenomena such as mountains, a fascination motivated (and disturbed) by their irreducible materiality. Nicolson accounts for the distinctive properties of Burnet's mountain sublime by alluding, once again, to the importance of an "actual View" of mountains: "Henry More never saw a mountain. Neither did Immanuel Kant."[80] Kant's image of "deformed mountain-masses" in the "Analytic of the Sublime,"

78. Burnet, *Sacred Theory of the Earth*, 105, 305–6.
79. Nicolson, *Mountain Gloom and Mountain Glory*, 15.
80. Ibid., 122.

even more than the "heaven-ascending mountain-masses" in his concluding remarks, are conventional mountains rather than real ones, but this is part of his point. In the former passage, Kant follows probably his clearest statement of the thesis that the "true sublime" is to be found in the judging mind, and not in natural objects, with a rhetorical question: "And who would want to call deformed mountain-masses, piled one over the other in wild disorder, with their pyramids of ice, or the dismal, raging ocean, etc., 'sublime' in the first place?" he inquires, with some degree of irritation.[81] While there may be some ironic distance in the later passage, he tolerates the "sacred shudder" there for ethical reasons. Here the mountains are merely deformed (or unformed, *ungestalt*) and disorderly, to emphasize their lack of "intrinsic" sublimity. This passage indicates the equivocal nature of the sublime as a locodescriptive mode; despite Kant's critical distance, it suggests the alienation the very idea of rude rocks is still capable of affording in 1790. The passage provides further evidence of the energy Kant finds it necessary to invest in negotiating rocks—these mountains require four times the amount of description given to the sea. The description caustically mimics earlier descriptions in the sublime mode, responding to tropes still active in "Mont Blanc," with its precipices and pyramids "piled" on top of each other (102–4). The unformed or "rude" quality and "wild disorder" are equally present in Shelley, but in Kant they signify the absence of the lawful order that aesthetic judgment mediates for practical reason. The mathematical sublime in this example can be "sublimed" alchemically out of the solution; but the stubborn precipitate of "unformed" rocks indicates that the fascination of matter itself weighs equally with the fascination of grandeur in these rejected tropes.

Through Shelley, Burnet, and Kant emerges one way of "reading" rocks. Within this hermeneutic circle, the salient signs are the "rude" forms of rocks, the apparent chaos of their arrangement, and the violence within or behind that chaos. The sense of wonder, though assigned a different value in each text, is essentially a descriptive response to the literal and figurative opacity of these signs or characters. This chapter concludes by considering two attempts to establish a text out of such characters. This more developed way of reading is still in the aesthetic mode, and the history it posits is largely untraceable. But the attempt to piece together a textual substance, however fragmentary, out of rocks, sets the stage in important ways for the geological and moral hermeneutics that operate alongside the sense of wonder in a letter by Dorothy Wordsworth and a passage from William's *Excursion*. Such hermeneutics make explicit a tension in rocky landscapes between

81. Kant, *Kritik der Urteilskraft*, 343 (B 95); *Critique of Judgment*, 113.

the appearance of utter chaos and theologically imprinted narrative, a tension widely reported by European naturalists in the eighteenth century.[82]

## The Illegible Rock Record

The earliest known manuscript of the "Excursion to the Top of Scawfell," incorporated into the third edition of *A Guide to the Lakes* in 1822, consists of extracts made by Dorothy Wordsworth from her own letter to William Johnson, dated 1818. Her brother made fewer changes in this text than in her account of Kirkstone Pass and the Ullswater excursion, which follows this one in the *Guide*. One especially vivid paragraph, describing "huge blocks & stones" at the summit of Scawfell (Sca Fell) Pike, is adopted almost verbatim from Dorothy's letter. The paragraph shares the journals' style of minute observation, and shares the image of rock adorned with lichens with one of her poems, "Grasmere—a Fragment." At the summit, she writes:

> Not a blade of grass was to be seen—hardly a cushion of moss, & that was parched & brown; and only growing rarely between the huge blocks & stones which cover the summit & lie in heaps all round to a great distance, like Skeletons or bones of the earth not wanted at the creation, & there left to be covered with never-dying lichens, which the Clouds and dews nourish; and adorn with colors of the most vivid and exquisite beauty, and endless in variety.

The paragraph concludes by emphasizing the desolation of the place, frequented neither by "the adventurous Traveller" (who seeks out another summit of Scawfell) nor by "the Shepherd . . . in quest of his sheep for on the *Pike* there is no food to tempt them" (*Prose*, 2:367).

The familiar emphasis on barrenness and chaos here carries two salient differences. One is the striking image "bones of the earth," and the other a counteremphasis on the one "softening circumstance," the beauty of the lichens. Through the former image, Dorothy Wordsworth reads a modicum of order into the chaos of these "huge blocks & stones." Her wonder is consequently more serene than Shelley's or Burnet's. The extraordinary stasis of these rocks and the neutral or faintly benign character of the agency they suggest soften their otherness almost to the point of legibility. Their desolation serves more clearly as a foil for the beautiful lichens, which achieve equal importance in the description. The origin the boulders seem to commemorate testifies less to cosmological violence than to a superabundance

---

82. Goethe's fragment "Über den Granit" (1784) is one remarkable example; *Geologische Schriften*, 325.

of material; no Deluge has interposed here to efface the traces of the Creation. The chronology they might offer is largely unreadable, but as "bones of the earth" they provide matter for speculation about origins. Dorothy's speculations may have been especially appealing to William in 1822 because of an encounter with geology reported by Theresa Kelley.[83] As Kelley relates, Wordsworth read an 1820 article on the stratigraphy of the Lake District arguing that the major peaks are composed of older rock that protrudes up through the surrounding limestone. She argues that this hypothesis, coupled with the recent discovery that Scawfell is in fact higher than Helvellyn, long considered the region's highest peak, confirmed for William the priority of the sublime and the sublime priority of Scawfell Pike. Dorothy seems to presume a preexisting stock of such "bones," of another order than the implied organic "flesh" of the earth; the ambiguity of the "bones" as both integral to the earth's body and as previously created, or even uncreated, produces the tension that makes this such a vivid description.

The lichens, in a certain sense, confirm the analogy to the body. Dorothy links them organically to the rocks by concluding her description with the phrase, "the beauty of some of these masses of stone" (*Prose*, 2:367). The "never-dying lichens," on the one hand, mimic the rocks in their quality of being the oldest and simplest form of plant (or protist) life. But as "accompaniments," they allow the rock to perform its mimesis of the body.[84] The rocks are not in fact dry bones, but carry a residue of organic matter that speaks of their original purpose. The otherness of the inorganic is sufficiently marked in these rocks by their composition (like those of Burnet) in "heaps," by their failure to sustain more complex vegetation, and hence by the absence of for-

83. Kelley, *Wordsworth's Revisionary Aesthetics*, 20–21. Kelley remarks that "the history of the ruin on Scawfell is so archaic that without these few physical remnants it would be lost to human memory. . . . They signify a past whose beginning cannot be determined" (37). Steven Goldsmith (personal correspondence) has suggested that rather than hinting at superabundance, these rocks might be seen as so recalcitrant as to be unformable, rejected because "uncreatable." It would follow that the lichens "adorning" these blocks provide an aesthetic cover masking this unassimilable materiality, a beautiful illusion somehow in tension with the raw material of the scene.

84. In her poem "Grasmere—a Fragment," Wordsworth describes a rock similarly adorned by vegetation (77–80):

> What need of flowers? The splendid moss
> Is gayer than an April mead;
> More rich its hues of various green,
> Orange, and gold, & glittering red.

This "stately Rock" is located at a lower elevation than those of the prose passage, and adorned with ferns and other vegetation, but some of the colors listed here are clearly those of lichens. The prose description strongly emphasizes the colors of the lichens in their "endless variety" because they stand out against the drab coloring of the rocks, marking the difference of organic and inorganic in yet another way.

age for the sheep. But they show the softening influence of weathering and vegetation, like the "gashes" in the earth analyzed in Uvedale Price's *Essay on the Picturesque*. Even the oldest materials, it seems, are integrated into cosmology by the organic processes of an ecosystem that assimilates excess material, processes that the rocks themselves make faintly legible. At the same time, the phrase "not wanted at the creation" speaks against such integration and signals the necessarily conjectural nature of any unifying reading of these rocky characters. William Wordsworth, in adopting this passage, would probably have noted its echo of Burnet's "skeleton of the earth." This echo helps to locate both Wordsworths in a larger pattern of speculation concerning the function of rocks as the original handwriting of a creator.[85]

A key passage from *The Excursion* more aggressively develops the metaphor of geological reading. This passage (III.50–158) employs a "mass of rock" to distinguish the aesthetic responses of the three principals: the narrator, the Wanderer, and the Solitary. The narrator introduces the scene as follows (50–54):

> Upon a semicirque of turf-clad ground,
> The hidden nook discovered to our view
> A mass of rock, resembling, as it lay
> . . . . . . . . . . . . . . . . . . . . . . . . . . . . .
> A stranded ship, with keel upturned.

His description of the rocks is chiefly limited to a series of similes, likening individual rocks to the ship, to pillars, and to an altar or "tablet" (61–65):

> Barren the tablet, yet thereon appeared
> A tall and shining holly, that had found
> A hospitable chink, and stood upright,
> As if inserted by some human hand
> In mockery.

The holly provides a striking contrast to Dorothy's lichens; here the vegetation is a single tree that wrests its nourishment out of a weak point in the rock's defenses. The lichens, "which the Clouds and dews nourish," integrate the rock into an economy of nutrition and generation, with a social quality of beauty. The lone holly, though also evergreen, accentuates the rock's barrenness. The suggestion of human agency ("as if inserted by some

85. Burnet, *Sacred Theory of the Earth*, 112. William holds up Burnet's Latin descriptions as models of style in at least two places, *The Excursion* III.112n. and *Guide*, 149 (letter on the Kendal and Windermere Railway). Concerning the idea of rocks as divine writing, see also Hartmut Boehme's discussion of *die Urschrift Gottes* ("Das Steinerne," 140).

human hand"), moreover, attaches an artificial aspect to this coexistence of organic and inorganic, and thus establishes the holly as an inscription on the otherwise "barren tablet."

For the Wanderer, however, the entire composition becomes a text whose theological tenor accords with the metaphor of an altar. There is a distinguishable trace of design in the bare rocks, even if the trace is faint and largely faded from the substance of nature (as distinct from the "rock record" seen elsewhere as the substance of human or natural history). The Wanderer speaks "reverend" words of praise on beholding this "cabinet for sages built" (80–91):

> Among these rocks and stones, methinks, I see
> More than the heedless impress that belongs
> To lonely nature's casual work: they bear
> A semblance strange of power intelligent,
> And of design not wholly worn away.
> . . . . . . . . . . . . . . . . . . . . . . . . . . . . .
> In these shows a chronicle survives
> Of purposes akin to those of Man,
> But wrought with mightier arm than now prevails.

The Solitary's opposite description of the same scene, especially in light of the Wanderer's subsequent commentary, becomes the focal point of the passage. The Wanderer's response deals in certainties; the vestiges of design are intrinsically there, a stable meaning of this passage from the "Bible of the Universe." The scene provides laboratory conditions for a display of the Solitary's "despondency," which it is the Wanderer's task in book IV to "correct." (The "argument" of book III notes the "contrast" between the "sensations" of the Wanderer and those of the Solitary "excited by the same objects.") The exemplary symptom here of a diseased moral constitution is the absence of design in the Solitary's reading of the rocks: for him, they are merely the radically indeterminate substance of nature, as capricious as his own conflicting readings. In this respect, however, his reading resembles that of the speaker of the Kirkstone Pass ode in the first stanza (124–33):

> The shapes before our eyes
> And their arrangement, doubtless must be deemed
> The sport of Nature, aided by blind Chance
> Rudely to mock the works of toiling Man.
> And hence, this upright shaft of unhewn stone,
> From Fancy, willing to set off her stores
> By sounding titles, hath acquired the name
> Of Pompey's pillar; that I gravely style

My Theban obelisk; and there, behold
A Druid cromlech!

As in the Kirkstone Pass ode, arbitrary and improvised readings of rude rocks are not enough, but must be followed by the moral reflection that is an established part of the protocol for the sublime. The Solitary explains that these notions occur to him when he indulges "the antiquarian humour," and then offers an alternative version that obtains "if the spirit be oppressed by sense / Of instability, revolt, decay" (137–38). In this latter mood, "these freaks of Nature" stimulate only "pity, and scorn, and melancholy pride" (142). In both cases, the Solitary's literacy is manifestly classical and secular. Rather than regarding the quasi-scriptural sense of the rocks, the Solitary self-consciously projects arbitrary readings onto them. But the persistent architectural comparisons (see also 143–52), even if they serve only to illustrate the vanity of human productions, may be deconstructed by means of moral "correction": these comparisons show that the Solitary, despite himself, perceives the vestiges of design in these rocks. The Wanderer's effusion incorporates moral reflection spontaneously into the aesthetic response; the otherness of rocks for him is sufficiently mitigated by their presence as illegible history. But the Solitary's failure to acknowledge that history, except unconsciously, indicates a moral deficiency; mere nature is not sufficient for a corrupted mind, and he requires the explicit moral narrative that a guide like the Wanderer can offer either through his glosses on the rocks or through direct homily. (The parallel structure of the ode makes it possible to imagine its first stanza as being spoken by the Solitary, and the remaining stanzas by the Wanderer.) Theresa Kelley points out the "healthy state" of mind required for the aesthetic education offered by Wordsworth's Lake District, comparing him to "Ann Radcliffe and other advocates of the doctrine of sensibility."[86] The history these rocks offer to a healthy taste also exemplifies *The Excursion*'s overarching principle of matter perfectly "fitted" to mind. The unstable history they offer (or refuse) to the Solitary as "unhewn stone" becomes confused in his "despondent" mind with images of worked stone, as he falsely conflates the pretechnological design of Nature with historical human artifice.

The conclusion of this passage in *The Excursion* allows us to distinguish both "readings," as aesthetic, from a species of practical reading that the Solitary ridicules while still reflecting on the scene. His satire of the geological specimen-hunter (III.173–89), with his "pocket-hammer" and "barbarous name[s]," describes an attempt to make the rock record fully legible,

---

86. Kelley, *Wordsworth's Revisionary Aesthetics*, 24.

and associates this attempt with the conversion of rude rocks into economic resources: should this geologist's or prospector's specimen be "haply inter-veined / with sparkling mineral," the Solitary continues, he "thinks himself enriched, / Wealthier, and doubtless wiser, than before!" (186–89). This clear hierarchy of aesthetic and practical reading responds to the increasing pragmatism and power of applied geology, demonstrated most impressively in William Smith's 1815 *Geological Map of England and Wales.* Wordsworth's polemical tone does not allow for the subtle gradation of such "purely" aes-thetic responses as the Wanderer's, or Shelley's "extatic wonder," into more explicitly economic responses to geological objects. The primitive material-ity of rock, in a concrete and technical sense, is central to geological discus-sions in Goethe, Humphry Davy, and James Hutton, as well as the poetry of Blake. Accounts like these bring the idea of primitive materiality into circu-lation in aesthetic discourses about rocks as resources to be contended with, physically and culturally. In these discourses, as in Wordsworth's imagined world of economic geology, geological otherness is articulated through the desire of domestication.

# 3

## Blake, Geology, and Primordial Substance

The idea of rock as primitive matter presented in "Mont Blanc" and other poems plays a central role in the wider field of the period's natural knowledge. The idea itself is as old as cosmology, and its continued currency also reflects the lingering influence of Thomas Burnet on both the literature and the emerging geology of Great Britain. The debate within late eighteenth-century earth science concerns both the agency responsible for forming rocks themselves and the chronology of the earth's formation: which rocks are oldest, or "original"? Is the absence of organic residue a sufficient criterion for distinguishing "primitive" from "secondary" rock? More fundamentally, can some rocks be regarded as "primordial substance"? A negative answer in this wide-ranging discussion comes from the surprising quarter of William Blake's poetry. While images of rocks, cliffs, and caves in Blake's later poetry are often associated with artistic production or with his Druid symbolism, they also constitute a symbolic intervention in the debate about primitive matter. Blake's "no," however, is ambiguous: rockiness is indeed the first or "primordial" state of materialization for many phenomena in his prophetic books, but materiality itself is adventitious.

Blake's friend George Cumberland referred to "the almost new branch of science, Geology," as late as 1826, and throughout his career Blake had reason to believe that ideas about primitive matter were the province of poetry and aesthetics rather than geology. On the other hand, James Hutton's *Theory of the Earth* (1795)—noted by historians for its newly rigorous science—

borrows some of its explanatory paradigms from the aesthetic sphere—to phrase this transaction, for the moment, in terms of the more recent paradigm of two cultures. Stratigraphy (in the absence of radio-carbon dating, for example) was both the primary source of evidence for historical geology and a subject of very general interest, attracting amateur naturalists, poets, and tourists. Blake's poetry shares the cosmological interest of geology, both registering and critiquing the wonder provoked by geological forms and processes. The geological accounts corroborate this wonder at what seems to be the primitive materiality of rocks, but they also compete to discern its physical and chemical principles. In the process, they analyze and question the very qualities provoking wonder—magnitude, duration, stability, inorganism. These accounts are often formally or structurally analogous to the images of rocks in Blake's narratives, though Blake has a different set of motives for his representations. This chapter compares Blake's poetry with the geology of James Hutton and others in order to argue that aesthetic categories in both discourses serve the explanation of nature and that both explanations of rocks are equally explanations of the material condition of things.

Blake's prophecies abound with images of rock. These images—in the form of British mountains and ancient British trilithons or cromlechs, as well as deterritorialized landscapes of "opake hardness"—are especially concentrated in *Jerusalem*, where they embody Albion's fallen condition, drawing on both geological and cultural registers of the primitive. Such negative images of a dehumanized cosmos stand opposed to the wonder evinced by geologists and other writers on landscape, but their negativity is hardly surprising. The critique of materiality and materialism is perhaps the most familiar philosophical dimension of Blake's thought. Even this obvious association with Blake's rocks, however, provides a vital piece of evidence for my larger argument that Romantic rocks stand for materiality in general. Blake eagerly embraces this metonymy even as he attacks materialism and resists the naive aesthetic response to geological objects. Readers tend to see the cosmogony of the later prophecies as a challenge to Newtonian space, but in many other systems from Descartes through Erasmus Darwin cosmology depends on geogony and geology, so that these also become factors in Blake's competing science. Especially in the Changes of Urizen episode, which occurs in four of the major prophecies, he adopts a geological model to articulate the genesis of fallen perception.

Blake's demonstrable exposure to geology was fairly limited and has been discussed briefly by several critics. David Worrall and Nelson Hilton have made strong cases for the importance of Erasmus Darwin's geophysics as a source of imagery in Blake, especially *The Book of Urizen.*[1] Other critics have established connections between Blake's poetry and designs and Thomas Burnet's *Sacred Theory of the Earth.* Darwin and Burnet are fruitful sources for the analysis of Blake's imagery, and both, especially Darwin's rough-and-ready brand of Vulcanism, also had a broader impact on the structure of Blake's cosmology. Blake probably had further exposure to geology from his friend George Cumberland. Only one sentence has been written concerning this connection, though it makes a strong claim:

> Blake makes no direct reference to fossils, although *Vala, or the Four Zoas* and *Jerusalem* are full of references to petrification which may, of course, be Biblical in their derivation, as in the story of Lot's wife. However, it would be absurd to suggest that he did not know of the speculation current at his time for, according to Gideon Mantell, his friend George Cumberland was "a celebrated geologist," and the subject must have been discussed by them.[2]

This chapter begins with a reading of what I will call Blake's "geological moment"—his first account of the Changes of Urizen—in *The Book of Urizen.* My frame of reference for this reading includes a survey of the established connections among Blake, Burnet, and Darwin, and also raises the unexplored question, did Cumberland's geology recognizably influence Blake?

The later sections of this chapter perform a different kind of work, approaching the culturally central issue of primitive matter from some of its manifold contexts rather than tracing a causal thread of influences on Blake's creative process. My two main sources for this longer discussion are *Jerusalem* and James Hutton's *Theory of the Earth* (1795). Hutton, though never cited by Blake and hardly mentioned in Blake scholarship, challenged the view of rock as primitive matter in his influential theory, much as Blake challenged the philosophical regime of solid surfaces in his negative images of petrifaction.[3] From a biographical point of view, Hutton's theory is an odd choice for situating Blake's geology. He was a deist and an avatar of the

---

1. Worrall, "William Blake and Erasmus Darwin's *Botanic Garden*," esp. 405–13; and Hilton, "The Spectre of Darwin."

2. Ruthven Todd, *Tracks in the Snow*, 16. See also Geoffrey Keynes, *Blake Studies*, 233; and G. E. Bentley Jr., *Bibliography of George Cumberland.*

3. The only references I have found are in the commentary of David Worrall in his edition of *The Urizen Books.* Two of these draw on Hutton's earlier work on the theory of matter (36, 142); in a third reference, Worrall uses Hutton's *Theory of the Earth* to illustrate the connections between anatomy and geology embodied in terms such as "veins" (37). Regrettably, Blake tells us too little about the eighth Son of Albion ("Hutton" or "Huttn") to warrant a direct connection. See *Jerusalem* 71:36–37. All references to Blake's illuminated books are by plate and line number, except as specified.

scientific rationalism that Blake attacked so vigorously. But Hutton's materialism has a political potential that would have appealed to Blake, just as other late Enlightenment thinkers, such as Paine and Priestley, held a strong if ambivalent attraction—political if not philosophical—for the poet. And Blake's view of geological process and the production of matter is cognate with the Plutonism of Hutton's theory. As Ruthven Todd points out in the above reference to Cumberland, Blake never mentions fossils—key evidence for Neptunists and diluvialists—and flood is a distant second to fire in his geological images. Both Blake and Hutton recognize, in their culture's preoccupation with rock as primitive or unformed matter, the need for a new account of solidification based on obscure and vastly powerful agencies.

## Geology as Anatomy

One argument from intellectual history can be adduced to corroborate Hutton's relevance for understanding Blake's system of the earth. In 1791 Blake engraved several plates for Erasmus Darwin's *Botanic Garden*, a poem with ample geological content influenced by Darwin's conversations and correspondence with Hutton over a period of twenty years.[4] Huttonian principles absorbed from Darwin's poem seem to have stimulated Blake's fascination with the idea of primitive matter, itself the product of a widely shared, polymathic literary culture. Geological ideas appear in Darwin's poem juxtaposed with incendiary political ideas "fueled" by the plutonic heat of his Huttonian geology. Desmond King-Hele points out that one of these passages must have been among the first Blake read as he set about his engraving assignment.[5] In this passage, an extended allegory of the "giant-form" of Liberty, rising up on Gallia's plains, interrupts a narrative in which the Gnomes form granite out of lava (*BG* I.ii.297–400). Darwin invokes the

4. Desmond King-Hele discusses the details of the Darwin-Hutton correspondence in *Erasmus Darwin and Romantic Poets*, 10–11. King-Hele also notes that "in the 1760s [Darwin] went on expeditions to the caves of the Peak District with John Whitehurst, collecting specimens and forming his own views on geological history. These views matured in the course of his correspondence with James Hutton in the 1780s. Geology is dominant among the notes to *The Botanic Garden*" (19). Whitehurst is the only contemporary British geologist cited by Hutton in his two-volume *Theory* (1:153). For Darwin's treatment of Whitehurst, see *BG* I.ii.17n., 36n.

5. King-Hele, *Erasmus Darwin and Romantic Poets*, 41. David Worrall points out allusions to Darwin's politically inflected Huttonian geology in Blake's *The French Revolution*, similarly concluding that he must have been engraving from page proofs of Darwin's poem while writing his own, ultimately printed several months before Darwin's in 1791 ("William Blake," 405). See also *BG* I.ii.297–400 and the plates accompanying Additional Note XXII.

same images to represent physical processes such as the "giant-power" of steam, in another personification that conflates natural and social forces (I.i.263). He is especially apt to harness geological forces to the causes of social justice—in moving from molten rock to metals, the geological narrative also touches on Spanish colonialism and the English slave trade and culminates in the historical allegory of a tyrant's army swallowed by an earthquake (I.ii.401–99). The scientific authorities cited in this section—including Hutton and Benjamin Franklin—also authorize the political use of science.[6] Darwin's poem, then, exposed Blake to Huttonian geology as well as providing a poetic model for harnessing the energy of the primitive—which Hutton shows to be an inconceivably powerful subterranean source of heat and pressure—to revolutionary politics.

Blake presents his first fully developed cosmogony in *The Book of Urizen* (1794). Though he had created his own Giant Form of Liberty, Gordred, as early as the American Revolutionary period, "we move beyond such sharing of commonplace matter and images," according to David Worrall, "when we come to the realm of poetized cosmogony."[7] Worrall's analysis of Blake's cosmogony and its Darwinian elements focuses especially on *Urizen;* he points out several echoes of Darwin in the poem's numerous passages describing the episodic formation of a physical universe. By virtue of Blakean inversion, the central episode in this successive and ongoing creation is an anatomical one, the geologic formation of a body-prison for Urizen, under the aegis of Los. This pivotal episode narrates the creation of a body for the self-styled creator of the physical cosmos. In the process, anatomy becomes closely identified with geology, an identification strengthened in the numerous recurrences of this passage (chapter IV[b]): it is incorporated almost verbatim into *The Four Zoas* (54–55), and occurs in abbreviated form in *Milton* and in several modified versions in *Jerusalem*.[8] The persistence of the Changes of Urizen suggests that this motif is for Blake the fundamental creation story. The errors of Urizen that lead to his Luciferian fall materialize as the body and bodily senses. Los binds these forms in place, constraining Urizen but also limiting his dissolution. The errors of perception, in their

---

6. *BG* I.ii.275n. and 398n. rely on Hutton's theory of continental uplift, and Darwin cites him directly in Additional Notes XXI, XXIII, and XXIV. Other significant Hutton references include those in I.ii.119n. and A.N. XVI, XIX.

7. Worrall, "William Blake," 406. "Gordred the giant" appears in "Gwin, King of Norway" and leads the people to a revolutionary victory over their despotic king. See *Complete Poetry and Prose*, ed. David V. Erdman, 417–20 (cited hereafter in the conventional form, e.g., E417–20).

8. I refer to *The Book of Urizen*, Copy G, ed. Kay Easson and Roger Easson, 12:24–15:27. I also refer (by MS page number) to the text of *The Four Zoas* in *Complete Poetry and Prose*, ed. Erdman.

hardness and opacity, become the substance of the material world perceived by the bound senses. The idea that matter itself is created out of a derangement of the faculties allows Blake to posit his doctrine of perception from *Marriage of Heaven and Hell* as an axiom of cosmology. Geological images—especially those associated with materialization and primitive matter—provide Blake with a topical idiom for his emphasis on the virtual identity of the physical world and the natural body.

As Eternal watchman over the fallen Urizen, Los becomes directly implicated in Urizen's "rebellion" against the Eternals. Urizen is "rent from his side," "rent from Eternity," and contemplated from above as Eternal Death: "What is this? Death. / Urizen is a clod of clay" (*Book of Urizen* 7:2–10). Both Worrall and Nelson Hilton point to a Darwinian model for this "clod of clay" "rent" from the Los-sun, namely the theory offered early in *The Botanic Garden* that the earth originated as a fragment ejected by a "solar volcano." In this theory, Darwin creatively embellishes ideas from Hutton and others while alluding to the explosive political energies that he links to the internal heat of the earth and other heavenly bodies.[9] Hilton makes Darwin's theory the model for a parallel scene, the birth of Enitharmon from Los's "bosom": "The sequence, in *The Book of Urizen*, of Los's bosom 'earthquaked with sighs' [13:49] followed by Urizen's being 'rent from his side' [7:4] points to the idea (popularized by Darwin) of the earth ejected from the sun, and the subsequent binding of Urizen in chains plays on the conceit, common from Thomson through Darwin, of gravity as a chain."[10] Hilton's inadvertent conflation (indicated in brackets) of these two passages as a "sequence" is illuminating because it reflects *Urizen*'s overarching representation of processes belonging to physics and even human biology as geological. The same could be said of the cave or "roof vast petrific / . . . like a womb" (5:28–29) that Urizen initially forms to seal his world off from Eternity. In a more recent commentary, Worrall persuasively matches this description of a womb-cave, supplied with cooling rivers of blood (5:31), to the full-page design (G17) depicting the birth of Enitharmon as a globe with veins and a polar icecap.[11]

Other rocky images, both visual and verbal, frame the entire narrative, from the hewn stone tablets of the title page to the rocky forms of the

---

9. *BG* I.ii.13–18; see further Additional Note XV; Worrall, "William Blake," 411–12; and Nelson Hilton, "Blake and the Perception of Science." According to Worrall, Darwin's theory "supplies Blake with images of solar and telluric earthquake and volcanic action clear and powerful enough to match his concept of 'Prophetic wrath,' and he uses them widely in his books of creation, *The Book of Urizen* and *The Book of Los*" (411).

10. Hilton, "Blake and the Perception of Science," 61.

11. *The Urizen Books*, ed. Worrall, 43.

poem's primitive landscapes. Plate 4, a full-page image of Urizen crouching passively in a cavity within the rock just the size of his huddled body, introduces a visual motif that recurs in *Jerusalem* and elsewhere. Los is depicted in such a form-fitting rock prison in plate 12, though he struggles to break free of it, a gesture confirmed by the rock fragments around him in plate 18 (visible only in Copy G) and by the parallel text of the *Book of Los*: having been "frozen amidst / The vast rock of eternity," Los breaks free, and "the vast solid"

> Crack'd across into numberless fragments
> The Prophetic wrath, strug'ling for vent
> Hurls apart.[12]

Worrall points out a connection between these two images (G12 and 18) established in another copy: "In Copy A, this pictured energy is politically focused because the figure is identifiable with the Los of Pl. 16 pushing asunder a trilithon-like structure, like Samson tottering Egyptian religion."[13] The suggestion is topical here because of the politicized geological agencies we have seen in Darwin. A pictorial analogue, the work of Blake's engraving master, reinforces the geological connection: James Basire's engraving after Raphael depicts the giant Enceladus, who causes earthquakes by pushing upward against the roof of his cavern from the same posture as that of Los in G12.[14] In discussing the fissures in the rock on this plate (G12/C9), Worrall refers directly to Hutton's geology: "The contemporary vocabulary of geology allow[s] these fissures to be 'veins' (as they are at 4:30) and for the whole of the Earth to be 'considered as an organized body.' "[15] *Urizen* Copy G features the strongest visual emphasis on rock (see esp. plates 4 and 18). Printed more than twenty years later than the other copies, it is hand-colored and not color-printed like the rest. It may also be relevant that the renewed work on *Urizen* falls in the period during which Blake's friend, George Cumberland, was publishing on geology (1815–25), and it is certainly relevant that this was the time when Blake was completing *Jerusalem*, a book teeming with visual and verbal images of rocks.

Mountains are the first physical feature mentioned in each version of

---

12. *Book of Los* 4:11–20, E92.

13. *The Urizen Books*, ed. Worrall, 37.

14. Thanks to Robert Essick for this suggestion. The image comes from Charles Rogers, *Collection of Prints*, 1:46–47. Although the image is dated 1767 by Basire, the book (published 1778) was evidently still in preparation during Blake's apprenticeship. In Basire's image the figure is seen from the front, but see also Blake's engraving "Earth," from *The Gates of Paradise* (E261).

15. *The Urizen Books*, ed. Worrall, 37.

Blake's prehistory of the Genesis chaos, including the initial survey (3:11) and the prelude to Urizen's speech.[16] The word "mountainous" introduces the ensuing vision of "ruinous fragments" and "hanging frowning cliffs" (5:7–10) produced by the separation of this chaos from Eternity, just as the mountainous postdiluvian earth is, for Thomas Burnet, a "heap of ruins." In response, Urizen encloses the rocky ruin in the "roof vast petrific" or Mundane Shell mentioned above. This is merely one in a series of attempts to reach the goal of "a solid without fluctuation" (4a:11) embraced in his speech. Unlike the originary word of Genesis, Urizen's first "words articulate" (4a:4) (shown as satirical by their consequences, but perhaps also by their deferral) recount a preverbal struggle with fire, winds, and waves to create the fourth element, earth, or "a wide world of solid obstruction" (14–23). On this foundation, Urizen institutes the social world through his laws (22–28; cf. plate 1), which translate "obstruction" into social terms. Several commentators have focused on the "solid without fluctuation" in its scientific dimension. Worrall, assuming that it refers to impenetrability, cites a work by James Hutton to illustrate that impenetrability, by the 1790s, had come to be seen as physically indemonstrable. This view complements Hutton's geological questioning of "primitive rock" as well as Blake's critique of materialist cosmogony. But Blake's is an ambivalent critique: he relies on an intimate knowledge of problems in natural philosophy, as Donald Ault and others have shown, and equally on natural history, especially the branch dealing with inorganic matter, for his anatomy of Eternal Death.[17]

Having healed his own wound, Los observes that

> the wrenching of Urizen heal'd not
> Cold, featureless, flesh or clay
> . . . . . . . . . . . . . . . . . . . . . .
> He lay in a dreamless night
> Till Los rouz'd his fires affrighted
> At the formless unmeasurable death.
> (9:4–9)

This passage alludes to the strong organic/inorganic distinction associated with the divisions of natural history as well as to the formative power

---

16. Urizen's speech, the bulk of chapter 3, is given on plate "a" (Copy C), following Copy G in the Eassons' facsimile (this plate is lacking from Copy G). The line cited here is thus 4a:5. However, I am following Erdman (E71, E804) and other critics in treating this speech as an integral part of the narrative. Cf. Angela Esterhammer, "Calling into Existence," 120–21.

17. *The Urizen Books*, ed. Worrall, 142. Ault cites Urizen's creation of a "solid without fluctuation" as the prototype for Blakean creations of condensed solids (*Visionary Physics*, 86). Blake's insistence that a void shrinks and condenses substance, according to Ault, stands opposed to the Newtonian idea that matter expands to fill a void.

of the earth's internal heat, with which Los (like Darwin's "giant-forms") and his fires are associated here. The watery images in plates 6 and 8 pose the problem of Urizen's disintegration, while the anatomically precise skeleton in the cave (plate 10) anticipates Los's Vulcanist "solution" for arresting Urizen's dissolution and restoring a modicum of "organization." As Los prepares to forge this skeleton (the subject matter of plates 12–15), several kinds of material already appear to be in circulation, above all the "flesh or clay" into which Urizen has degenerated. The "clod of clay" observed by the Eternals then represents the endpoint of Urizen's initial trajectory, rather than a point of origin, as in Genesis 2:7. Blake also derives from geology and natural history a formidable vocabulary to describe the "unorganiz'd" state of Urizen: he is "rifted with direful changes" (9:6); he is a "surging sulphureous / Perturbed Immortal mad raging / In whirlwinds & pitch & nitre" (10:3–5); and his sleep is "like a dark waste stretching chang'able / By earthquakes riv'n, belching sullen fires" (12:3–4). This vocabulary suggests a plutonic origin for the earth that embodies the state of Urizen's mind. The formulation "flesh or clay" expresses the identity of these raw materials. The uniform softness or fluidity of this "flesh or clay," the "sulphureous fluid," pitch, niter, and so on, generates a need for the definite hardness of solid rock or bone. Los's technique of forging a solid chain to bind these changes out of molten iron and brass (12:29–30) parallels the state of the material. The whole account presents the body as a geological formation, a process defined rather than created by Los, seen in plate 12 as athletic geologist. Urizen's geological activity, on the other hand, his furious creation of mountains in the early part of the narrative, can only culminate in his own rocky materialization.[18]

The bones and organs form under Los's watchful eye as geological features forming out of a primordial plutonic chaos. The process commences with the skull ("a roof shaggy wild inclos'd / In an orb. his fountain of thought"), followed by the spine,

> Like the linked infernal chain:
> A vast Spine writh'd in torment
> Upon the winds; shooting pain'd
> Ribs, like a bending cavern

18. According to Stuart Peterfreund, Urizen refuses to take definite form just as Newton's God is "utterly void of all body and bodily figure" and as Newton himself refuses to "form" hypotheses (*Blake in a Newtonian World*, 29). Peterfreund interprets the performance of Los as likewise allegorizing a Newtonian creator who refuses responsibility for his creation, pretending it is "natural" and "out there"—and thus becoming what he beholds (30–31).

>And bones of solidness, froze
>Over all his nerves of joy.[19]

A later passage amplifies the trauma of solidification, figured here as the solidification of rock and ice: "Nerves change[d] into Marrow: / And hardening Bones began" (25:23–24). On the anatomical level, no change expresses more rigorously the confinement of sensation: nerves are the superlatively sensitive, soft organs of unbounded "joy" or "prolific delight." The trauma of anatomical hardening occurs throughout *Jerusalem*: "the interiors of Albions fibres & nerves were hidden / From Los. astonishd he beheld only the petrified surfaces!" (32[46]:4–5). As with Newtonian space, the opaque surface obscures and confines transparent depths; solid surfaces are the structural principle of both vegetating consciousness and the fallen nature it beholds (hence Erin's prophetic exclamation, "Remove from Albion these terrible surfaces!"; 49:76).

Absolute sensitivity appears only as chaos (Los's "formless unmeasurable death") from the deluded natural perspective. To restore a kind of order, then, the "surgeing sulphureous fluid" of Urizen's "prolific delight" must harden into a spine and ribs, just as molten rock solidifies into cavern walls. In the six subsequent "Ages" and "states of dismal woe" that make up the Changes of Urizen, the heart and arteries, eyes, ears, nostrils, digestive system, and extremities form on a succession of more or less geological models (plate 13; plate 15:1–19). The red-hot fluid of the heart and arteries emerging from the "cavern" of the ribs suggests an eruption of lava; the eyeballs are confined to "caves" in the skull; the ears are simply "petrified"; and the stomach is a "Hungry Cavern," producing a "channeld Throat." A striking passage in *Jerusalem* doubles this movement in a different context and produces a verb to express it. The narrator here describes the consequences of Albion's flight from the Divine Vision for the "Inhabitants of Albion," who

---

19. 12:33–41. The parallel text in *The Book of Los* is of interest here (4:54–57, E93):

>The Lungs heave incessant, dull and heavy
>For as yet were all other parts formless
>Shiv'ring: clinging around like a cloud
>Dim & glutinous as the white Polypus.

The image of Urizen's spine has a surprising Darwinian analogue in a *Botanic Garden* passage on the crocodile, first pointed out by D. C. Leonard, "Erasmus Darwin and William Blake." Darwin describes how the crocodile "burst into life" following a sequence somewhat reminiscent of Blake's in the Changes of Urizen (*BG* I.iv.425–28):

>First in translucent lymph with cobweb-threads
>The Brain's fine floating tissue swells, and spreads;
>Nerve after nerve the glistening spine descends,
>The red Heart dances . . .

See also Hilton, "Spectre of Darwin," 39–40.

> Feel their Brain cut round beneath the temples shrieking
> Bonifying into a Scull, the Marrow exuding in dismal pain
> They flee over the rocks bonifying.
>
> (58:6–9)

The image of the skeleton on this plate also links it to the Changes of Urizen sequence, especially to *Book of Urizen* 11.

The coinage "bonify" ("not found in the *OED* in this sense," Paley remarks dryly) expresses on a lexical level the conjunction between the Urizenic body and the geological earth. "Bonify" clearly echoes another Blakean coinage, "stonify" (which could be taken as a more literal variant of "astonish").[20] The two processes are identical in the Changes of Urizen. Blake interprets geology as anatomy in this episode, in accordance with the principles of human-centered cosmology; but the anatomy is significantly shaped by geological terms and concepts. By emphasizing plutonic processes such as earthquakes and volcanism (*Urizen* 12) and associating them with cosmogony, Blake creates a world akin to Hutton's dynamic earth. Blake depends on geological associations to ground his creation myth on a notion of primitive matter as generally understood. Urizen's creation of a world of solid surfaces draws on the recognizable properties of rocks as raw material—hardness, opacity, resistance, massiveness. These tropes allow Blake to figure that world as a product of corrupted human faculties rather than the "foundation" of human and other organic life. The Darwin-Hutton version of primitive matter clearly also held a political charge (especially in the 1790s) that increased its appeal for Blake, despite his philosophical ambivalence toward it.

The influence of Thomas Burnet and George Cumberland—both diluvialists—on Blake's cosmogony is geologically opposed to that of Darwin and Hutton and lacks its political energies. Burnet's *Sacred Theory of the Earth* (1684)—a verbal and visual presence in Blake's work, and still widely admired in his time—was originally held to be dangerously heterodox, if on theological rather than political grounds.[21] When Blake's friend George Cumberland embraced the same premises over a century later, however, his science was seen as deeply conservative, and the theology accompanying it

---

20. *Jerusalem*, p. 223. Both occur for the first time in *The Four Zoas* (57:2; cf. 4:69); cf. "stonied" (106:22). For a complete list of "stone" and its variants, see David Erdman's *Concordance*, 2:1806–8. The considerably more common "rock" and its derivatives occupy 2:1575–80.

21. Martin K. Nurmi, "Negative Sources in Blake," 312–18; Vincent de Luca, *Words of Eternity*, 153–63; and Morton Paley, "Blake and Burnet's *Sacred Theory*"; see also *Jerusalem*, p. 201. I assume with de Luca that Blake depends on nature and empirical natural history in such a way that "he would regard his images as reflecting accurate science" (*Words*, 154), even if that dependence is "negative" in Nurmi's sense.

suggests a politics equally conservative.[22] Burnet's and Cumberland's theories of the earth, based on the agency of water on the earth's surface, are clearly opposed to those of Darwin and Hutton, based on internal heat. By Blake's time the contest between these two schools of thought had taken shape as the politicized Neptunist-Plutonist debate, detailed elsewhere in these pages. Cumberland's preference for Neptunism and its embedded conservative ideology, when he became a serious geologist well into his fifties, is hard to reconcile with the aesthetics and politics of his earlier work. Blake responded with great enthusiasm to such works as *Thoughts on Outline* (1796), for which he engraved several plates.[23] He also read or knew of Cumberland's picturesque travel narrative, *An Attempt to Describe Hafod* (1796). In addition to a probable reference to the proprietor of Hafod in *Jerusalem*, there are several ironic echoes of picturesque travel writing à la Cumberland in the poem's geographic catalogues, particularly in their emphasis on the primitive.[24] It seems less likely that Blake read many of the numerous scientific essays Cumberland produced in his fifties and sixties. But the title of Cumberland's geological chef d'oeuvre suggests a derivation from the earlier picturesque interest in primitive landscape: *Reliquiae Conservatae, from the Primitive Materials of Our Present Globe* (1826). Blake would have objected to the theology behind this late work, and though they continued to correspond until Blake's death, Cumberland's move to Bristol well before this time must have diminished his influence on the poet. It is also tempting to object—especially in light of Ruthven Todd's unsupported in-

22. Cumberland, though a conservative geologist, was Blake's contemporary and lifelong friend and originally shared more of his political sympathies. (See, e.g., Bentley, *Bibliography of George Cumberland*, xviii, xxii.) The despotic God of Cumberland's geology probably would have appeared to Blake a logical consequence of his earlier deism: "Man has . . . often mistaken his proper offices and forgotten his origin . . . until called back to self knowledge and humility by the exhibition of divine power and justice. . . . The most striking example on record is the universal Deluge, of which the surface of the earth bears testimony, and Geology, or the study of its exterior, comes now in aid of that early tradition and record" (George Cumberland, *Reliquiae Conservatae*, iii–iv).

23. See his letters of December 23, 1796, and July 2, 1800 (E700, 706–7), as well as Robert Essick and Morton Paley, "'Dear Generous Cumberland.'" See further Geoffrey Keynes, "George Cumberland and William Blake," in *Blake Studies*, 230–52. G. E. Bentley's introduction to Cumberland's novel sheds some light on his apparent retreat from radicalism (*Captive of the Castle of Sennaar*, xliv–xlix).

24. Blake's references to Snowdon, Plinlimmon, and the "horrid Chasm" of "Derby Peak," discussed further below, all fall under this category. *An Attempt to Describe Hafod* was in part a testimony to Cumberland's friendship with Hafod's proprietor, Thomas Johnes, praising the "savagely grand" landscapes of his estate. Blake may have engraved the map for this volume. Morton Paley ("Thomas Johnes, 'Ancient Guardian'") has suggested that Johnes is the basis for Blake's "Hereford, Ancient Guardian of Wales" (*Jerusalem* 46[41]:3–4), whose "mountain palaces" suggest a familiarity on Blake's part with Cumberland's description of Johnes' actual mansion.

sistence that Blake and Cumberland "must" have discussed geology—that Blake was simply not much interested in natural history.

Blake's relationship to science is complex, however. It may be impossible to determine whether Cumberland discussed his geological interests with Blake on occasional visits to London or sent Blake copies of his geological writings, as he did copies of his other writings. But the *Monthly Magazine*, which printed Cumberland's earliest pieces on geology, provides a literary locus of connection. A polymathic literary culture, in which periodicals such as the *Monthly* covered all the arts and sciences and much besides, was one avenue of scientific influence on Blake. He is especially likely to have seen the *Monthly* on occasion because he had professional connections to its publisher and himself wrote letters to the editor of the magazine. The *Monthly* also regularly published Cumberland's letters to the editor over a period of twenty-five years.[25] His letters on geology were especially long—a series written in 1815 is equivalent to a substantial essay (installments were published in three consecutive issues). This essay chides "Huttonians" and "Neptunists" alike for "los[ing] sight of the stable foundation of tradition and revelation, *the veracity of Moses*," and articulates a new theory of the Deluge that will vindicate biblical literalism as a scientific practice.[26] The tenor of these letters reflects geology's progress as a discipline: Cumberland airs his theories about the Deluge in this general forum while his empirical papers on fossil anatomy are being read at the Geological Society and published in its *Transactions*. Cumberland's insistence that "primitive" meant "from a former creation" alienated professional geologists such as John Farey (who attacked Cumberland's "mosaical or bible geolog[y]" in a letter to the *Monthly*), but need not have alienated Blake.[27] Whether or not Blake

25. See Bentley, *Bibliography of George Cumberland*, 57–71; one of these letters mentions Blake (57). In *Blake Records*, Bentley describes Blake's extensive interactions with Richard Phillips, the publisher of the *Monthly*, who also published Hayley's *Ballads* (with Blake's designs) in 1805 (156–62). Around the same time, Cumberland was urging Blake to send him a manuscript on his "new method of engraving," offering to "prepare it for the Press" and have it published by Phillips or by William Nicholson, who published Cumberland's pieces in his *Journal of Natural Philosophy, Chemistry, and the Arts* (211–12).

26. *Monthly Magazine* 40 (August 1815): 18. Cumberland develops a chaos in attempting to reconcile Genesis and geology that surprisingly resembles Urizen's chaos (*Book of Urizen* 3:14–17; cf. 19:34) of twenty years earlier: "there is nothing in the word chaos to imply that its component parts were not made up of fragments of a former creation, where vegetables, fish, reptiles, and certain large quadrupeds, might be the natural inhabitants" (19). The other two installments of the article appear in the September and October issues, with a postscript in December.

27. Cumberland became an honorary member of the Geological Society in 1811, thanks to his donations of specimens, and published six papers in the society's *Transactions* between 1817 and 1824. But his role in this increasingly professional milieu seems to have been the relatively marginal one of a provincial fact-gatherer. For the dispute between Farey and Cumberland, see *Monthly Magazine* 52–54 (1821–23).

actually read Cumberland's geological letters, he surely became aware of Cumberland's new interest by around 1815 and as a consequence paid renewed attention to geology. It seems clear, however, that he found richer poetic resources and greater political energy in the Plutonist geology that his old friend opposed.

The web of acquaintances linking Blake and Cumberland includes several other figures who shed light on Blake's relationship to science. William Nicholson's *Introduction to Natural Philosophy* (for which Blake engraved a plate in 1782) shows how the period's physics and chemistry relate to Blake's concern with matter and materialism. Nicholson defines matter unsurprisingly in terms of its typical properties, beginning with extension; more surprisingly, he concedes that "we are totally ignorant of the substratum in which these properties are united. The essence of matter is unknown to us." This post- if not anti-Newtonian position helps to contextualize Blake's treatment of Newton and his interest in matter, which was at the same time the "best established part of Natural Philosophy" and the most unstable. The account of matter, as a whole, is more complex than Blake's critiques of mechanism might lead us to expect. Echoing Joseph Priestley and anticipating Hutton, Nicholson states: "It is not in our power to determine, whether extension or impenetrability be essentially necessary to existence."[28] Priestley's influence on Blake, both scientific and philosophical, has been traced specifically to his *Disquisitions on Matter and Spirit*.[29] Thomas Taylor, another common connection of Blake and Cumberland, satirizes the Jacobin-rationalist inversion of metaphysical hierarchies that Blake picks up from Priestley and elsewhere.[30] Priestley, Nicholson, and Cumberland—all loosely connected via the publishing world—are likely to

28. Nicholson, *Introduction to Natural Philosophy*, 1:7–8, xi, 17. Nicholson's privileging of demonstration (1:4–6)—all the more because he claims to be writing a textbook or epitome—illuminates Blake's use of that word. Blake in turn is apt to connect demonstration with geological condensation. As Albion declares, "I therefore. condense them into solid rocks. stedfast! / A foundation and certainty and demonstrative truth" (*Jerusalem* 28:9–10). On the Blake-Nicholson connection, see Christopher Heppner, "Another 'New' Blake Engraving." For details on Cumberland's articles in *Nicholson's Journal*, see Bentley, *Bibliography of George Cumberland*, 58–61.

29. This work—also published by Joseph Johnson—appeared in 1777. Nelson Hilton, "Blake and the Perception of Science," traces the idea that force is prior to matter (56)—a cornerstone of "Romantic science"—from a 1727 critique of Newton through Priestley (58) to a culmination in *Jerusalem* 98 (61). See further Mary Lynn Johnson, "Blake, Democritus," 116.

30. Along with attacks on Paine and Wollstonecraft, Taylor's *Vindication of the Rights of Brutes* contains a satirically distorted scientific narrative alluded to in Blake's *Visions of the Daughters of Albion* (5:8–9/E48; cf. Taylor 81–83 and 76 ff.). Taylor concludes his satire with a prolegomenon to a wholesale vindication of the rights "of vegetables, minerals, and even the most apparently contemptible clod of earth" (103). It is tempting to speculate that the apparently absurd idea of mineral rights—clods of clay's manifest lack of any rights—might have contributed to Blake's adopting petrifaction as his favored image of dehumanization and op-

have impressed Blake with the privilege attached to natural philosophy. Newton is, after all, Blake's privileged representative of scientific discourse. As noted above in the analysis of the Changes of Urizen, Blake's account of void and solid and of condensation has more to do with Newtonian science than with geology. The prevalence of rock imagery in such passages is due as much to the evident power such imagery had in Blake's culture to metonymize abstract materiality as to any interest in geology per se. This power motivates my reading of Blake's science. Other critics, including Donald Ault and Stuart Peterfreund, have opened important dimensions of Blake's poetic construction of nature by concentrating on his use of Newton. Ault explains that Blake's descriptions of solidification serve his competition with Newton as a system-builder and mythmaker, suggesting that the transformations of thoughts and feelings into rocks in *Jerusalem* are subversively anti-Newtonian.[31]

Though geology and physics combine to reinforce Blake's account of materiality, his use of geology differs in some important ways from his relationship to Newtonian science. Geology was barely in embryo when Blake was born, and still an "almost new branch of science" (as Cumberland put it in his *Reliquiae*) even in 1826, the year before his death. Rocks and mountains must have appealed to the later Blake because they represented a territory not yet claimed by one established "system." Geological vocabulary was more general and more literary than the specialized vocabulary (e.g., "vortex," "fluxion") that Blake used in passages satirizing natural philosophy. But he knew enough about geology to draw on the scientific associations beginning to cluster around rocks and mountains. This attitude certainly does not amount to an endorsement of geology or a geological sublime, but it represents an artistic opportunism much less conflicted than his ambivalent use of Newtonian (and Cartesian) physics. Ault's surprising analogy between Blake and Thomas Kuhn illuminates this difference: Blake views Newtonian science in quasi-Kuhnian terms as the product of a paradigm shift, an exclusionary redefinition of a field of inquiry.[32] Mark Greenberg's pithy observation on Blake's science makes a similar point: " 'Science' becomes for

pression. The fullest discussion of Blake's probable allusion to Taylor is Nelson Hilton, "An Original Story," 94.

31. In Newton, condensation and contraction are necessarily "operations within the physical realm itself" (Ault, *Visionary Physics*, 88). But Blake's point is that this only seems distinct from the mental: "Blake provides an explanation . . . for the very existence of substances in the external world. Blake even draws on the implications of the Newtonian doctrine, by making the solid substances, created by condensation and contraction of mental entities, seem to be the only reality, a reality which is not transformable out of its current solid, heavy state." See further *Visionary Physics*, 30, 48–49; and Peterfreund, *Blake in a Newtonian World*.

32. Ault, *Visionary Physics*, 51.

Blake a kind of demonic synecdoche—the whole word symbolizing only a part of its former meaning."[33] Greenberg and others define Blake's corrective use of science with reference to Los's redemptive explanation of the "systems" of "Demonstrative Science" as "Giving a body to Falshood that it may be cast off forever" (*Jerusalem* 12:12–14). Geology, as preparadigm science, is particularly suitable for the poetic work of Los. Blake's history of petrific bodies and sense organs, primitive, alien landscapes, and environmental upheaval is, like geology, a new kind of natural history. This history intersects with early geology to the extent that both require an account of materialization and build it on the common perception that rocks embody the primitive.

## Primitive Matter as Scientific Topos

The geological category of primitive rock has a rich cultural history. Humphry Davy's account of "the distinctions between the primitive and the secondary rocks," in his 1805 lectures on geology, relies on the relatively stable distinction between rocks containing no fossils ("igneous" and "metamorphic" in our terminology) and fossiliferous sedimentary rocks, respectively. By this definition, "All substances that contain organic remains that bear evidence of having been produced since the existence of living beings on the globe are excluded from the series of primitive masses" (*Lectures*, 60). Davy also touches on several points raised in the meditations of such literary figures as Goethe and Dorothy Wordsworth. In one of the narratives incorporated by William Wordsworth into his *Guide to the Lakes*, she refers to scattered stone blocks on Scawfell Pike as "bones of the earth" (here the metaphor also recalls the Changes of Urizen). Similarly, Goethe responds to a mass of granite, on the summit that forms the setting for his "Über den Granit" (1784), as a remnant of the oldest materials of the creation, the *Urwelt*. As evidence for this feeling, Goethe points out that the highest and the lowest places of the earth are composed of this primitive rock.[34] Davy's definition also emphasizes this distribution, and his popular lectures share the poets' interest in formal properties, in primitive rock as primal object.

---

33. Greenberg, "Blake's 'Science,'" 125.
34. Goethe, "Über den Granit," in *Geologische und mineralogische Schriften*, 323. For a translation of the essay (actually a fragment from a projected novel), see *Goethe's Collected Works*, 12:131–34. Goethe reviews the scientific record and points to contemporary debates about granite, but insists that primitive rock is worth studying because it is sublime and ultimately unknowable, and hence that geology itself should be formed on a more metaphysical basis. For an extended discussion of Goethe's geology, see Helmut Hölder, "Goethe als Geologe."

These lectures relied heavily on specimens, which the audience saw and admired and probably sometimes handled.[35] Like Goethe, Davy refers to ancient Egyptian sculpture made of granite, "wholly unimpaired by time," concluding that "no rock is grander in form nor more sublime in structure" (61–62). He draws on Jean-André de Luc for his final definition, noting that "from his infancy [de Luc] had been accustomed to the contemplation of the grandest and most elevated of the mountain chains in Europe" (53). Davy's definition emphasizes the importance of primitive rock to geology as a science now constituting itself as a historical one: "When the word 'primitive' was first applied . . . it signified that matter of our globe as yet unchanged by any known natural operations. In this sense, its meaning was definite and it was sufficient for all the purposes of science. In every system some primordial state of things must be allowed. And where we can perceive no certain indication of a prior arrangement, there it is reasonable . . . to fix the foundations of our science" (68–69).

James Hutton (1726–97) would have agreed with both Goethe and Davy on the unknowable origin of plutonic rocks, but he nonetheless rejects the idea of "primitive" mountains. Though analysis and induction predominate over wonder in Hutton's *Theory of the Earth* (1788/1795), it comes close to realizing the rigorous science founded on aesthetic response that Goethe envisions in his fragment on granite. He decides the questions that Goethe leaves as undecidable—the method is empirical and not metaphysical; heat, not water, is responsible for the fusion of granite; and rocks are purely architectonic principle, not ruins—but preserves the aesthetic moment by calling up aesthetic response as evidence for the magnitude of the plutonic forces he is positing. Hutton's is an aesthetic response to the power of nature, coming from a sensibility as close to Goethe as to Davy. As Dennis Dean has pointed out, Hutton's attempt to disentangle "natural operations" from divine intervention suggests deism, also evoked by Goethe's image of granite as "eternal altar."[36] Hutton shares with Davy his reliance on aesthetic categories and the principle of an unknowable origin. But these shared premises lead to widely varying conclusions on the subject of primitive rocks. While Davy sees the unknowable origin of these rocks as a limit on the scope of geological science, Hutton wants to make it an enabling condition, a necessary ground for inference. Hutton rejects entirely the standard definition of

35. The manuscripts of Davy's geology lectures, explicitly aimed at a "general audience," indicate at what points he showed specimens, often several times per page. The painting serving as the frontispiece to this book further illustrates the visual aesthetic of these lectures (see also *Lectures*, xxxiii).

36. Goethe, "Über den Granit," 324; cf. 322. Dean presents deism as a key aspect of Hutton's intellectual context, the Scottish Enlightenment; *James Hutton*, 2, 5–6, and passim.

primitive or primary rocks still espoused by Davy. He finds legible traces of
their formation in the chemical composition of these materials (an approach
rejected in Davy's critique of Hutton; *Lectures*, 57). To make a case against
the established notion of primary or originally solid rock, Hutton must re-
examine both the concept and the substance of primitive materiality, establishing an origin for solidity itself. As in earlier theory, observable rocks and
landforms become a textual surface signifying the operation of immense elemental forces. But in Hutton, the text becomes more legible. By inquiring
minutely into the nature of primitive materiality, Hutton makes the earth's
material more available as a comprehensible natural resource; but at the
same time, he imagines more vividly, on the basis of his new knowledge, the
immensity of the forces at work in generating this material.

Hutton offers empirical evidence for his contention that the so-called
primitive rocks do contain organic residue, thus upsetting the strong or-
ganic/inorganic distinction used to support the standard definition of primary rocks as original matter. He takes his evidence from observations
made in the Lake District, the "alpine schistus country of Cumberland"
(*TE*, 1:330). This "schist" or slate provided probably the best-known English example of primary rock and primitive mountains. Hutton reports on
finding a specimen of this rock that appeared to contain fossils, a report cor-
roborated, he says, by his landlord (an innkeeper on the shore of Windermere), who "had seen evident impressions of marine objects . . . in the slate
of those mountains" (331). Hutton's specimen, having been "ground and
polished," is now "most evidently full of fragments of entrochi. The schistus
mountains of Cumberland were," he concludes, "as perfect primitive mountains as any upon the earth, before this observation; now they have no claim
upon that score, no more than any limestone formed of shells." This narrative concludes Hutton's long examination of naturalists' records of the Alps,
implying that observed fossil evidence proves that even the Alps are not really primitive mountains. Hutton's test would not hold if applied rigorously
to the central peaks of either mountain range, which are volcanic and metamorphic, with igneous intrusions; but had he known this, it would only have
strengthened the point that the vast majority of alpine rocks have been ob-
servably altered and so are not "original."[37]

The inorganic components of primary rock can be chemically analyzed, a

---

37. The rock Hutton refers to as both "slate" and "schist" may be what is now called the
Skiddaw Group of metasedimentary rocks, chiefly slate; slate is now distinguished from schist
as the product of a lower-grade metamorphism. He may also mean the volcanic rock of which
the peaks between Ambleside and Keswick are formed, also referred to as "slate" in the period.
Either one of these would naturally have been known for containing no fossils. Windermere,
however, is located south of the main peaks of the Lake District; the Silurian rock here is

fact complicating the category of primitive matter even in the absence of organic residue. Davy essentially reiterates what Goethe had observed in 1784, that the elements of granite could be identified, but their fusion could not be explained on the basis of any observed process or agency. But Hutton focuses on the crucial implication that these component minerals must have existed in a prior, simpler state. A chapter titled "The Supposition of Primitive Mountains Refuted" contains his most concerted reply to the objection that *primitive* mountains cannot have had an igneous origin, as he argues all rocks did. Here he particularly addresses the problem of granite, with its special claim to primary status. Hutton builds his solution—that all solid rock was once molten—on the observation that the mineral components of granite occur in varying proportions. Such chemical analysis proves that no substance, including granite, is simple enough to be original (*TE*, 1:312). "No part should be considered as original," Hutton argues, "in relation to the globe, or as primitive, in relation to second causes, *i.e.*, physical operations by which those parts should have been formed" (311). Granite is not older, but it is more durable than other rock, and that is why the highest mountains, at the center of things, are made of it (315). Its formlessness, moreover, does not merely disprove its formation by marine deposition (as other theories maintained), but proves instead that it was once molten, or "subterranean lava" (317–18). This account disputes not the relative, but the absolute age of "primary" material; all materials have originated in the same way, but at different times (323). For Blake, too, no rock is original, but in his case this is because "everything is human" (*Jerusalem* 34[38]:48); Blake's omission of fossils could, then, be explained by the supposition that all rock is fossilized human anatomy, as in the Changes of Urizen.[38]

Primitive materiality is thus displaced from the solid surface of landforms to the depths of a dynamic earth, where rock comes into being under the influence of heat and pressure as magma (though not yet so called), in which solid particles are recycled. This is the primitive, though not original, state of matter. The appeal to an *Urwelt* is for Hutton little better than superstition, and the idea of primitive mountains a myth: "this form of discussion, with regard to a physical subject, is but a mere concession of our ignorance" (*TE*, 1:387). This kind of rhetoric shows the side of Hutton least amenable to Romanticism. But in his conjectures about subterranean heat and pres-

---

younger shale, unmetamorphosed, and hence—the whole having been a seabed—contains some fossils. See John Whittow, *Geology and Scenery in Britain*, 199–208.

38. Thanks to Steven Goldsmith for this suggestion. Hutton's case against original rocks could be taken as a veiled polemic against diluvialism, likewise contested by Blake. The fact that the "physical operations" altering all rocks are partly unknown does not justify, for Hutton, the invocation of supernatural agency. See further Stephen Jay Gould, *Time's Arrow, Time's Cycle*, chap. 3.

sure, however plausible and reasonable, the "occult cause" he seeks to banish is replaced by another kind of occult cause. Hutton operates as fully as Whitehurst or Davy in the mode of "aesthetic geology," vesting his account of the natural agency responsible for forming rocks with all the trappings of the sublime. In the process, he develops a concept of primitive matter built on a fluid state that precedes solidity but dissociated from the origin and its theological connotations. If molten rather than solid rock is the archetype of primitive, unformed matter, then such matter is the locus of even greater indeterminacy than the various inscrutable stones and rude rocks thus far considered. It is entirely a natural production involving unfathomable amounts of energy that cannot be harnessed for human applications.

I use the term "occult" not to impugn Hutton's scientific authority, but to point out the surprising affinities between his necessarily conjectural description of this energy and the languages of Romanticism, his willingness to write on subjects "beyond the reach of our faculties." Besides the general echoes of discourses of reading and the sublime, Hutton's paradigm of original subterranean fluidity displays a particular affinity to Blake's descriptions of his printing process. Blake famously refers to this process as "printing in the infernal method, by corrosives, which in Hell are salutary and medicinal, melting apparent surfaces away, and displaying the infinite which was hid."[39] The process and its materials are allegorized in the next plate of *The Marriage of Heaven and Hell* as a universal dissemination of knowledge in the form of "living fluids," molten metals. In this sense, Hutton's new way of reading the landscape can also be seen as a way of writing, as a way of harnessing an empirical infinite to an expanded or "improved" representation. As much as Blake later insisted on the ideality of phenomena, the emphasis here is on the phenomenality of the ideal, on the role of the corrected senses, and it seems likely that he is indebted to natural history for this notion of a reality with hidden depths.

Solid rocks and landforms are a textual surface that becomes legible as the myth of the primitive is set aside. "We must read the transactions of times past," Hutton writes, "in the present state of natural bodies" (*TE*, 1:373). Unlike earlier geological readers, Hutton makes the textual metaphor explicit and extends it beyond (or below) the sedimentary strata that were already recognized as a rock record. His method of reading arises in answer to a question that highlights Hutton's distinctively radical premises: "How shall we acquire the knowledge of a system calculated for millions, not of years only, nor of the ages of man, but of the races of men, and the successions of empires?" (372). Hutton's contemporaries simply

---

39. *The Marriage of Heaven and Hell* 14:13–16 (E39). I am indebted in this paragraph to suggestions from Leo Damrosch and Steven Goldsmith.

rejected this vast timescale, which gradually became conceivable in the course of the nineteenth century. Yet he merely begins by observing, like others before him, that "it is impossible" dislocated strata "could have originally been formed . . . in their present state and position" (128). Hutton takes the established premise that nothing deformed can be original and turns it to his advantage by reading deformation as a recurring process immanent in the natural world: what seems "another power, which had introduced apparent confusion," becomes "instructive, with regard to what had been transacted at a former period of time" (129). He insists that this "power" can be accounted for by inference from observed phenomena, as ordinary miners do when they affirm that "there is but one place from whence these minerals have come; this is, the bowels of the earth, the place of power and expansion" (130). Veins of metal contain matter that is foreign to the earth's surface, and so must be considered as "the continuation of that mineral region, which lies necessarily out of all possible reach of our examination" (130–31). Hutton reads these veins or intrusions as signs of a limited dislocation of the strata, which in turn implies the possibility of "violent fracture and unlimited dislocation," characteristic of earthquakes (133). Earthquakes, too, provide a plainly legible text in Hutton's reading: "when fire bursts forth from the bottom of the sea . . . there is nobody but must see in this a power" capable of erecting the continents themselves (139–40).[40]

The sublime depths corresponding to Hutton's surface text of deformation are the site of inconceivable heat and pressure whose representation requires aesthetic categories. "The place of mineral operations is not on the surface of the earth," Hutton proclaims, "and we are not to limit nature with our imbecility, or estimate the powers of nature by the measure of our own" (*TE*, 1:94; see Fig. 3.1). All consolidation is the result of fusion, an operation "which has been transacted at great depths of the earth, places to which all access is denied to mortal eyes" (98). Hutton argues by analogy that unregistered degrees of heat and pressure at literally "unfathomable" depths beneath the sea can melt any mineral substance (which Davy denies outright). Natural catastrophes provide the only visible index "that these operations of the globe remain . . . in the fulness of their power" (141). Hutton mobilizes the high style conventional for describing natural disasters—reminiscent of the sublime of the Hebrew Bible analyzed by Robert Lowth—to gesture toward the magnitude of force only shadowed forth by these events: "when

---

40. Volume 2 of the *Theory* is a sustained reading of the surface of the earth that supplies causes for its various features from among three constant operations: stratification, deformation, and weathering (2:4). The burden of this reading is to appeal convincingly to the signified of subterranean forces and "amazing power" established in volume 1.

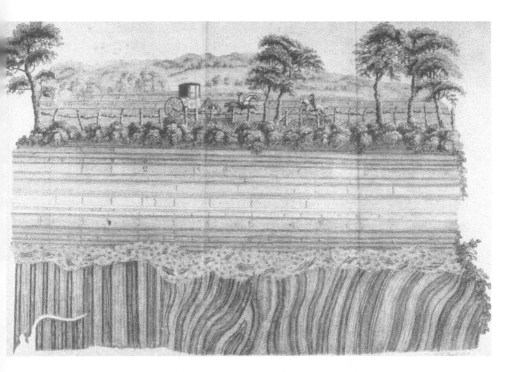

3.1 "Jedburgh Unconformity," from James Hutton, *Theory of the Earth* (1795), vol. 1, plate 3. Engraving by D. B. Pynt from a drawing by John Clerk of Eldin. Courtesy of the Linda Hall Library of Science, Engineering, and Technology, Kansas City, Mo.

fire bursts forth from the bottom of the sea, and when the land is heaved up and down, so as to demolish cities in an instant, and split asunder rocks and solid mountains . . ." Hutton makes use of aesthetic categories to adumbrate the otherwise unknowable processes of rock formation, which Davy merely alludes to as "the operation of unknown powers" (*Lectures*, 96). For Hutton, the sublimity of natural catastrophes entails a universal process and agency much wider in scope than isolated catastrophic occurrences. If the force of an isolated earthquake is sublime, then it becomes conceivable that generalized subterranean heat and pressure could be responsible for the formation of the earth itself. In other words, the premises of Plutonism—that all rock is capable of fusion, that subterranean heat and pressure are constantly renovating the earth's material from below, that landforms evolve over a vast period—are partly consequences of the wonder prompted by rocks and natural catastrophes, and an analysis of this wonder.

Hutton's scientific reasoning and his aesthetic response interact most visibly in his account of volcanoes. He first lays out the argument that "subterraneous fire" both consolidates the strata and raises entire landmasses above the sea (*TE*, 1:121–25). After positing the necessary subterranean forces, Hutton cites descriptions of earthquakes and volcanoes as the best available illustrations of his principle. He pauses after a spectacular description of Etna to reinforce the legitimacy of his statement of the geological problem and the necessity of a Plutonist solution: "Has the globe within it such an active power as fits it for the renovation of that part of its constitution which may be subject to decay? Are those powerful operations of fire, or subterraneous heat, which so often have filled us with terror and astonishment, to be considered as having always been?" (143). The terror and astonishment here constitute the best evidence that these operations are continual and universal—that is, the magnitude or intensity of this general aesthetic response must stand in for the unattainable empirical proof that these operations "have always been." Hutton invokes aesthetic response to supply the gap created by cognitive indeterminacy.[41] The philosophical necessity of a plutonic principle and the aesthetic impact of its surface text are more than adequate, as these rhetorical questions insinuate, to motivate the inferences about otherwise unknowable processes. As "a spiracle to the subterranean furnace," the volcano is an integral and purposive part of nature's system for the formation of solid matter and not mere "accident" or "in [itself] an end, for which nature had exerted such amazing power" (146–47).

Hutton's notion of primitive matter as fluid differs greatly from the various prevailing accounts of solid primitive rock. His *Theory*, accordingly, was widely attacked for decades after its publication—not a surprising response to a theory that attempts to pull the solid ground from under our feet (among other controversial gestures). But Hutton's Plutonism still bears out my larger argument that aesthetically motivated inquiry into rocks generated widely shared speculation about the otherness of the physical. Here, as in Wordsworth, Shelley, and various writers on landscape, that otherness takes the form of a primitive materiality thrown into relief by aesthetic response and a resistance to cognitive operations. The idea of rock as the primitive form of matter has intuitive appeal for Hutton, as a geologist who sees it as the basic material of the earth; but he constructs a genetic account in which the primitive state of that primitive matter is a fluid one conditioned by the "amazing power" of subterranean heat and pressure. In this state, physical substance is not merely (like the trope of "rude rocks") in-

---

41. The point here, once again, is not to deconstruct Hutton's claim to scientific truth, but to identify the degree to which literary language constitutes that truth. The larger problem of geological time famously requires such recourse to "myth and metaphor," as Gould puts it. Hutton is confronted with one of the earliest versions of the problem of expanding the scope of our perception (again in a somewhat Blakean way) from a scale of 6,000 years to an indefinitely larger one of "deep time"—for us, 4.5 billion years.

scrutable but empirically unknowable; its indeterminacy is not merely formal but absolute; its force and its magnitude are greater than anything apparent on the surface of the earth; and its radical otherness finds expression in a defiance of human "imbecility" and inaccessibility to human technologies.

Universal mineral fusion becomes an axiom for Hutton because—ironically, in view of the outcry against his theory—no theory is needed to remove solidity as an essential quality of matter. As in the case of frozen water, present solidity naturally implies past fusion; given that the basic types of minerals are insoluble in water (*TE*, 1:48), past fusion equally implies original heat (39–40). The axiomatic nature of liquefaction in Hutton recalls Locke's treatment of solidity in the *Essay Concerning Human Understanding*. Locke concludes his chapter on solidity with the following observation: "If any one asks me, *What this Solidity is*, I send him to his Senses to inform him: Let him put a Flint, or a Foot-ball between his Hands; and then endeavor to join them, and he will know." Such tactile examples convince us, Locke argues, that solidity is "the *Idea* most . . . essential to Body, so as no where else to be found or imagin'd, but only in matter"; solidity allows us to think of bodies as filling space or having extension, and ideas of space and of the universe are negative forms of it.[42] It is not necessary to embrace Lockean epistemology in order to see that an idea of the physical world as a tangible place outside the body arises from physical contact with bodies substantial enough to feel. Rocks act as representative bodies because they display the qualities required for such physical knowledge, but little else that is conducive to cognitive determinacy. The sense that they stand for materiality in general therefore comes frequently from aesthetic response. From this perspective, they are primitive matter in the sense of being inorganic, merely matter, but also full of the mystery of that alien state. Natural history also designates rocks as primitive matter because they are the oldest things. Hutton only changes their state, but retains the premise, satisfying the requirements of natural history.

Of all the great poetry of the Romantic era, Blake's deals in the most general way with solidity and its relation to materiality, largely because of his preoccupation with Bacon, Newton, Locke, and their scientific legacy. *Jerusalem* develops this interest more fully than the earlier prophecies and makes it more central to Blake's cosmogony. Following Blake's critique of Locke, the construction of the material world might be termed a regime of

---

42. Locke, *Essay Concerning Human Understanding*, II.iv, 126–27, 123. Boswell's anecdote of Johnson kicking the stone in order to "refute" Berkeley amplifies the link here between solidity and materiality (*Life of Johnson*, 333).

materiality; the measurable dimensions of a Lockean-Newtonian spatial universe famously restrict human vision for Blake. This "opake" epistemology intersects with the ontology of Generation to create a philosophical double-bind that makes the vegetative state of human faculties seem both original and permanent. Blake crosses the divide within natural history to reveal this permanent generation as a paradox, creating a world of "caverns rooting downwards . . . / . . . rock and stone in ever painful throes of vegetation."[43] The living elements of this scenario imply a prior state, the fiery state of Blake's Eternity. This narrative—the solid, rocky state considered the primitive form of matter must be preceded by a fluid one, out of which the limited solid form congeals—strikingly resembles Hutton's. Blake's interest in this morphology is not, of course, geological; but his readiness to adopt rock as a consistent symbol for a restricted physical universe suggests a deeper connection between ideas about rock and ideas about materiality that underlies both poetry and geology. Moreover, Blake's project of converting solid back to fluid is conceptually aligned with Hutton's project, despite broad differences of form and content. Both projects are revolutionary in the sense that they must overturn what were seen as natural structures of human perception, embodied in the polarity of fluid and solid, active and inactive, living and dead. Hutton's molten rock is "primordial substance" without solidity. But his endeavor to establish an origin for fundamental qualities of the physical universe otherwise considered original with rock (such as stability and duration), his challenge to the notion of primitive solidity, places even greater stress on the category of primordial substance.

## "On These Rocks"

Blake, in this context, can be seen as posing another kind of challenge to such categories and offering another explanation for the origin of solid substance. Blake's rocks provide an example of the way in which poetic constructions of aesthetic response resemble and compete with scientific explanation. The aesthetic response to nature seldom seems as immediate in Blake as in the other Romantics—though the letters from Felpham and occasional descriptive passages such as those on the lark and wild thyme in *Milton* show a capacity for such response.[44] Description commonly entails explanation in Blake because rather than dramatizing a response to recognizable natural phenomena, he interrogates those phenomena as features of

43. *The Four Zoas* 74:12–13/E351.
44. *Milton*, E130–31; cf. 136.

a cosmology. So, for example, the cliffs of Albion (by which Blake seems to mean chiefly the chalk cliffs of the south coast) become the excrescence of a diseased mind fashioned forth in order to bind perceptions within a measurable compass and, in social terms, to bind the affections with moral law. These cliffs, then (as in the first plate of *Jerusalem*, chap. 1), do not only represent, but become such confining or condensing agents in the prescriptive cosmology of Newton and Locke (as Blake sees them). The project of *Jerusalem* is to liberate Albion from this rocky state, and Blake relies on natural history's strict division of the organic from the inorganic to show this liberation (95:2–4):

> Albion mov'd
> Upon the Rock . . . in pain he mov'd
> His stony members, he saw England. Ah! shall the Dead
>    live again.

Like Hutton, Blake sees solid rock as a "condensation" demanding an explanation of materiality. Blake's explanation, however, serves his famous challenge to mechanist philosophy, his project of reclaiming and redeeming the physical through art.[45] This challenge generates another important paradox in the cosmology of *Jerusalem*: the "starry Wheels" of the geometric cosmos open onto a space that is simultaneously "enlarg'd without dimension, terrible." Mechanist space is the premise of the chaotic void; but the forge of Los redeems this paradoxical identity of overdefinite form and formlessness: "It / Became a Limit. a rocky hardness without form."[46] The sheer fermenting chaos of vegetable incarnation demands the uniform surface and solidity of rocky condensation. "The rock: the stone: the metal: / Of Vegetative Nature" (73:20–21), which conclude the sequence of creations from "Los's Furnaces," mark a "Limit of Opakeness" (27) because they represent unformed matter (the "rocky hardness without form" above), the point beyond which even generated forms cannot degenerate. The image of "Los's Furnaces" clearly preserves the Vulcanist/Plutonist association (coming from Hutton

---

45. As Morton D. Paley writes in *The Continuing City*, "it is with such unpromising material [as physical nature] . . . that Los must work to create a structure for reality, making possible regeneration and redemption" (195). Whereas Hutton takes us beneath the surface of the earth to explain the continuity of the ground under our feet (cf. Locke, *Essay Concerning Human Understanding*, 123), Blake takes us to a place before topography, before resistance and opacity.

46. *Jerusalem* 5:4–5, 73:22–23. Paley points out a possible allusion to Genesis 1:2 in line 23 (ibid., p. 252). Los, in his Vulcanic guise, forges the limits of opacity (Satan, rocky hardness) and contraction (Adam, fallen perception), which prevent the human world and capacities from disintegrating any further. Such limits are a pervasive concern in Blake; see also *The Four Zoas* 56:17–21, *Marriage of Heaven and Hell* 4:9, and the Changes of Urizen.

via Darwin) between the earth's internal heat and the production of matter. Hutton's notion of a textual surface of landforms, with their subtext of occult causes and plutonic deformation, finds several analogues in Blake, beginning with his iconic chalk cliffs. The slab of rock that functions pictorially as a cliff also appears in vertical close-up as a cabinet specimen of platy limestone or as a stack of stone tablets, of which the top one is inscribed with the text at hand (4:1–34). From any perspective, "natural" and "artificial" are deliberately hybridized in this image, which becomes a figure for the particular meaning of "solid rock" in Blake's lexicon: a textual surface into which Albion (or, later, his Sons and Daughters) condenses living substance. This inaugural "wall of words" performs the synthesis, as noted by Vincent de Luca, between the natural sublime and the densely inscribed plates (or "verbal outcroppings") of *Jerusalem*.[47] The design with its embedded text is most important for being a chunk of England, the beginning of an epic poem set "on these rocks," as William Hayley advised in his *Essay on Epic Poetry*—an epic in which Albion is "a Rocky fragment from Eternity hurld" (54:6) and "London is a stone of [Jerusalem's] ruins" (29[43]:19).[48]

This complex image, incorporating the text, embodies the "hard restricting condensation" (*Jerusalem* 73:21) already thematized in the text as the poem begins: Albion refuses the Saviour's appeal, extolling demonstrative certainty over "indefinite" faith and confining humanity within the English hills and mountains (4:26). "My mountains are my own," he proclaims (28), with a probable allusion to the Mosaic tablets: "here will I build my Laws of Moral Virtue" (30). Blake goes on to name these mountains, creating the first of *Jerusalem*'s many catalogues of place-names. Morton Paley points out that "what may at first appear a farrago of place names is actually carefully chosen to bring out a rich texture of meaning," and in this case Plinlimmon and the Malvern and Cheviot Hills all mark internal borders within the island, signifying Albion's fragmentation.[49] Geological condensation becomes clearer in a later speech of Albion, his condemnation of the "ornamental" Edenic landscape (28:9–11):

> These hills & valleys are accursed witnesses of Sin
> I therefore. condense them into solid rocks. stedfast!
> A foundation and certainty and demonstrative truth.

47. De Luca, "Blake's Wall of Words," 218. These images also evoke Blake's time in Sussex, near the cliffs celebrated in Charlotte Smith's *Beachy Head* (1807).

48. Hayley (1782) quoted in Susan Matthews, "*Jerusalem* and Nationalism," 89. On the design, see also *Jerusalem*, p. 135 and Erdman, *Illuminated Blake*, 284; cf. *Book of Urizen* 1, *America* 1.

49. Paley, *Continuing City*, 197–98.

The world of solid surfaces is conducive to both the mechanist regime of certainty and the moral one of sin. In this respect, the explanation of solidity becomes an explanation of matter in its restrictive sense. The condensed form becomes a limit for the "unnatural consanguinities" of vegetation— again, from the enlightened perspective of Blake's cosmology, an effectively benign "limit of opacity." (The complementary "limit of contraction," embodied in Adam or "clay," also has a particular geological resonance.) Albion's response here, which generates the explanation of rock and solidity, is an aesthetic response (or perhaps an antitype of aesthetic response) to the original landscape. "Every ornament of perfection . . . / In all the Garden of Eden" provokes only "remembrance of jealousy" and original sin in Albion, who has become "punisher & judge" (*Jerusalem* 28:1–4). This response is akin to Burnet's "horror" in the Alps, a reading of mountains as "ruins" of a perfect original globe, wreckage left behind by the Deluge, and hence as "remembrances" of sin. Albion, in a parody of natural history, transforms Eden itself into such a landscape, stifling "ornament" or beauty with a stern Burkean sublime.

The Huttonian textual metaphor is also appropriate to the rocks in this second passage. By condensing the Edenic (or human) landscape into rock, Albion makes it legible. He posits his own subtext of moral deformation, accounting for landforms on the basis of moral agency. Through Albion, however, Blake also offers his own occult causes (analogous to Hutton's "subterraneous fire") for the solidity and opacity of landforms. They appear "unhumanized" because of imposed epistemological limits attaching to the qualities through which rocks are generally understood, especially their inorganic quality as mere matter. Albion's speech here alludes to one such reductive explanation, the reading of topographic deformation as a register of human moral deformity, common in medieval cosmology but still implicit in the diluvialist geology of Blake's time—George Cumberland's being one relevant example. The story of a Garden of Eden corrupted by original sin is itself a textual maneuver of the kind that Albion is making. The mechanism responsible for rock formation in the terms of this allegory is the heroic, if misguided, effort on Albion's part to confine his environment to the scope of his "fallen" capacities. Such mechanisms resemble Hutton's "subterraneous fire" inasmuch as they aesthetically trump competing explanations by means of their sublime style, just as Hutton ropes in natural catastrophes in the service of his argument. Ultimately, any sort of natural explanation becomes unnecessary if we remember, as Albion cannot, that "Mountains are also men; everything is human. Mighty! Sublime!" (*Jerusalem* 34[38]:48).

Albion's most spectacular exercise in condensation results in the solidification of space itself. At the culmination of another episode in the long series of Albion's repudiations of Eden, he rebuffs the "kindest violence" of the Zoas, who have been organized by Los "in love sublime" to bear Albion back into Eden and unity (*Jerusalem* 44:8–12):

> And all the Gate of Los, clouded with clouds redounding from
> Albions dread Wheels, stretching out spaces immense between
> That every little particle of light & air, became Opake
> Black & immense, a Rock of difficulty & a Cliff
> Of black despair; that the immortal Wings labourd against.

This passage documents the origin of the physical universe. Blake proceeds to describe the transformation of the English Channel—from "waves of pearl" into a "boundless Ocean" filled with "clouds & rocks," but at the same time a "horrid Void" (44:15–17)—and to explain the material world now created: "Such is the nature of the Ulro" (21). The first result of Albion's defensive condensation of "light & air" is clearly physical space, the "Opake / Black & immense" sky of the astronomers, inimical to the transparent depths of Eternal vision. Because of the connotations of opacity both physical and metaphysical (resistance to the faculties), rock becomes the symbol for all material being in its Newtonian form, with solid particles suspended in a void that makes objects seem impenetrable. The rocks here and in scores of similar images are without a geographical place, like those in *The Book of Urizen*, making it easy for a Newtonian reader such as Donald Ault to ignore the rocky nature of Blake's opacity.[50] But in *Jerusalem* Blake aggressively *re*territorializes rocks as well with his profusion of place-names, underscoring the importance of geology and geography for his account of materiality. This complementary emphasis may be anti-Newtonian inasmuch as it problematizes the universality of physical laws. Within the poem's geographic allegory, rock expresses perfectly the opposition between the material world and the "human" universe in Blake's particular sense: "the Cities & Villages of Albion became Rock & Sand Unhumanized" (63:18).

The relationship of the rocky cosmos to vegetable culture becomes most explicit in chapter 3, in another account of the production of the material world, this time by the Daughters of Albion (*Jerusalem* 67:3–15):

50. Ault argues that Albion here *causes* Newtonian opacity by resisting his recuperation, that Blake's cosmology moves into a world obeying Newton's laws in order to expose those laws as constructions (*Visionary Physics*, 83–84). Here and elsewhere the geological nature of Blake's opacity fruitfully complicates Ault's readings (cf. 78–80).

>     They drew out from the Rocky Stones
> Fibres of Life to Weave for every Female is a Golden Loom
> The Rocks are opake hardnesses covering all Vegetated things.
> . . . . . . . . . . . . . . . . . . . . . . . . . . . . . . . . . . . . . . . . . . . . .
> They cut the Fibres from the Rocks groaning in pain they Weave;
> Calling the Rocks Atomic Origins of Existence: denying Eternity
> By the Atheistical Epicurean Philosophy of Albions Tree
> Such are the Feminine and Masculine when separated from Man
> They call the Rocks Parents of Men. & adore the frowning Chaos

Nelson Hilton notes that "rock" at this time could mean "distaff" and links the Daughters to the Three Fates, a link that does much to explain how weaving, for Blake, becomes such a central image for the production of materiality.[51] Rock in this double sense is both process and material. As the source of the "Fibres of Life," it provides the bodily machinery for the involuntary functions or "Vegetative powers" (5:39) which the Daughters set in motion. As the physical form of bodies in general, "rock" is the opaque surface "covering" the processes of life. Geological associations predominate over textile ones in the last four lines of the above passage. They are crucial to the irony in Blake's observation that the most barren substances—both "primitive rocks" and the materialists' abstract atoms (12–13)—assume generative power when regarded as building blocks of the cosmos, and hence "Parents of Men" (15). "Frowning" is also a conventional epithet for rocks, even in Blake's descriptions (see also 66:13). Blake here invokes a complex materialism that probably derives from both Lucretius and British empiricism, while "Albions Tree" points to the "Moral Virtue" (28:14) that is a corollary of materialism. Lucretius would be relevant as both an Epicurean philosopher and the patron saint of natural philosophy in verse. More broadly, the passage illustrates the "ideology of the natural," in Stuart Peterfreund's term.[52]

Rock is above all the basic and generic substance *worked on* by spiritual agencies in *Jerusalem,* so that the primitive materiality it represents becomes

---

51. Hilton, *Literal Imagination,* 109–10. See also idem, "Sweet Science of Atmospheres." In its "covering" role (67:5), rock is the stuff of Vala's veil, whose evolution is aptly traced by Paley: "If the Veil is . . . as Damon defines it, 'the film of matter which covers all reality,' then it is this that Los, the laborer in the rough basement of English, must use as building material" (*Continuing City,* 195). Paley points out that the veil is transformed, after its traumatic proliferation in chapter 1 (21:50–24:13), into the "beautiful Mundane Shell" (59:7), "the natural world as re-ordered in our perceptions by the imagination" (196).

52. In Peterfreund's view, Blake critiques scientific discourse for granting matter autonomy—as he does in this passage—insisting that otherness is always constructed rather than natural (*Blake in a Newtonian World,* 165). Lucretius' *De Rerum Natura* was frequently cited by scientific and topographical poets throughout the eighteenth century.

implicated in social constructions of the body and of history. It requires work because of the "Energy" confined in it; "primordial substance" itself, as in Hutton, is fluid. These relationships are especially clear in Los's work at his forge, the norm contrasted with deviant productions of materiality such as those of the Daughters of Albion above or that of Hand, the son of Albion (and spiritual father of Newton, Locke, and Bacon), another black-smith (8:44–9:4):

> The mighty Hand
> Condens'd his Emanations into hard opake substances;
> ∙ ∙ ∙ ∙ ∙ ∙ ∙ ∙ ∙ ∙ ∙ ∙ ∙ ∙ ∙ ∙ ∙ ∙ ∙ ∙ ∙ ∙ ∙ ∙ ∙ ∙ ∙ ∙ ∙ ∙ ∙ ∙ ∙ ∙ ∙ ∙ ∙ ∙ ∙ ∙ ∙ ∙ ∙ ∙ ∙ ∙
> He siez'd the bars of condens'd thoughts, to forge them.

Los combats this corrupted creation by melting down the torments of the Sons and Daughters themselves into "the gold, the silver / . . . & every precious stone" (23–24), what Paley suggests are "the materials for building the New Jerusalem."[53] The "condensation" of psychic phenomena rather than physical things in this passage suggests the body, the seat of the passions, as the most important condensation of all. As in the Changes of Urizen, the association between stone and bone readily brings anatomy to mind, and anatomy often parallels geography as well as geology. The confinement of the sense organs, for example, parallels the geographic shrinking of the Polypus, a negative form of the World of Generation woven from the Rocks by the Daughters of Albion (67:31–43). In another symbolic register, the Daughters' weaving is equivalent to their druidic torture of human victims "under the knife of flint," described so vividly in the preceding plates (65–66). Blake's account of Druid rituals merits particular attention here because it weaves together ideas about rocks both natural and artificial (Stonehenge), English history and geography, and male and female bodies: "the Twelve Daughters naked upon the Twelve Stones / Themselves condensing to Rocks & into the Ribs of a Man" (68:24–25). Throughout this passage (65:63–69:5), Blake invokes the mythical-historical model of human sacrifice as ostensibly practiced by the Druids. The Daughters' sexualized torment of male victims also parodies Los's efforts to establish a "Limit of Contraction" through anatomy (see esp. 66:30–38), as in the Changes of Urizen or Los's experiment on Reuben earlier in the poem (plates 34–36). In one prominent design, three female figures draw forth the bowels of Albion, whose body is painted blue and decorated with moons and stars, as in seventeenth-century accounts of Druid customs.[54]

53. *Jerusalem*, p. 143; cf. 14:17–24.
54. Ibid., 25. Paley cites John Speed's *History of Great Britain* (1611) as the source for these customs (ibid., p. 169; see further *Continuing City*, 99–104), and David Worrall offers a richly

Druids share with rocks the fascination of the primitive, augmented for Blake by the analogy that he perceives between the Druids' culture and his own. Blake's contemporaries worship "unhewn Demonstrations" in a state religion that resembles the rituals evoked by the "unhewn stones" of Stonehenge, another false Jerusalem (*Jerusalem* 66:1–3). Stonehenge, in turn, is an object of current interest because of the general eighteenth-century fascination with ruins. Because such "ruins of time" are seen as cognate with the "ruins of nature," the seemingly more benign institutions of tourism and landscape aesthetics fall under the same suspicion as science and religion. *Jerusalem*'s numerous allusions to the Peak District, England's most fashionable "primitive" landscape, must be seen in this light as implicating the fashion for the primitive along with druidic barbarities ancient and modern. Blake's Druid rocks have many resonances beyond aesthetics and geology, however, and previous accounts have addressed their social and political implications.[55] The negative charge of druidism for Blake comes from his central project of reclaiming primitive Christianity from more conservative thinkers who valorized the Druids. The immense potential of the "primitive" as a cultural force clearly has a positive attraction for Blake from early on, to judge from his discussions of Poetic Genius circa 1790; his use of the term in *Jerusalem*, in an address "To the Jews" that seems to mock the debates about primitive Christianity, is more equivocal.[56] His ambivalent fascination with rocks in this poem is partly due to the fact that their primitive status is less problematic and can potentially be used to authenticate his counterclaims about the cultural primitive.

The work of William Stukeley, one of the best-known eighteenth-century authors on the Druids, helps to establish the context of Blake's references to Druids and Druid rocks. Stukeley was a syncretic mythographer, like Jacob Bryant after him, but many strands are woven into his discourse about myth, including biblical exegesis and deistic theology as well as the archaeology and natural history of the stones themselves. Blake's poetry demonstrates his interest in these subjects and provides internal evidence that he knew Stukeley's treatises, *Stonehenge: A Temple Restored to the British Druids* (1740) and *Abury, a Temple of the British Druids* (1743). Stukeley's plan of the temple at Avebury in its original "serpent" form is the probable source for a very similar temple in the background of *Jerusalem* 100 (and for several related images in Blake). For Stukeley, these structures are Gothic

---

textured account of Blake's use of this and several similar sources in "Blake's *Jerusalem* and Visionary History."

55. Along with Worrall, "Blake's Derbyshire," several more recent works have helped to situate Blake's Druids historically, including de Luca, *Words of Eternity*, chaps. 5–6; and Jon Mee, *Dangerous Enthusiasm*, chap. 2. See also Todd, *Tracks in the Snow*, chap. 2.

56. But see *A Vision of the Last Judgment*, E559.

cathedrals before their time, and he contends that British Christianity can be traced, via the Druids, to the time of Abraham. Jon Mee has pointed out how politically topical this legitimating narrative was in the 1790s, arguing that Blake aligned himself against it in favor of a radical discourse of "primitive literature."[57] Blake seems to engage directly with Stukeley in his account of the building of Stonehenge, which alludes to the structure's real astronomical function as well as positioning it metonymically within the geography of Blake's Mundane Shell (66:1–8):

> In awful pomp & gold, in all the precious unhewn stones of Eden
> They build a stupendous Building on the Plain of Salisbury; with chains
> Of rocks round London Stone: of Reasonings: of unhewn Demonstrations
> In labyrinthine arches. (Mighty Urizen the Architect.) thro which
> The Heavens might revolve & Eternity be bound in their chain.
> Labour unparallelld! a wondrous rocky World of cruel destiny
> Rocks piled on rocks reaching the stars: stretching from pole to pole.
> The Building is Natural Religion & its Altars Natural Morality.

"Natural Religion" forms the central link to Stukeley in this passage, which occurs in a chapter addressed "To the Deists" and prefaced by a critique of eighteenth-century doctrines of natural religion (*Jerusalem* 52). Blake hints at the cruelty of the prehistoric "nature religion" reconstructed by Stukeley and links this religion with current "worship" of rocks and landscape. Stukeley argues that Stonehenge and the other "Druid" temples provide historical proofs of the thesis of natural religion: the Druids, famed for their theological sophistication, must have arrived at a notion of the Son of God "by the mere strength of natural reason. . . . nought else could induce men to make such . . . stupendous productions of labour and art."[58] Blake's first and last references to Stonehenge in *Jerusalem* clarify its relation to "natural" reason and "Natural Morality." Early in the poem it is a sign of sexual guilt: "We reared mighty Stones: we danced naked around them" (24:4, see also page 168). Near the end its altars are central to the confession of Brittannia, who professes to have "murdered" Albion in "Dreams of Chastity & Moral Law": "In Stone-henge . . . / I have Slain him in my Sleep

57. Mee, *Dangerous Enthusiasm*, 76. Mee compares Blake's attitude with James MacPherson's rejection of the Druids as "cunning and ambitious priests" (89). Specifically, he argues that Blake inverts Stukeley's argument for continuity between Druid wisdom and Anglican authority, suggesting instead that Christian priestcraft perpetuates druidic barbarity (93; cf. 7). Mee usefully quotes a 1792 progovernment poem, "Stone Henge," that draws on Stukeley (95).

58. Stukeley, *Abury, a Temple of the British Druids*, 89. Stukeley published actively in natural philosophy as well.

with the Knife of the Druid" (94:23–25). The design, with its enormous trilithons in the background, offers an unusually precise match for the text of this plate as well as developing a visual pattern unfolding toward the end of the poem (trilithons also appear on 100, 92, 70, and 69). These images, perhaps especially plate 70 (and its counterpart in *Milton*, plate 4), participate in the poetic and pictorial tradition of sublime landscape that Blake evokes with the stock phrase "rocks piled on rocks" (66:7).[59] Such phrases occur widely in the descriptive poetry of James Thomson and his followers, and are of a piece with the aesthetic sensibility informing Stukeley's geography and geology of the Druid temples. Stukeley begins his account of Stonehenge by describing the aesthetic qualities of its component stones, which have led tourists to break off pieces under the assumption that they were factitious. Stukeley assures us that the stones are real, insisting that the Druids always preferred unquarried ("unhewn") stones for their temples. The aesthetics is informed by a geological theory that the stones were flung to the surface of the earth when the planet was still fluid.[60]

Stukeley's demystification parallels Blake's translation of "Fable or Allegory" on the history of ancient Britain into "Vision."[61] As required by his geology, Stukeley provides the enormous rocks with a local habitation (Marlborough Downs) and a name ("grey weathers"); he also explains the "mortaise and tenon" technology used for their construction, suggesting that the same technology was used for the "rocking stones" in Derbyshire.[62] Blake's account of "rocking stones" (*Jerusalem* 90:58–66) is likewise closely related to his account of Stonehenge; he gives other locations (including the Hebrides) for his rocking stones, but elsewhere he refers many times to the Peak District in Derbyshire, whose scenic gorges and hot springs (the "wonders of the Peak") had attracted tourists from the late seventeenth century. He links it to Stonehenge in one of his more concise surveys of British geography in the poem (57:5–7):

> The Great Voice of the Atlantic howled over the Druid Altars:
> Weeping over his Children in Stone-henge in Malden & Colchester.
> Round the Rocky Peak of Derbyshire London Stone & Rosamonds Bower.

59. Erasmus Darwin (*BG* I.ii.531) is one in a long line of eighteenth-century poets to use the phrase. Cf. James Beattie, *The Minstrel* II.56–57. See de Luca, *Words of Eternity*, 174, for an analysis of plate 70, situating it in relation to the sublime of history and Stukeley in particular (171–73). See also Worrall, "Blake's *Jerusalem* and Visionary History," 207, on the related design of plate 69. De Luca also provides a vivid survey of the field of syncretic mythology in which he situates both Stukeley and Blake (179–92).

60. Stukeley, *Stonehenge*, 5–6; cf. *Abury*, 15–17.

61. *A Vision of the Last Judgment*, E554.

62. Stukeley, *Stonehenge*, 5, 49.

In the following plate Stonehenge is identified as raw material for the much vaster global geography of the Mundane Shell, raising some of the numerous issues that critics have pointed out within the poem's geography.[63] Blake's local geography, however, is shaped by what I have called the "geological moment," his reading of anatomy (English geography being Albion's anatomy) in terms of geology. Stukeley gives us one point of reference for Blake's conjunction of Stonehenge with the Peak and other sites of the geological primitive. The presence of geology in Stukeley seems less coincidental when he is compared with John Whitehurst, the Derbyshire geologist, who concludes his geological treatise by claiming that his motivation is to vindicate "the great antiquity of arts and civilization" (*Inquiry*, 272–73). According to Paolo Rossi, mythographers' efforts to reconstruct the historical basis of myth contributed to the rise of geology, because they needed evidence of earth history to corroborate what Whitehurst calls the "antiquity of civilization." If the "romantic caverns" of Derbyshire offer a primitive record of the planet's first "subterraneous convulsions" (*Inquiry*, 63–66), then any historical evidence predating the origin of these landforms proves that "civilization"—of a kind compatible with eighteenth-century European ideas—must date from before the Flood, if not from Creation itself.[64]

Blake's account of "primitive" sites such as the Peak, the Hebrides, Stonehenge, and Snowdon is oppositional to the extent that their primitive character, for him, only confirms the antiquity of barbarism. But he must rely on the same geography for his own vision of "British Antiquities," including Albion as "Patriarch of the Atlantic."[65] David Worrall argues that *Jerusalem* 100, depicting Blake's "giant forms" with the Stukeleyan serpent-temple in the background, revises the myth that Stonehenge was transported by giants from Africa to Ireland.[66] Competing with other revisionists, including Stukeley, Blake instead has Urthona, Los, and Enitharmon building Jerusalem in Atlantis, farther west even than Ireland. From the beginning of the poem (4:28), we know that the British mountains are Albion's "hard restricting condensations," or "mountains of Moral / Virtue," as Blake later

63. See, for example, S. Matthews, "*Jerusalem* and Nationalism," 96, 98–99.

64. Rossi, *Dark Abyss of Time*, 16–17. Cf. Stukeley, *Abury*, 54–55.

65. *Descriptive Catalogue*, E542–43.

66. Worrall, "Blake's *Jerusalem* and Visionary History," 209–10. This myth also lends itself to Stukeley's theory that Stonehenge was built by Phoenician colonists—the Druids—who later went to Ireland (*Stonehenge*, 49). Worrall points out that Blake takes Stonehenge and the other Druid temples as being central to British history but pointing beyond it, as portals to Atlantis.

puts it (31[45]:19–20). The long, prophetic speech of Erin, midway through the poem, confirms this diagnosis and pleads for a cure (49:76–50:2):

> Remove from Albion these terrible Surfaces
> And let wild seas & rocks close up Jerusalem away from
> The Atlantic Mountains where Giants dwelt in Intellect;
> Now given to stony Druids.

Mountains here are both symptoms of the disease and promises of a cure, geographic signs of a restoration of the full scope of human territory and capacities. The design of plate 50, which concludes the speech and the chapter, concentrates on the symptoms: Blake performs a spectacular condensation of England, shrunk to a miniaturized white grass-topped rock and flanked by the sun and moon of a condensed cosmos. Albion's son, the three-headed giant Hand, dominates the composition, kneeling in front of the tiny island as three male figures emerge (Athenalike) from a cavity in his chest. A later narrative explains that Hand gives birth in this fashion to Newton, Locke, and Bacon (70:1–15). It is tempting to allegorize the image as "England dominated by materialism."

Erin's long speech, with its Christian conclusion, is known for being one of the most hopeful moments before the poem's end, hinting at political revolution as well as other elements of a potential cure for Albion's rocky disease. Albion's "terrible surfaces" are the mountains embodying "laws of chastity & abhorrence" (49:26) and other forms of repression detailed in Erin's jeremiad. But she also offers a topography of liberation, featuring a poetic circumnavigation of Ireland (4–5), twin Emanations hovering over England and France (48), and a recollection of Albion's Atlantic legacy as "the Mountain of Giants," now "witherd" but one day to be restored to its lost continent (6–7; cf. 19–20).[67] The complex, multivalent roles of these various rocks make her instructions to remove Albion's surfaces somewhat perplexing. The existing surfaces are terrible, and Erin warns that Jerusalem must not be confined within them lest she "consume under Albion's curse," condemned by laws of Moral Virtue to play the part of Vala, the "proud Virgin-Harlot! Mother of war!" (50:14–17). Yet the "wild seas and rocks" provide a saving refuge for Jerusalem, and the mountains of Atlantis, though "now given to stony Druids," raise the possibility of restoration. As a visionary locus one station closer to Atlantis, Ireland is closer to political ferment,

---

67. David Erdman points out that English radicals around 1810 pinned their revolutionary hopes on Ireland; *Blake*, 482–83.

for Blake. He also seems to single out Welsh mountains—Plinlimmon and Snowdon—as retaining some Atlantic sensitivity, which might be associated with Welsh/Celtic cultural nationalism.[68] British mountains in general, however, remain the scene of Albion's curse—especially in the first narrative plates of chapters 1 and 2, with their echoes of Burnet, discussed above. The Miltonic "pits of bitumen" mentioned in a geographic speech of Los (43[38]:62) evoke the economic dimensions of Albion's curse. Partly because of its important role in mining and industry, the rocky landscape of Derbyshire becomes the center of Blake's topography of oppression.

Albion's curse is localized in Derbyshire. Although he utters several related curses at various points in the poem's complicated chronology, Blake stages it most dramatically as a dying curse toward the end of chapter 1 (after which Albion is asleep or dead, in the poem's primary present, until plate 95). Blake's narrator first sets the scene—"These were his last words, relapsing! / Hoarse from his rocks, from caverns of Derbyshire & Wales / And Scotland"—and then the curse materializes in Albion's speech: "Lo here is Valas Veil whole. for a Law. a Terror & a Curse!" (23:26–28, 32). Although the text inevitably mentions other places besides Derbyshire—geographical dispersion being the omnipresent sign of fragmentation—the elaborate full-page design encompassing the text can be clearly identified as a Peak District image. As Paley describes it, "the design is conceived as a mountain lying under the text" (*Jerusalem*, p. 166). The peak, with a winged female figure (Jerusalem) half embedded in it, appears between the first two portions of text, with a cross-section of the mountain and its "roots" appearing in the middle and lower portions of the plate, respectively. These two portions feature massive tree roots interspersed with caverns, recalling the "caverns rooting downward" of *The Four Zoas* (74:12). The caverns, however, are occupied by primitive human figures ("Niebelung-like men," in Paley's words) who crouch in their tightly "restricting" spaces, much like Los and Urizen in the *Book of Urizen* images discussed earlier. Like

---

68. They look on one another: the mountain calls out to the mountain:
   Plinlimmon shrunk away: Snowdon trembled: the mountains
   Of Wales & Scotland beheld the descending War.
   (66:58–60; cf. 4:29)

Paley's gloss alludes to the special status of Snowdon for Welsh and Briton nationalism (*Jerusalem*, p. 239); see also idem, *Continuing City*, 55–57; and Mee, *Dangerous Enthusiasm*, 110–11. George Cumberland celebrates the Welsh mountains in *An Attempt to Describe Hafod* as well as in *A Poem on the Landscapes of Great Britain* (1793). Bentley (*Bibliography of George Cumberland*) also credits him with an untraced 1804 manuscript "Tour of Plinlimmon."

Jerusalem herself, these men are embedded in the mountain, anticipating Erin's warning about a Jerusalem confined by Moral Virtue.

These primitive figures inhabit a social context associated more strongly with economic than with moral or sexual repression. Worrall first proposed these connections, identifying Blake's image as an impression of the Devil's Arse (also known as the Peak Cavern, near Castleton), a large cavern counted as one of the Wonders of the Peak (i.e., the Peak District). He suggests that Blake might have seen one or more travel narratives describing "a great many" poor people living in the mouth of the cave.[69] Because the cave and the district were so frequently visited throughout the eighteenth century, this picturesque circumstance became a topos in a long sequence of travel narratives dating from 1636 to the early nineteenth century. Worrall mentions only two of these accounts, but numerous others also provide relevant details (see further chapter 6, below). The French geologist Barthelémy Faujas de Saint-Fond reports that the Devil's Arse has been "regarded from all time as the chief of the seven wonders of Derbyshire." Much of this account is devoted to the primitive—both in its geological form as "toadstone," whose origin Saint-Fond debated with Whitehurst, and in its social form: in his Peak District chapters, he comments several times on the abject condition of the miners there. In one description that is practically of a piece with Blake's image, he refers to families working the limekilns near the caves as "moles" and "troglodytes," reporting that they "scoop out their dwellings among heaps of cinders and lime-refuse."[70] This description illuminates the point of Blake's design: not just nature but also social reality conformed to the expectations of travelers who went to mountainous places in search of the primitive. Through much of the century, the livelihood of the poor in this rugged district depended almost wholly on tourism and on mining (cf. *Jerusalem* 94:7–8), which in turn motivated the geological inquiry; after 1770, the Peak's abundant water power also made it an important industrial center.[71]

69. Worrall, "Blake's Derbyshire," 34. Worrall cites two sources on the Peak that Blake may have known, arguing that he took their fanciful descriptions of shapes in the Devil's Arse and another Derbyshire cave as allegories of Albion reposing on his couch.

70. Faujas de Saint-Fond, *Journey through England and Scotland*, 2:315, 288–89.

71. Richard Arkwright's cotton mills near Matlock were among the most famous early factories in the Peak. Geologists including Whitehurst, Cumberland, and William Smith are very explicit in justifying geology as a tool for the "improvement" of landscape. The picturesque travel narrative, too, celebrates the ethos of improvement—for a relevant instance see Cumberland's *Attempt to Describe Hafod*, iii. Blake's perceptions about rural poverty accompany Los on his sorrowful journey among the London poor (31[45]:3–43), an urban companion text to the *Jerusalem* 23 design in which Los sees "every Minute Particular of Albion degraded &

While British topography, as Albion's anatomy, is gendered male, Derbyshire has the singular status in *Jerusalem* of being linked to female anatomy. This association, too, has a strong geological basis, both in the ancient tradition of representing the earth as a womb and in more contemporary expressions such as "horrid chasm." John Whitehurst, for example, cites the "horrid chasms" of the Peak District as evidence for the cataclysmic origin of its landforms (*Inquiry*, 190). In Burnet's earlier, less technical idiom, we encounter a semieroticized "vast and prodigious Cavity, naked and gaping at the Sun." Blake's "horrid chasm" is explicitly female:

> Derby Peak yawnd a horrid Chasm at the Cries of Gwendolen. & at
> The stamping feet of Ragan upon the flaming Treddles of her Loom
> That drop with crimson gore.
>
> (64:35–37)

As it was for many geologists, the "horrid chasm" here is a sign of original rupture, which in Blake's case has the concrete referent of the vagina and the abstract one of Albion's division from Jerusalem. The interposing "chasm"—rendered, as usual, in terms of both geography and anatomy—is a patriarchal conception of femininity taking a variety of forms in the course of the narrative (62:30–65:4).[72] Another geological moment in this sequence is when "All the Daughters of Albion became One before Los: even Vala!" whose anatomy concludes with her genitals: "in her Loins Earthquake. / And Fire" (64:6, 10–11). This feminine geography becomes manifest when Los realizes that his vision—a distorted vision of femininity occupying the place of a lost Jerusalem—has assumed historical reality (63:40–41). The geological form and negative space of the "chasm" are another product of the Daughters of Albion's weaving, the production of a historical world of bodies petrified by repression and bloody wars (linked to the vaginal blood of the loom-chasm as "crimson gore"). In Blake's view, divided sexuality is "hermaphroditic," as aptly expressed by the paradox "Derby Peak yawnd a horrid chasm." Here Vala merges with the Spectre of Albion ("Thou art / thyself female, a Male: a breeder of Seed") to become "a dark Hermaphrodite" (64:31), and the ensuing narrative on the building of Stonehenge develops this idea (65–69). As I pointed out earlier, the Daughters' and Sons'

---

murderd" (7) and sees that "every Universal Form. was become barren mountains of Moral / Virtue" (19–20).

72. This narrative follows a pivotal scene in which Jesus explains to Jerusalem that she will endure a time of "prison & judgment" (62:24) but will be nourished by faith "tho Valas cloud hide thee" (28); in other words, Jerusalem's femininity will be understood in terms of the negative models provided by the war goddess and Whore of Babylon, but she will be vindicated in the end. See further Paley, *Continuing City*, 178–96.

interspersed operations of weaving and building are parallel images for the same process of materialization. This entire twofold operation is also "hermaphroditic" in the sense that the separate activity of both sexes produces the identical result of human sacrifice.

*Jerusalem's* embedded travelogue records two further Wonders of the Peak, Mam Tor and Dovedale, again associated with Gwendolen and female sexuality. Blake juxtaposes Dovedale—the scenic limestone gorge made famous by Izaak Walton—with Mam Tor, a major double peak above Castleton, to create a geographic *blason* for Gwendolen (82:45–47):

> She drew aside her Veil from Mam-Tor to Dovedale
> Discovering her own perfect beauty to the Daughters of Albion
> And Hyle a winding Worm beneath.

This is a highly philosophical striptease: Gwendolen demonstrates that under the Veil of Vala ("the film of matter," in Damon's phrase) lie only natural causes and natural features, a topography of mountains and valleys mirroring female anatomy. The twin peaks and the gorge, of course, stand for the breasts and vagina.[73] Materialization here takes the anatomical form of gestation. Hyle (the Greek word for "matter") turns out to be "a winding Worm . . . / & not a weeping Infant" as Gwendolen had claimed to her sisters (8, 37). This revelation concludes yet another version of the foundational myth I have called Blake's "geological moment": dwindling human faculties are given a "limit of contraction" in the form of anatomy understood as geology. Gwendolen gives such a form to Hyle by means of a process that parallels both the Changes of Urizen and the conception of Orc, upon which Enitharmon "felt a Worm within her womb" (*Book of Urizen* 19:20). This iteration of the myth is remarkable for being staged as a love story and set in Derbyshire: "Hyle on East Moor in rocky Derbyshire, rav'd to the Moon / For Gwendolen" (*Jerusalem* 80:66–67). Gwendolen encloses

---

73. See *Jerusalem*, p. 269. As the worm in her womb, Hyle is the opposite of the white dove or Holy Spirit we might expect to find taking refuge in Dovedale; likewise, the barren bosom of Mam Tor is not a source either of milk or of the gentility implied by "ma'am." (Etymological thanks to one of the anonymous readers for Cornell University Press.) The position of Gwendolen's audience "on Skiddaws top" (82:16) marks this passage as a possible parody of picturesque travelogues, since Skiddaw, one of the highest peaks in the Lake District, features prominently in such narratives along with the Peak landforms personified by Gwendolen. Skiddaw's claim to "primitive" status was specifically disputed by James Hutton. Blake's probable allusion to the Welsh country house celebrated in Cumberland's picturesque narrative, *An Attempt to Describe Hafod*, also suggests parody: the key phrase here is "stupendous works" (*Jerusalem* 46[41]:3–4), which refers on three later occasions to entirely demonic structures (58:48–50, 89:22, 91:32). Paley points out a "parody of sublimity" in Gwendolen's striptease (*Continuing City*, 65; cf. 134–35).

Hyle's soft organs in definite forms, beginning with the heart and the images of love lyric: "she took up in bitter tears his anguished heart" (67–77).
She triumphs over the results, observing that "the mighty Hyle is become a
weeping infant" (82:8) and enjoining her sisters to follow her example:
"The fury of Man exhaust in War! Woman permanent remain" (82:35).
Hyle, appropriately, displays a dizzying variety of "primitive" forms here as
a Son of Albion pining for his Emanation, a captive warrior sacrificed according to Druid ritual, an infant or worm, and a personification of matter
or materialism.

## The Production of Matter

I began this chapter with accounts of primitive matter implicitly gendered
male, including those in *The Book of Urizen*, Hutton's *Theory of the Earth*,
and especially Darwin's allegorical version of Hutton. Other accounts, such
as Burnet's and Whitehurst's, feminize the earth and earth processes, as do
many of the mythical cosmologies that form an ultimate common source for
Blake and the geologists. Blake makes explicit an essential connection between the concepts of materiality and gender, which partly motivates his critique of materiality. His ironic exclamation, "They call the Rocks Parents of
Men" (*Jerusalem* 67:15), thus not only expresses his critique of scientific and
aesthetic materialism, but also literally describes a petrified reproductive
sexuality. The "Rocky Law" of sexual reproduction materializes in the act of
intercourse itself (30[44]:35–37):

> The Male enters magnificent between [the Female's] Cherubim:
> And becomes One with her mingling condensing in Self-love
> The Rocky Law of Condemnation & double Generation, & Death.

Condensation, as we have seen, is the most frequent image in *Jerusalem* for
a restrictive materialization that collapses reality into geology. The idea of
primitive matter and its association with rocks are immensely useful to
Blake as a critical tool, enabling him to point out that even "primitive" matter is always imagined in human, gendered terms, that materiality itself is
inscribed with concepts such as gender. Judith Butler, in her more recent
exploration of the intimate connection between gender and materiality, un-
cannily echoes Blake in a modern theoretical register: "We may seek to return to matter as prior to discourse . . . only to discover that 'matter' is fully
sedimented with discourses on sex and sexuality that prefigure and constrain

the uses to which that term can be put."[74] In Blake's cultural lexicon, geology provides the apparatus for such a critique. The geological concept of rock as primitive matter implies a World of Generation, in Blake's terms, forms of life produced by the physical environment and sexual reproduction. If Generation threatens to terminate in Eternal Death—another Blakean paradox (e.g., *The Four Zoas* 106:7–8, 117:5–6)—this threat exists not merely because of a commonplace sense that matter is dead, but specifically because of an increasingly dominant paradigm of matter as inorganic and geological. Blake's image of sexual reproduction as a "Rocky Law" refers not only to the morality of sex—sexuality subordinated to reproduction (*Jerusalem* 69:43–44)—but also to the ontology of gender: if all bodies are built up on a basis of primitive matter, then gender must be natural and not the construction Blake holds it to be.[75]

Blake draws on many other registers besides sexuality to debunk the myth of the primitive, the myth that any physical state is original. The abstract materiality earlier figured forth in Urizen's rocky body of a world not only becomes more explicitly gendered in *Jerusalem*, but is reterritorialized in specific "primitive" landscapes as well. The idea of rocks as primitive matter, then, also serves as a critical tool for the analysis of economic issues. Blake's mapping of Derbyshire, for example, reveals the economic contours informing the "primitive" construction of that landscape. In his two longest poems, Blake struggles to produce a unified body for Albion, a geographic body with physical as well as social and political aspects. As a physical entity subject to geological processes, it closely parallels Hutton's anatomy of the earth: just as Hutton posits an original, uniformly fluid (molten) state for all the earth's material—even for the "primitive" peaks of the Alps and other sublime mountains—so Blake posits a uniform, original body in which all things bodily were once integrated. While this body and its disintegration take abstract forms in *The Book of Urizen*, drawing especially on Darwin's allegorized geology, geological bodies are more localized in *Jerusalem*. Blake

---

74. Butler, *Bodies That Matter*, 29. Note the geological metaphor, "sedimented." Butler goes on to point out that because of this sedimentation, "'materiality' designates a certain effect of power or, rather, *is* power in its formative or constituting effects" (34–35). On the surface, this resembles very closely many of Blake's claims that materiality is a kind of regime. Butler also makes the Blakean-sounding claim that materiality is contaminated because "its status as contingently constituted through discourse is erased, concealed, covered over" (251).

75. As Butler puts it, "we have barely begun to discern the history of sexual difference encoded in the history of matter" (ibid., 54). For her, this history might include representations like Blake's, in which materiality is sometimes "constituted through an exclusion and degradation of the feminine" (30). See also Suzanne Vesely, "Daughters of Eighteenth-Century Science."

here incorporates geographically specific images of rocks and mountains in order to harness their rich cultural significance. Geology itself continues to run the same gamut from abstract, theoretical discussion of earth processes to empirical descriptions of specific formations and outcrops. Though never deeply immersed in geology, Blake evidently paid attention to the revealing relationships between abstract and concrete descriptions of rocks and landforms. In *The Botanic Garden*, the empirical illustration of scientific principles—whether geological or botanical—is always a cultural story, often animated by economic motives. Blake's friend Cumberland is even more explicit about such motives, recommending "the knowledge of the mode by which the form of this globe has been destroyed by the flood, in order that we may know where to find the valuable materials that the waters robbed the old continents of, such as mould, clays, marls, sand, peat, gravel, and the debris of ores, &c."[76]

Though still cherished today for his iconoclasm and individuality, Blake shared at least one quality with his great literary contemporaries: his absorption in "the history of his own times" and his broad and vigorous literacy within the voluminous cultural production of those times. Geology emerged during the same transitional epoch of cultural production, partly because images of rocks and mountains had become part of a broad literary vernacular fueled by the various "booms" of the early industrial age, including tourism and mining. Blake made this widespread fascination with rocks and mountains a target of philosophical critique, but he also noticed the industrial subtext, the fact that "primitive" or unspoiled profiles sometimes became a code designating landscapes as suitable for industrial exploitation, as was the case with the Peak District and its "raving streams," or water power.[77] The immense proportions and geological properties that mark "primitive" rocks as sublime also mark them as raw materials. This paradoxical entanglement of aesthetics and economics is more explicitly present in Erasmus Darwin and in later poets and geologists such as Shelley and William Smith. It is vividly suggested as well by Blake's representations of the Peak District and by Hutton's speculative geology. Blake politicizes geology in *Jerusalem* by aggressively naming and historicizing the many rocks and landforms featuring in the narrative; earlier, in *Urizen*, the images of

---

76. George Cumberland, "Mr. Cumberland on Proper Objects of Geology," *Monthly Magazine* 40 (1815): 131. In Darwin see, e.g., *BG* II.ii.85–104.

77. De Luca complicates Blake's critique of rock- and nature-worship by pointing out that despite the "delusive pretensions" of "displays of the material sublime," Blake "is sufficiently fascinated to write about them obsessively, often in an exclamatory tone that seems as much designed to astonish the reader as to critique the premises of astonishment"; *Words of Eternity*, 166.

rock are more nearly identical with the deterritorialized raw materials of Plutonism.

A deep source of parallels between Romantic poetry and geology lies in the creation myths that both are engaged in rewriting. Both Blake and Hutton are especially close to myth in their insistence that appearances demand to be explained by means of vast and powerful agencies not yet fully understood (or even anticipated). For Hutton, the "amazing power" in the "bowels of the earth" is the agency behind all geological processes, the dynamism of the earth's core, which we are still unable to describe empirically. In Blake's cosmology, human creativity and artistic process— often figured as vast, obscure Plutonic forces—make up the central transformative agency of the planet. In his famous parable of the "Printing house in Hell," this agency is attributed to "Unnam'd forms" much akin to geology's "unknown powers" (*Marriage of Heaven and Hell* 15:15–21):

> In the fourth chamber were Lions of flaming fire
> raging around & melting the metals into living fluids.
> In the fifth chamber were Unnam'd forms, which
> cast the metals into the expanse.
> There they were receiv'd by Men who occupied
> the sixth chamber, and took the forms of books &
> were arranged in libraries.

# Interchapter: Literary Landscapes and Mineral Resources

## The History of the Earth and the Economy of Improvement

William Blake and George Cumberland, in spite of their artistic and personal alliance, represent the two opposing faces of their period's Januslike idea of the earth's material. Though he is critical of them, Blake internalizes literary *topoi* that emphasize the otherness of rock, filling his poetry with sublime scenes of desolate mountains and imposing ruins. Cumberland, in his writings on geology, takes the contrary approach of emphasizing the domestication of nature. In a passage quoted near the end of the previous chapter, Cumberland justifies geology on the basis that it teaches us "where to find . . . valuable materials." The numerous images of alien, sublime, incomprehensible rocky landscapes anatomized in the first half of this book, concluding with those of Blake, are created in a context of other discourses on rocks that aspire to more control over the knowledge and use of geology and its "valuable materials." The second half of the book addresses the economic context of aesthetic discourse and cultural practices surrounding geological objects. My coinage "aesthetic geology" is intended as a counterpart to "economic geology," a term still widely used today. This interchapter maps the contours of early nineteenth-century economic geology to create a transition between "romantic rocks" and "aesthetic geology." Mapping provides an apposite metaphor because both economic and aesthetic interest necessarily attach to specific local landscapes, many of which are featured in the following chapters. The present map also has a sociological aspect, introducing applied science through the careers of William Smith, Humphry Davy, and Mary Anning. Geological fieldwork helped to uncover a conceptual terrain that still occupies mapmakers, a terrain featuring such concepts as "deep time," the "rock record," and "natural resources."

Economic geology cements an association between natural history and morality originating in the culture of "sensibility," the stylized sensitivity to nature that linked natural knowledge and literary form in the decades preceding Romanticism. Moral profit becomes identified with economic profit as the language of the aesthetic is applied to the "improvement" of land, a primary objective in the applied science of both Smith and Davy. The sequential form of the rock record, which is read almost like a literary narrative, strengthens this identification of the moral, economic, and aesthetic. The appearance of *topoi* from natural history in imaginative literature reflects the increasing tendency toward the practical within geology, but also shapes the vocabulary that is used to represent scientific discovery and economic progress, and conditions the sovereignty over nature that these seem to imply. William Smith's claim that stratigraphy "must lead to accurate ideas of all the surface of the earth," as both "intelligible and useful," illustrates the "sprawling subjectivity" of Romanticism, but also anticipates Alexander von Humboldt's discovery of geologically identical formations in both hemispheres.[1] The planet's geological uniformity promises the triumph of European economic geology as globally profitable science—a promise seemingly confirmed by the eventual Englishing of whole periods of the planet's history (e.g., "Cambrian," "Devonian"). The remainder of this book focuses on the physical and cultural factors linking scenic beauty and mineral wealth in early industrial Britain.

Rachel Laudan, Roy Porter, and other historians of geology have examined early geology's connections with industrialization. The distinction between primitive and secondary rocks discussed at length in chapter 3 originates in eighteenth-century treatises on mining and derives from practical methods of classification used in the mines.[2] Roy Porter argues that early nineteenth-century geology is still "isolated from economics" by "cultural mediations," such as the gradual incorporation (by "establishment geology") of the empirical tendency of provincial science.[3] Yet as Porter acknowledges elsewhere, industrialization depends on a broad mastery of nature: "as man

---

1. Smith, *Memoir*, 6. Smith suspected as much, but evidently had not read either Humboldt's 1801 geological sketch "prov[ing]" the identity of the formations of the two hemispheres" or its successors. See Humboldt, *Personal Narrative*, 1:285. In her reading of Humboldt, Mary Louise Pratt points out some connections between the global claims of imperialist science and the Romantic sublime (*Imperial Eyes*, 111–43). For a view of Humboldt in context, see Fig. I.1.

2. Laudan, *From Mineralogy to Geology*, 58.

3. Porter, "Industrial Revolution and Rise of Geology," 343.

conquered it, it became more friendly." This war on nature is waged on both cultural and economic fronts. On the one hand, historical geology's new emphasis on the progress of the earth as a system motivates the appearance of "truth in the rocks." Porter also emphasizes the Romantic prestige of tourism and natural scenery, expressed in Humphry Davy's quest for "solace and knowledge together in the rocks."[4] David Allen makes the stronger claim that the fashion for the sublime directly caused the rise of geology: "the ultimate, avoided part of nature was now to be examined." But Allen also chronicles the rise of a "geological lobby," an accumulation of political power by a group of scientist-bureaucrats in the second quarter of the nineteenth century that reflects the concentration of economic interest around the field.[5] The influence of Romantic literary culture on geology is in some respects more obvious than geology's impact on cultural production. Historians of geology refer to "shared norms . . . and practices" and a "shared . . . cultural environment" to explain the relationship of geology to Romanticism, in some cases making the rise of geology a component of Romanticism.[6] In the following chapters, however, I hope to show how economic factors conditioning the history of geology become significant in literature as well.

Applied geology brought economic incentives to the problem of determining the age of rocks, also an important factor in both the geological sublime and foundational scientific debates about "primitive" and "secondary" rocks. The issue of geological time also links the history of geology to other branches of inquiry, including cosmology and the philosophy of science and of history. Stephen Jay Gould, for example, reads the works of Burnet, Hutton, and Charles Lyell as both geology and philosophy of history, arguing that the geologists' "visions" of time's metaphorical arrow and cycles "fueled the discovery of deep time as surely as any observation of rocks and outcrops."[7] The subtitle of Paolo Rossi's *Dark Abyss of Time* makes this connection explicit: *The History of the Earth and the History of Nations from Hooke to Vico*. The assimilation of natural history to human history is one of the Enlightenment's most important legacies for Romanticism. Rossi's emphasis on this assimilation—particularly associated with the morality and practice

---

4. Porter, *Making of Geology*, 219, 220–21, 142.

5. Allen, *Naturalist in Britain*, 47 (cf. 51), 52–53. Rachel Laudan cites the growth of the money economy and the cultivation of state-run industries, such as mining and porcelain manufacture, as fostering geology in Germany, where the situation is admittedly clearer; *From Mineralogy to Geology*, 47–48.

6. Nicolaas Rupke, "The Apocalyptic Denominator," 30; Porter, *Making of Geology*, 6.

7. Gould, *Time's Arrow, Time's Cycle*, 7–8. Scholarly work on the history *in* geology has tended to focus on the earliest phase of geology; see also Paolo Rossi, *Dark Abyss of Time*; and Rhoda Rappaport, *When Geologists Were Historians*.

of improvement in William Smith's geology and the scientific poetry of Erasmus Darwin and Percy Shelley—helps to counterpoise my earlier argument that "deep time" emerged as part of a larger emphasis on the alien, nonhuman quality of the earth's material. Gould and Rossi share an interest in the human–natural history analogy; Rossi is particularly helpful here because his argument diverges into a "search for the historical and cultural contexts" that he proposes as "an essential and constitutive part of any scientific discourse."[8] Rossi traces local instances over a century in which fossils were increasingly treated and read as documents equivalent to historical records. In my next chapter, tracing the evolution of a "rock record," I suggest that this secular history appears in nature in proportion as matter seems to resist cognitive ordering. Both of these *topoi* become increasingly prominent toward the end of the eighteenth century and into the nineteenth. In the following chapters I shall argue that they are interdependent: the higher the stakes in the mastery of nature, the more salient its real and idealized resistance.

Rossi's work also speaks to the paradox of otherness apprehended by means of human structures. He argues that a shift from theology to history enables early geology to identify the alien quality expressed as "the dark abyss of time." But intelligibility is not lost with the vanishing of sacred history and its human timescale. Because recognition of the "dark abyss of time" coincides with reflection on the "boundless antiquity of 'nations,'" the otherness of an unlimited time is expressed, paradoxically, in human terms. Just as fossil species have disappeared, "entire cultures" may have been swept away, and fossils may be found to "document" deliberately obscure ancient myths of human origins. Rossi reads Thomas Burnet's geology of "ruins" in the same way: the ruins of nature are not only symbolic of but in some sense coeval with the ruins of time. In the preceding chapter I suggested that Blake consciously inherits this project, along with contemporaries such as the geologist and chronologist John Whitehurst, whose motive was to vindicate "the great antiquity of arts and civilization" (*Inquiry*, 272–73). However abyssal, the earth's material becomes legible and thus historical in Rossi's rendering. Rossi reads the Romantic period debate about geological time as a recapitulation of the "great dichotomy of deism and materialism," which he suggests is "reopened" by Hutton's *Theory of the Earth*.[9] As Rossi points out, the French Revolution reactivated the moral sensitivity to materialism. Materialism also acquires a heightened charge in the wake of the Industrial Revolution, especially clear in the animated matter of a rock record narrativized at the instigation of the coal industry. The

8. Rossi, *Dark Abyss of Time*, xii.
9. Ibid., viii–ix, x, 15, 17, 36, 118.

Romantic period discourse on the rock record differs fundamentally from the Enlightenment discourse traced by Rossi in the heightened urgency behind the later claim to mastery over nature, in spite or because of nature's heightened otherness. The ability of diluvialist geology—as practiced by William Smith and by A. G. Werner in Germany—to deliver concrete assistance to mining operations is a new criterion operating through the growth of industry. (Fig. I.1 schematizes the theoretical distance between Werner and Hutton.) The inherited narrative form of this science contributed to its new applications. Rossi shows that the history of the earth continued to rely on the history of nations as a fictive organizing structure through the nineteenth century; it was fully absorbed into the truth-claims of geology only after 200 years. This is an instructive paradigm for the interdependence of Romantic literature and geology.[10]

Making preparadigm earth science the focal point of literary and cultural analysis complicates the relationship between industrialization and the rise of geology. Arguing for a more central role for tourism in the history of geology, Richard Hamblyn suggests that applied geology, in particular, was oriented as much toward tourism as toward industry. Hamblyn points out that mineral dealers bought their stock from "miners, quarrymen, and canal builders, who . . . were equally as alive to the non-industrial demand for mineralogical deposits as they were to the needs of coal-fired technology." Emphasizing the simultaneously scenic, industrial, and scientific interest of Coalbrookdale, the Duke of Bridgewater's Canal, and other "sites of environmental disruption," Hamblyn posits an "economic coupling of industrial mineralogy with the leisure economy of movement between an itinerary of prospects." Thus tourism both ratified industrial sites and "funded the earth sciences" through a souvenir and specimen trade that required field research independent of "mineral genera produced elsewhere." The " 'low' cultural practioners" pursuing this research "name[d] the field agenda for science" as they "named . . . the agenda for tourism." Their publications made mineralogy more accessible and contested as "undemocratic" the "obscurantism of 'high' published mineralogy" and geology.[11] While tourism altered the socioeconomic and geological contours of landscape, scientific specialization can also be seen as anticipating the division of labor characteristic of industrial capitalism. Stephen Toulmin has argued that the "bureaucratic rationalization" (Max Weber's term) of science dates back to the late sixteenth century and introduces a form of productivity cognate with that of capital-

---

10. Despite Hutton's anticipation of "deep time," geology until 1900 continued to depend on the analogy to history, Rossi argues (ibid., 120).

11. Hamblyn, "Landscape and Contours of Knowledge," 98, 109–11, 94, 29, 45.

| Geologists | Scale | Region | Description |
|---|---|---|---|
| WHISTON, Theory of the Earth, 1725. BUFFON, Theorie de la Terre. LEIBNITZ, Protogæa, 1768. DESCARTES. | 100 — 95 | Plutonic Region. | All Rocks affected by heat. The Earth struck off from the Sun by a Comet. |
| BOUÉ, Essai sur l'Ecosse, 1822. | 90 | | All Rocks of Chemical origin igneous. |
| HUTTON, Theory of the Earth, (Ed. Trans. v. i.) PLAYFAIR. Illustrations, 1820. Sir J. HALL, Edin. Trans. vol. vi. 1806. Sir G. MACKENZIE, Travels in Iceland, 1810. | 85 | | All the older rocks either fused or softened by heat. Metallic Veins injected from below. |
| Sir H. DAVY, On Cavities in Rock Crystal, 1822. MACCULLOCH, Various papers in Geol. Trans. from 1814 to 1817. KNIGHT, Theory of the Earth, 1820. BRIESLAC, Journal de Physique, vol. xciii. | 80 — 75 | | Granitic Rocks igneous. Granitic Veins injected from below. Some Granite and Sienites igneous. |
| FAUJAS ST FOND, Essais Geologiques. HUMBOLDT, Travels and Memoirs. SPALLANZANI, Sur les Isles Ponces. Sir W. HAMILTON, Memoirs, &c. DOLOMIEU, Voyage aux Isles de Lipare, 1783. SAUSSURE, Voyages dans les Alpes, 1787. W. WATSON, Section of Derbyshire, &c. WHITEHURST, Theory of the Earth, 1786. | 70 — 65 — 60 | Volcanic Region. | All Trap Rocks igneous. |
| CORDIER, Sur les Substances Minerales dites en masse, 1815. VON BUCH, Travels, Memoirs, &c. | 55 — 50 | | Augite Rocks igneous. |
| BUCKLAND, Memoirs. CONYBEARE, Geology of England, 1822. SEDGWICK. HENSLOW. | 45 | | Flœtz-trap Rocks igneous. Whin-dikes injected in a fluid state from below. Some Flœtz-traps igneous; others aqueous. |
| DOLOMIEU, Journal de Phys. vol. xxxvii. 1790. SAUSSURE, Journal de Phys. (an. 2.) 1794. DAUBUISSON, Ib. 1804, Sur Volcans d'Auvergne. DAUBENY, Edin. Philos. Journ. 1821, On the Volcanoes of Auvergne. | 40 — 35 | | |
| DAUBUISSON, on the Basalts of Saxony, 1803. DELUC, Treatise on Geology, 1809. KLAPROTH, Beiträge, vol. iii. JAMESON, Edinburgh Philos. Journal, 1819. RICHARDSON, On the Giant's Causeway, Ph. Tr. | 30 — 25 | Neptunian Region. | Igneous origin of any Trap Rocks questioned. Whin-dikes cotemporaneous with the rocks they traverse. |
| MACKNIGHT, Wernerian Memoirs, 1811. JAMESON, Geognosy, 1808. MURRAY, Comparative View, 1802. MOHS, Memoirs, &c. KIRWAN, Geological Essays, 1795. WALKER, Lectures, 1794. WERNER, Theory of Veins, 1791. | 20 — 15 — 10 | | All Rocks (except the Volcanic) deposited from aqueous solution. Metallic Veins poured in from above. |
| LAMARCK, Hydrogeologie. DEMAILLET, Telliamed. | 5 | | Secondary Rocks secreted by animals and vegetables, and formed out of Water. |

I.1 "Geological Thermometer," by William Buckland, *Edinburgh Philosophical Journal* (1822). Courtesy of the Linda Hall Library of Science, Engineering, and Technology, Kansas City, Mo. This thermometer classifies Werner (11°) and his student Humboldt along the Plutonist-Neptunist axis. Notice how far Humboldt has moved in the Plutonic direction (69°) by the time of his South American travels. The thermometer registers all the geologists I have discussed except Smith.

ism. Geology and other modern disciplines depend on the "professional exclusion" of cosmology and the broader outlook of natural theology.[12] The culture of applied geology thus makes visible the emergence of new social classes—a gentlemanly scientific class, a class of professional guides and mineralogists, an industrial-managerial class—whose increasing specialization will ultimately lead them away from the shared vocabularies of landscape aesthetics and natural theology.

## Applied Science and the Competition over Nature

William "Strata" Smith (1769–1839) employs a number of the human-natural analogies cited by Rossi in articulating a new kind of earth history, a "fossil record" that integrates a narrative of biological succession with the sequence of the strata. The precision and accuracy of Smith's fossil record allowed him to construct a geological map that "changed the world," as a popular recent biography seeks to persuade us.[13] There are literary reasons, too, for Smith's contemporary reputation: his prose is more accessible than Hutton's (for example) and succeeds in integrating the aesthetics and morality of "improvement," the widely accepted equation between the development of land and moral progress. His geological observations, arising from the unique opportunities provided by his career as a surveyor, shed light on the interaction of geology and economic history. Though Smith's maps and system of fossil classification were much used by other economic geologists and "improvers" of land, his scientific career was hampered by his status as a self-educated, provincial practitioner. Martin Rudwick has analyzed Smith's belated recognition by the metropolitan Geological Society in 1831. This recognition was a strictly political matter; Smith, as a professional field geologist, had been studiously ignored by the society since its founding by gentleman landowners in 1807. Rudwick vividly shows how the change of attitude served a certain faction of gentlemen who now wanted to privilege fieldwork.[14] In this instance, the "landed interest" finally recognized the impetus provided to provincial applied science by the "commercial interest"—Smith made his discoveries while employed by the Somerset Coal Canal Company, and David Allen points out that he delayed the publication of his

---

12. Toulmin, *Return to Cosmology,* 229–30, 234.
13. Simon Winchester, *The Map That Changed the World.*
14. Rudwick, *Great Devonian Controversy,* 63–68. Hugh Torrens suggests that even in the society's early years, it was Smith's collection of fossils, viewed by its members in 1808 with ostensible indifference, that prompted the inclusion of an item concerning fossils in the *Geological Inquiries;* "Arthur Aikin's Mineralogical Survey," 119.

maps and sections because of his apprehension that they contained vital commercial secrets.[15]

Humphry Davy's stellar scientific career illustrates more vividly the scientific impact of the competition between the landed and commercial interests, as well as the literary affiliations of metropolitan scientific culture. After a short but highly productive tenure at Thomas Beddoes' Pneumatic Institution in Bristol, Davy was offered a post at the newly formed Royal Institution in London. As Morris Berman has shown, the founders and Managers of the Royal Institution represented the commercial interest, fostering the evolution of applied science as an instrument to facilitate the improvement of industry and agriculture using mercantile capital. It was the first scientific institution associated with acquired, rather than inherited, wealth.[16] Previous scholarly work on Davy has focused on his chemistry, which was the chief component of his multifaceted career. The political implications of Davy's chemistry have been explored by David Knight and Jan Golinski, among others.[17] His agricultural chemistry has particular implications for land use and natural resources that overlap with the implications of his geology and link Davy to a growing class of applied scientists—such as William Smith—who operated outside the gentlemanly learned societies of metropolitan scientific culture. Davy's social ascent from the provincial bourgeoisie to the presidency of such a learned society was the product of a transitional phase, during which the learned societies gradually became more professionalized. Rudwick and James Secord, among other historians, have illuminated the transition from the gentlemanly learned society to scientific professionalism.[18] This transition provides a unifying social context for the intellectual production of figures as diverse as Davy and Smith. Both naturalists extolled the history of the earth as a morally and aesthetically uplifting project as well as a valuable instrument for maximizing profits from a landed estate, whether acquired or inherited. Both these impulses inform

15. Allen, *Naturalist in Britain*, 50.

16. According to Morris Berman, the institution still owed its origins to the landed aristocracy, but "it was organized by a select group of the upper class with a very different ideology of science in mind," and thus became "the opening wedge in a major ideological shift" (*Social Change and Scientific Organization*, xxi). Berman argues that the Royal Institution (RI) was in the vanguard of middle-class ascendancy: "The RI became an illustration of how the leadership of the nation might be transferred to another social class," the "group of 'economic men'" using science to "improve their economic standing" who increasingly made up the institution's membership and managerial board (xxii, see further xxiii–iv and chaps. 1–2).

17. Knight, *Humphry Davy;* and Golinski, *Science as Public Culture*, 191–203; cf. 176–87.

18. Methodological differences between historians of nineteenth-century versus earlier geology seem to reflect this evolution of geology's concerns. Secord, for example, favors the study of geology's "social locations," rejecting the traditional view of early earth science as "an essentially historical enterprise concerned with process and time"; *Controversy in Victorian Geology*, 4.

the poetry and aesthetics examined alongside applied science in the following chapters.

Davy (1778–1829) is a uniquely representative figure for the period: he was a naturalist and a poet, a popular champion of interest in the natural world, a medium of access for commerce and industry to the ferment of ideas, and a key player in the production of Wordsworth's *Lyrical Ballads*. Davy occupies a unique historical position as both a great Romantic thinker and (in the words of one scholar) a "malleable orator and experimenter" hired by "improving landowners . . . to assist them in social control as a 'Mr. Fixit,' and to improve the farming, tanning, and mining industries on which their incomes depended."[19] His literary contributions might seem insignificant by contrast with the chemical discoveries and practical inventions that earned him his knighthood and tenure as president of the Royal Society, but any literary history must take note of his sponsorship of Coleridge's 1808 lectures on Shakespeare and the lively exchange with Wordsworth that culminated (as Roger Sharrock has shown) in the introduction of the "Poet" and the "Man of Science" in the 1802 revision of the *Preface* to *Lyrical Ballads*. Other critics have discussed Davy's relationship to Coleridge, Wordsworth, and the Shelleys.[20] Part of my purpose in rehearsing Davy's Romantic pedigree is to point out the intrinsic importance of Davy's economically inflected aesthetics—whether manifested in his early, "pantheistic" manuscript poems, his geology lectures, or other writings—for any general model of Romantic aesthetics, which is thus linked to the cultural legitimation of applied science. The brilliant young lecturer who writes of physical "laws" as "the most sublime objects" (*Lectures*, 12) is also the aspiring poet who seeks "on Newtonian wings sublime to soar / through the bright regions of the starry sky."[21]

Davy's work embodies what might seem to us a paradox: a sense of scientific practice as aesthetic experience—which we associate with "pure" science—accompanies his vocal and at times mercenary advocacy of *applied* science. "The study of . . . the laws," he writes, "by which dead and inorganic matter are rendered subservient to the purpose of living beings, affords some of the most sublime objects which even the most insensible mind can scarcely consider without deep feeling" (*Lectures*, 12). Here, scientific law becomes a hinge between utility ("subservient") and the aesthetic ("sub-

19. David M. Knight, "Commissions, Creativity, Esteem," 47.
20. See Roger Sharrock, "The Poet and the Chemist"; and Rom Harré, "Davy and Coleridge." Coleridge's letters of July 28, 1800, and February 3, 1801, provide an especially vivid portrait of their relationship. On Davy and Wordsworth, see Catherine Ross, " 'Twin Labourers' "; on Davy and the Shelleys, see Anne Mellor, "*Frankenstein*"; and Carl Grabo, *Newton among Poets*, esp. chap. 7.
21. This poem of Davy's is quoted in Sharrock, "The Poet and the Chemist," 59. Much of Davy's poetry is included in *Fragmentary Remains*, edited by his brother John.

lime"). This reasoning offers itself as a counterpart to aesthetic explanations of rocks as sublime because unfathomable and alien. Davy moves toward a vision of rocks as demystified commodities, but this emerging style of explanation depends heavily on the one it is displacing. Aesthetic explanation continually embellishes and at times legitimates the rhetoric of appropriation and mastery in Davy's lectures. The "laws" in question are associated with the applied chemistry Davy wants to use in order to systematize geology, a scientific practice linked, in turn, to his institutional setting. The "interests of the human species" that Davy presents as being served by his vision of science are always in part the economic interests of the landowners who control the Royal Institution and are the most influential sector of Davy's audience. Morris Berman points out, for example, that the 1805 *Lectures* sometimes refer for their examples to the estates and resources controlled by the powerful members of Davy's audience.[22] Davy's aesthetic language about nature cannot be reduced to economic interest, of course, but it helps to identify social contradictions that inform Romantic thinking about the earth's material.

The founding of the Geological Society in 1807 marks a crucial moment in the history of geology, perhaps most of all because the society retained its place as geology's central institution by incorporating and consolidating other geological endeavors through the early Victorian period, when the science became fully established. Rudwick and Secord have documented two phases of the society's transformation, which was partly a change in its social agenda and composition. The society's *Transactions* initially included some contributions by "practical men" such as John Williams, but gentlemen—whether academics like William Buckland or amateurs like George Cumberland—dominate its pages. The inaugural volume (1811) announces the project of a comprehensive geological map of "the British territory," deliberately disregarding William Smith, whose map was already in progress. Smith became the first to realize such a project, working independently, in 1815. Rudwick's account shows that Smith was eventually "adopted" by the society, and Secord describes the related compromise by which the society's conflict with the Geological Survey was eventually resolved. Despite these social conflicts, the economic interests of the "landed" and "commercial" classes can be difficult to disentangle. Paul Weindling has shown that "as far as mineral history motivated the Geological Society, it represents an eco-

22. Berman, *Social Change and Scientific Organization*, 65. The prospecting instructions he delivers in such instances also provide a substantive affinity with William Smith: "The only real indications of metallic substances . . . are those founded upon a knowledge of the directions of different strata, and of the substances usually found in them, and of their relations to each other"; *Lectures*, 106.

nomically-oriented factor in its foundation."[23] George Bellas Greenough, its first president, learned much of his geology from Davy, and Weindling describes in generic terms the fieldwork they did together in the summer of 1801: "mineralogical and stratigraphic observations had featured in technological tours, which were an important link between economic and scientific concerns. Besides offering picturesque views, tours were a chance to survey natural productions both in their native state and in mining and metallurgy."[24] The following panegyric locates a similar convergence in the provincial culture surrounding Smith's geology:

> Then, hail, ye patriots, who can prize renown
> On peaceful plains, where sylvan honors crown!
> Let morning air invigorate your pow'rs,
> And plans of increase mark your passing hours.
> Waste long-neglected acres be your care,
> And barren wilds with fruitful fields compare.
> The depth and kind of surface-soil explore,
> How marked with fatness, or how tinged with ore.[25]

Mary Anning (1799–1847) was another provincial field geologist whose work was incorporated and symbolically adopted by the Geological Society. Anning supplied valuable fossil specimens and anatomical knowledge to its members and achieved a measure of formal recognition when her death was noted at a meeting of the society at which a speaker lamented the passing of "this hand-maid of geological science."[26] Anning's career testifies to yet another economic application of a knowledge of fossils and minerals. The fashion for fossil-collecting and the birth of scientific paleontology generated a specimen industry quite distinct from the economic geology concerned with the improvement of land. While Anning's gender made it easier for the Geological Society to co-opt her science, her knowledge commanded respect and raises the possibility that a number of other, less distin-

23. Weindling, "Geological Controversy," 250. Suggesting that collectors of mineral specimens served as "middlemen" communicating the advantages of scientific discovery to industrialists, Weindling concludes that "if [the] pursuit [of economic geology] was a sustained cultural response to remedy the lack of applied science [or its suppression by protectionists], then the very efforts to promote utility mean the science should be viewed as a product of industrialization" (252).

24. Ibid., 250. On Greenough and Davy's relationship, see also Robert Siegfried and Robert H. Dott's introduction to Davy's *Lectures* (xvii, xxiii).

25. Quoted in Hugh Torrens, "Geological Communication," 237. William Matthews, a coal merchant of Bath, published these lines in 1800. A footnote to the last quoted line describes Smith's work and comments that the "publick, it is hoped, will ere long be benefited by a publication of his important discoveries."

26. W. D. Lang, "Mary Anning," 81.

guished women practitioners have been "written out" of early nineteenth-century geology.[27] Mary Anning eventually became the most celebrated prospector and dealer in fossils, sought out by tourists and collectors as well as the most influential geologists of the day, including Buckland, Henry de la Beche, and William Conybeare.[28] When they published articles based on her finds, however, they routinely failed even to mention her name until 1829, when she had already become famous in her own right. George Cumberland, who attempted to set the record straight in a letter to the *Bristol Mirror*, also felt compelled to speculate on the sexual ambiguity of Anning's position as an unmarried, working-class woman. The independence that prompted Cumberland's curiosity also helped to create the legend that surrounds her life. Anning made her first major discovery, a complete Ichthyosaur skeleton, at age eleven, only a year after her father's death. This discovery has been celebrated in over a dozen children's books, while her discovery of the first complete Plesiosaurus skeleton in 1824 cemented her reputation and netted her £100.[29]

One point of intersection between applied geology and literary culture lies in the moral dimension of "improvement." Anning's career, like Smith's and Davy's, shows the economic impetus to empirical rigor. Economic and cultural factors combine to permit the discovery of new fossil characters illuminating the rock record. These discoveries in turn appear to increase the stock of amalgamated wealth—economic, scientific, aesthetic, and moral—secreted in the rocks. Anning's contributions to academic paleontology, and the moral interest of her life as a working-class orphan who made such a career for herself, also illustrate the pedagogical function of the rock record, the moral profit that accompanies economic gain and aesthetic experience. The public notices of Anning that began appearing in the early 1820s seem to take their pattern from the literary ethos of sensibility: "This lady, devoting herself to Science, explored the frowning and precipitous cliffs there, when the furious spring-tide conspired with the howling tempest to overthrow them, and rescued from the gaping ocean, sometimes at the peril of her life, the few specimens which originated all the fact and ingenious theo-

27. Simon J. Knell vividly sketches the economics of early Victorian fossil-collecting, emphasizing the number of rural poor employed by connoisseurs and middlemen in fossil-rich districts; *Culture of English Geology*, 3–7, 193–221.

28. W. D. Lang details Anning's relations with these geologists ("Mary Anning," 75–77). For her business, Anning depended mainly on the Lyme tourist trade immortalized in Jane Austen's *Persuasion* (see also Lang, 66–67).

29. Lang, "Mary Anning," 74–75. According to Hugh Torrens, there were at least a dozen children's books on Anning between 1925 and 1995, and two more appeared in 1999 ("Mary Anning," 274–77). Torrens also quotes Cumberland's speculations on Anning (261).

ries of those persons."[30] The idea of such a pedagogical function can be traced to the Age of Sensibility with its "cult of nature," which influenced Wordsworth's "Nature" (the "soul / Of all my moral being"; "Tintern Abbey" 110–11) as much as the rhetoric of natural history. The naturalizing of Ann Radcliffe's heroines would have been one of the most widely known literary prototypes for a practice that was becoming increasingly common in the education of girls. The anecdote of Anning's exertions on behalf of "Science" depends on the contrast between the "peril" she underwent and the relative comfort in which Radcliffe's characters—and genteel young Englishwomen—pursued this study. Anning's exertions seem to testify to a correspondingly extraordinary virtue, though the professional nature of her pursuit is bracketed by the comparison. In *The Mysteries of Udolpho* (1794), Radcliffe capitalizes on the rock record in a scene in which the safety, and perhaps the virtue, of one of her heroines appears to be threatened, describing

> the mineral and fossil substances, found in the depths of these mountains,— the veins of marble and granite, with which they abounded, the strata of shells, discovered near their summits, many thousand fathom above the level of the sea, and at a vast distance from its present shore . . . the tremendous chasms and caverns of the rocks, the grotesque form of the mountains, and the various phenomena, that seem to stamp upon the world the history of the deluge.[31]

Radcliffe's heroines scrutinize and narrativize landforms on their travels to derive the sort of moral education that must be acquired by *reading*. This literary genealogy must have been especially present to commentators describing a young female naturalist whose adventures helped significantly to elucidate the rock record.

While it sheds light on the formation of masculinist modern science, the narrative of Mary Anning's career also shows how moral and aesthetic discourses were used to legitimate increasingly utilitarian readings of the rock

---

30. Lang ("Mary Anning," 77) quotes this excerpt from a traveler's diary. Since Anning herself did not publish, her fame was largely due to the circulation of such anecdotes, which attracted visitors from both Britain and the Continent.

31. Radcliffe, *Mysteries of Udolpho*, 602. The analogy to Anning's geological reading is not perfect, since in this scene the Count de Villefort interprets the scene for his daughter Blanche while positioned strategically between Blanche and her suitor as the party takes shelter from a storm in the mountains. Elsewhere, however, Radcliffe asserts that "susceptibility to the grandeur of nature" is the medium for "frankness" and "simplicity" in both sexes (34).

record. This new inflection in turn becomes a part of the literary legacy of the rock record as it is mobilized by Wordsworth, among others.[32] Wordsworth's conservationist program is never incompatible with poetic domestication—with the practice of inscription, for example, that is modeled on the history inscribed in rocks by natural agencies.[33] Rocks and landforms become poetic resources, spaces for inscription, and records of an aesthetic history, as well as moral textbooks. By the early Victorian period, Wordsworth's domestication of landscape was providing a new aesthetic framework for science and technology. John Wyatt argues, for example, that Victorian geologists "confirmed their faith by reading *The Excursion*," which Wyatt sees, along with *A Guide to the Lakes*, as "a major apologia for an orderly world with a benevolent intention." Wyatt persuasively points out ordering principles shared by the poetry and the geology, but he limits the affinity too narrowly to purely theoretical or "poetic" geology. He argues that geology appeals to Wordsworth because it "has ascended to a rank above mere technicality," concluding in a section titled "The Noble Science of Geology" that "the function of science is not to aid the economy but to elevate the human spirit." Tellingly, however, Wyatt focuses on Greenough, who ushered in the economic shift within mainstream geology in the 1810s and 1820s. He quotes the following passage to illustrate Greenough's philosophical affinity with Wordsworth: "Geology in its comprehensive sense is consequently a sublime and difficult science, but fortunately for its progress it is susceptible of divisions into many different departments, several of which are capable of being extended by mere observation."[34] The passage is from the Geological Society's *Geological Inquiries*, regarded by historians of geology as a cynical and "condescending" attempt by "establishment geology" to co-opt the fieldwork done by provincial geologists and "practical

32. Theresa Kelley points out that "like Ann Radcliffe and other advocates of the doctrine of sensibility, Wordsworth assumes that aesthetic response is an index of moral feeling" (*Wordsworth's Revisionary Aesthetics*, 24). Science becomes increasingly compatible with sensibility in the later Wordsworth (see, for example, *The Excursion* I.253–79). On the gendering of modern science, see Barbara Gates and Ann Shteir's excellent introduction to their collection *Natural Eloquence*.

33. Wordsworth's three groups of "inscription" poems (1800, 1808–11, 1818) are chiefly commemorative in function, but their largest implication is to claim inanimate nature in toto as a textual space for inscription. The second volume of *Lyrical Ballads* (1800) includes a set of related "Poems on the Naming of Places," and a single inscription poem of 1813 ("Written with a slate pencil on . . . Black Comb") is particularly interesting for its fusion of geological and national history.

34. Wyatt, *Wordsworth and the Geologists*, 111–12, 110, 179 (cf. 191–92). Wyatt also gives a good account of Wordsworth's geological sources (59) and the "availability of scientific culture" in general (64).

men" excluded from the society.[35] The *Inquiries* initially met with very little success, because those in possession of the desired facts knew the economic value both of "mere observations" and of geological order or legibility. Without disputing the humanist connection between Wordsworth and the geologists, it must also be recognized that the shared project of domesticating the natural world is economically motivated, leading Wordsworth to appropriate geological concepts like the rock record in the service of his own vision of utility.

Wordsworth's own theory of geomorphology owes as much to the ideology of improvement as to theories of the earth and their "ennobling" successors. Just as a model of the Alps brings "delight to the imagination" as well as "the more substantial pleasure" of a scientific comprehension of topographical relations (*Prose*, 2:170), the tone set here establishes science as an imperative distinct from aesthetics: "Sublimity is the result of Nature's first great dealings with the superficies of the earth; but the general tendency of her subsequent operations is toward the production of beauty, by a multiplicity of symmetrical parts uniting in a consistent whole" (181). However much Wordsworth's geology in the *Guide* may be informed by aesthetic categories, he seems to read the rock record quite deliberately as a "Man of Science," in order to establish a distinct prehistoric narrative from which the aesthetic itself may be derived. The ensuing transition, leading from the scientifically oriented material introduced in the 1820 edition back to his original discussion of the geometry of shorelines, is also striking: "But, checking these *intrusive calculations*, let us rather be content with appearances as they are, and pursue in imagination the wandering shores" (182, emphasis added). The rock record has prompted a digressive calculation of the landscape's potential for improvement, a speculation about the human relationship to the inscribed trajectory of increasing utility and beauty. Wordsworth became increasingly liable to such digressions, to judge from the 1833 poem "Steamboats, Viaducts, and Railways," in which "Nature doth embrace / Her lawful offspring in Man's art" (10–11) even as it "mars" her "loveliness." Throughout *A Guide to the Lakes*, Wordsworth adopts the terms of both improving landlords and applied scientists—to some extent, perhaps, in spite of himself—because he is competing against both to establish his vision of a British landscape compatible with "healthy" taste and traditional modes of production.

Jealously attuned to the economics of landscape, Wordsworth senses a connection between the tradition of landscape aesthetics and applied sci-

---

35. Weindling, "Geological Controversy," 261; and Roy Porter, "Industrial Revolution and Rise of Geology," 341. See also Allen, *Naturalist in Britain*, 59.

ence, the new face of improvement. This connection is one basis for the modern concept of natural resources, which arises where sensations of wonder or sublimity are connected with technical mastery. William Chambers seems a likely target for Wordsworth's skepticism about "a practice which by a strange abuse of terms has been denominated Ornamental Gardening" (*Guide*, 189n.).[36] Chambers was much ridiculed for his "Chinese" gardening in nationalist terms that anticipate Wordsworth's attacks on excess and exoticism. Chambers defended himself against such attacks by suggesting that his methods had scientific value: "As the principal parts of this supernatural Gardening consists [*sic*] in a display of many surprizing phenomena, and extraordinary effects, produced by air, fire, water, motion, light, and gravitation, they may be considered as a collection of philosophical experiments, exhibited in a better manner, upon a larger scale, and more forcibly than is common."[37] One of Chambers' "experiments" closes the gap between eighteenth-century aesthetics and nineteenth-century applied science in a startling way. This experiment originates in the "scenes of terror" ostensibly created in Chinese gardens: "To add to both the horror and sublimity of these scenes, they sometimes conceal in cavities, on the summits of these mountains, founderies, lime-kilns, and glass-works; which send forth large volumes of flame, and continued columns of thick smoke, that give to these mountains the appearance of volcanoes."[38] Humphry Davy's lectures on volcanism featured a similar spectacle on an appropriately smaller scale. While Davy's model cannot duplicate the economic productivity that Chambers so thoughtfully incorporates, it concentrates as much exoticism and aesthetic effect as the life-size spectacle. A member of the audience at one lecture recorded his recollections of the event: "I remember with delight the beautiful illustration of his theory exhibited in an artificial volcano constructed in the theatre of the Royal Institution. . . . Red hot lava was seen flowing down its sides, from a crater in miniature—mimic lightenings played around, and in the instant of dramatic illusion, the tumultuous applause and continued cheering of the audience might almost have been regarded as the shouts of alarmed fugitives of Herculaneum or Pompeii."[39]

36. Wordsworth deleted this polemical description after the second edition, but de Selincourt includes it with other variants in his appendix (*Guide*, 189; cf. 69). In the same passage Wordsworth indicates that "ornamental gardening" came into vogue about 1770, that is, two years before the publication of Chambers' *Dissertation*.

37. Chambers, *Explanatory Discourse*, 132.

38. Chambers, *Dissertation on Oriental Gardening*, 37.

39. John Ayrton Paris, *Life of Sir Humphry Davy*, 218–19. Davy may also be drawing directly on the London stage. Two popular musical pantomimes, *The Volcano* (1799) and *Harlequin Teague* (1782), incorporated the theme of volcanism or (in the former case) actually staged volcanic eruptions.

Wordsworth more quietly stages the "production of beauty" by the "secondary agents of nature." He probably did not have Chambers' artificial volcano in mind, but in turning his attention to landforms he probably did think of Davy, perhaps even in terms similar to those used by Davy in the first of his geology lectures, also quoted above: "The imagery of a mountain country, which is the very theatre of the science, is in almost all cases highly impressive and delightful, but a new and higher species of enjoyment arises . . . when the arrangement . . . and its subserviency to the purposes of life are considered" (*Lectures*, 13). Much of this introduction to geology is adapted and elaborated from Davy's introduction to his course of chemistry lectures in 1802, attended by Coleridge and Wordsworth. Other scholars have noted the caution with which Wordsworth greets Davy's notion of the "man of science" as "a master, active with his own instruments."[40] My point here is to suggest another ground of suspicion on Wordsworth's part: Davy's representation of nature relied on a landscape aesthetics tainted by economic motives. Just as Davy's 1805 lectures represent an extension of his 1802 challenge to Wordsworth, so *A Guide to the Lakes* can be considered an extension of Wordsworth's rebuttal. It could only help Wordsworth's case if Davy's aesthetics appeared closer to someone like Chambers than to the program outlined in the *Guide*.

Davy recommends scientific pursuits to his audience at the Royal Institution by means of a curious blend of economics and aesthetics: through scientific study, he says, "nature arises subdued by artificial means, not impoverished or deformed, but enriched, and made more beautiful" (*Lectures*, 7). This formula establishes the aesthetic provenance of the concept of natural resources. Davy conjoins a sovereignty over nature ("subdued by artificial means," i.e., scientific knowledge and economic development) with its conventional admiration ("enriched and made more beautiful"). This is a form of the conjunction I described in earlier chapters as a merging of aesthetic response ("wonder") into economic relations ("sovereignty") and of aesthetic objects into resources. Though otherwise very different, Wordsworth's *Guide* and Chambers' *Dissertation* share a practical slant, a focus on intervention in landscape, that makes it possible to develop these economic relations more fully. Davy's applied geology analyzes and marshals landforms in a way that is often similar to both texts, drawing some aesthetic values from Wordsworth, but closer to Chambers in its emphasis on technical "invention" (*Lectures*, 7). To put it another way, each of these writers has a very different use for landscape and natural resources. Chambers' appropri-

---

40. Davy, "Discourse Introductory to a Course of Lectures on Chemistry," in *Collected Works*, 2:319. See further Sharrock, "The Poet and the Chemist"; and Ross, "'Twin Labourers.'"

ation of "Chinese" gardening technique and Chinese *nature* compensates for the lack of mysterious commodities inaccessible to British trade, technology, and science. Wordsworth's and Davy's reactions to such aesthetics, in turn, make visible the global narrative of exotic beauty and natural superabundance latent in the nativist discourse of Romanticism and its concept of natural resources.

## Aesthetic Geology

The economic geology producing such concepts as the "rock record" and "natural resources" is never divorced from aesthetic geology. The aesthetics associated with applied geology provides a useful guide to the common ground of the two Romantic-period discourses now known by the mutually exclusive names of literature and geology. Turning from the symbolic register of geological otherness, and the environmental concepts which accompany it, the remaining chapters of this book examine the literary genres and geographical places in which literature and geology cohabited most flagrantly. Chapters 5 and 6 in particular contribute to the cultural history of Romantic landscapes less renowned than the Alps or the Lake District but equally important for the evolution of the aesthetics of landscape and the science of landforms. At the same time, "aesthetic geology" is designed as an intervention in the history of the disciplines, a space between literature and science that helps to contextualize modern disciplinary boundaries. The cultural history of places powerfully illuminates the individuation of particular discourses about them, as in the earlier examples of Lyme Regis, the site of Jane Austen's "romantic rocks" and Mary Anning's astonishing fossils; and Middleton Dale in Derbyshire, where Thomas Whately used stratigraphic correlation to articulate his model for "romantic" scenes within the landscape garden and John Whitehurst adduced the same evidence for his theory of the earth.

These are romantic places in more than one sense. Thus far I have explored the places of Romanticism as traditionally understood, through the converging and diverging lenses of the poets usually associated with it and their geologically minded contemporaries. Numerous questions remain, however, questions about the geographic and stylistic senses of "romantic" that refer to cultural territory often considered peripheral to both Romanticism and geology: the worlds of fashionable tourism and grinding poverty operating next to each other in the Peak District and parts of Scotland and Ireland; the discourse and practices of popular science equally scorned by Wordsworth and self-consciously modern geologists; the rich periodical archive of poems about places; the complex material culture surrounding specimen collection; and the period's large body of travel narrative, much of it still unincorporated by literary history. This material provides the best available map of the cultural territory through which Wordsworth, Blake, Shelley, Hutton, Davy, and others traveled when they created their eventually famous images of rocks and landforms. Two locations on this cultural map are especially rich in particulars: the complex generic locus of Erasmus Darwin's *Economy of Vegetation* and the geographical locus of the Peak District in Derbyshire, both romantic places in the sense that they incorporate the five elements listed above. The extensive geological and topographical passages in Darwin's poem, like many of the prose and verse descriptions of the Peak, illustrate the transfer of a style of description from geology to literature as geology began to separate itself in the later eighteenth century. These relationships in turn provide material for case studies in discipline-formation and the broader relations of literature and science.

The relations between literature and geology from about 1770 to 1820 embody crucial aspects of the more complex and gradual shift from "natural history" to "science."[41] Because of the unique importance of landscape in this period, what we now term "geology" called forth great volumes of poetry and natural history that shared many of the same philosophical, rhetorical, and commercial concerns. This unified commerce in print and ideas certainly became less unified over time, and its chronological dimension—illustrated by my initial comparison between Whitehurst's and Austen's uses of the term "romantic rocks" (1778–1818)—is a more explicit focus of the last two chapters. What appears in this comparison to be a transfer of style from geology to the novel is more accurately described as an incipient shift in the definitions of the disciplines themselves. Chapters 5 and 6 consider the relations among aesthetics, earth science, and poetry, as well as the relation, often thematized in topographical poetry, between poetry and science in general. The encyclopedic nature of other prose on landscape, such as tours and guidebooks, helps to illustrate the importance of place for all these forms of writing. By the early to middle nineteenth century, what we consider literary forms, especially descriptive poetry, absorbed the spectacular style of description associated with aesthetic geology as modern geology became increasingly empirical and positivistic. The proliferation of topographical poems about rocky and mountainous places (generating the refrain "not a mountain rears its head unsung"), in conjunction with the close kinship between aesthetics and early geology, suggests that interest in rocks and landforms initially arose from aesthetic experience and organized itself around aesthetic categories. Literary descriptions of places carried on a kind

---

41. For a discussion of this epistemic shift and review of some relevant literature, see my introduction to *Romantic Science*.

of scientific practice eventually rejected by science as the study of places became more systematic. As the earlier mode of geological description, itself informed by descriptive poetry, became naturalized in literary forms, generic boundaries between poetry and science, as well as description in verse and in prose, were newly configured.

Place is crucial to the materiality of aesthetic experience. The local nature of all forms of geological description is one constant underlying the epistemological shift described above. The local emphasis on observation and description informs geological representations of materiality, which often derive their force from the particularity of individual landforms and landscapes. My last chapter attends expressly to this issue, focusing on three "romantic" places in the Peak District. The Peak District, among geologically remarkable places in the British Isles, is the earliest to develop a tradition of local literature, a genre of spectacular description later echoed in more remote places such as the Lake District, Staffa, and the Giant's Causeway. One of the longest of these local poems, William Drummond's *Giants' Causeway*, is also the most geologically remarkable. Drummond published his poem in 1811, late enough to display the old, spectacular style of geology in a new, distinctly literary form, as well as an impressive command of geological issues. Here the description of a place clearly serves as the hinge by which geology opens onto poetry; the poetry engages science, but not overtechnically, more in the manner of Tennyson than of Erasmus Darwin. Wordsworth's discussion of the primitive and secondary "operations of nature" helps to bear out this stylistic development. By embracing a terse and functional style of description when dealing with the subject matter claimed by modern geology (especially in the 1820 revisions to *A Guide to the Lakes*), he acknowledges the new classification as "poetic" of a more metaphorical, allusive, and spectacular style of description formerly shared by geologists and poets.

The stylistic component of aesthetic geology illuminates the generic relation between poetry and two increasingly distinct areas involving the physical study of rocks: geology and topography. The division between poetry and science becomes increasingly fundamental in the decades after Wordsworth's exploratory discussion in his 1802 *Preface*. Topographical poetry, with its interest in physical landscapes, is perhaps the poetic genre to which geology remains most relevant and in which geological content begins to acquire distinctly poetic form. Drummond's *Giants' Causeway* is one topographical poem that provides striking evidence of geology's becoming naturalized in poetry. This substantial work (over 2,000 lines) aptly demonstrates the importance of particular places as points of common interest among geology, poetry, and other topographical literature. The evidence comes as much from the verse as from Drummond's elaborate preface and copious scientific and historical notes, but the decorous placement of these notes at the end of the text, unlike Darwin's, signals a growing awareness that scientific content is not always compatible with poetic decorum. This is one of many tokens in such discourse of a major epistemic shift—a late stage in the transition from "letters" to "science" as the cultural rubric for study of the natural world. The major poets who have figured prominently in previous chapters certainly participated in the cultural moment of aesthetic geology, but scientific poetry and the topographical genres are its most concentrated and characteristic forms.

Much popular poetry throughout the eighteenth century and into the nineteenth was more assertively local in its interests than even Wordsworth's generally was. *The Giants' Causeway* demonstrates the ability of the established genre of topographical poetry to accommodate the proliferation of data while articulating a new boundary between poetry and science. Geology, antiquarianism, and travel narratives had made available a great deal of new information about places since the days of John Denham's *Cooper's-Hill*. The innumerable topographical poems produced during the intervening period often abandoned physical description almost entirely in favor of moralizing, or made their descriptions so derivative as to lose, ironically, their local character. The rigid dominion of conventions within the genre leads Robert Aubin, in his exhaustive study of topographical poetry, to argue that it was basically moribund by the last decade of the eighteenth century.[42] Drummond is not free of the obeisance to convention: morning, noon, and night revolve through the poem with stifling regularity; and his couplets, elaborate apostrophes, and other "gaudy and inane phraseology" seem old-fashioned from a post-Romantic vantage. But his consistent focus on the physical landscape, resulting in sustained original descriptions, allows even an apparently exhausted literary form to convey the intrinsic aesthetic interest of the Giant's Causeway, and to define that interest as distinct from those of science and antiquarianism. Drummond's poem preserves aesthetic geology, the style rejected by a more consciously positivistic geology, in a real sense, because it incorporates serious geological study into a literary formulation of the aesthetic experience of a particular rock formation.

While Drummond's poem is an interesting case, it represents a terminal point for my analysis of aesthetic geology, which is chiefly occupied with earlier stages of the process by which geology sloughed off the old style of description as it became a modern science. *The Giants' Causeway* offers an early instance of self-consciously poetic science, that is, poetry investigating the natural world that nonetheless acknowledges the prior claim of "natural

42. Aubin, *Topographical Poetry in XVIII-Century England*, 102.

philosophy" (in Drummond's lexicon). The difficulty of fixing these developments in time accounts for my focus on a transitional period rather than on a decisive movement from one paradigm to another. Geology was always literary, but its appeal to aesthetic response, at times studiously cultivated, came to be seen as unscientific. The new catastrophism, the diluvialism of the 1810s and 1820s, studiously avoided such effusions, despite the proven literary potential of geological catastrophe. James Parkinson's *Organic Remains of a Former World*, for example, despite its dramatic title and apparent literary potential, consists mainly of dry anatomical recitative, leaving Shelley to exploit its literary potential in the fossil archive of *Prometheus Unbound.* Though published eight years after Drummond's poem, Shelley's puts forward a stronger claim to the status of natural knowledge. Shelley's fossil archive incorporates economic concerns closely related to those of William Smith's paleontology and thus illuminates the aesthetic investments of modern empirical science as well. All in all, geological discourse provides a literary index of a massive epistemic and socioeconomic shift. The economics of geological discourse and practices, examined alongside the aesthetics of romantic rocks, offer a way to recontextualize the aesthetic/practical distinction. The prehistory of modern geology is ultimately a story of changing relations between the aesthetic and the practical. As a long-term result of these changes, it is now more difficult than ever, as William Cronon has pointed out, to recognize "wild" nature in the depleted industrial landscape.[43]

Despite my increased focus on emergent modern geology, old-fashioned theorists like Hutton and Whitehurst remain important throughout the book. Geologists do not repudiate aesthetic discourse entirely or instantaneously, but the older generation's style yields more readily to the concept of aesthetic geology. Moreover, the definitions of science and literature themselves shift in the course of these changes, and the repudiation of the aesthetic helps to *constitute* two newly separated discourses. Hutton's and Whitehurst's conformity to an older, more "literary" school of natural history is made especially clear by their adherence to a fiery (Plutonist) origin, though ironically this explanation proved—as far as we know—to be the correct one. The impact of the older style, which borrowed from descriptive poetry and aesthetic theory and helped to shape them in turn, appears especially in descriptions of natural catastrophes—as in the volcanic

43. I believe Cronon is wrong to suggest that Wordsworth and the Romantics set this dualism in motion ("Trouble with Wilderness," 73–74). Early industrial thinkers were aware that "landscape" and "natural resources" are identical. Our alienation from nature has more to do with applying the Romantic preference for "wilderness" to an American ideal that was always incompatible with intensive land use. See also James McKusick, *Green Writing*, 7–11.

episodes quoted above. Vesuvius, which erupted "with obliging frequency" from the mid-eighteenth century well into the nineteenth, provided the focus for a great deal of aestheticizing representation, from illustrated natural history folios to public spectacles including the pantomime, fireworks shows, the *eidophysikon*, and the panorama.[44] Such nonliterary forms influenced Erasmus Darwin's portrayal of geology and other sciences in *The Botanic Garden*, which draws broadly on visual culture as well as the scientific associations of earlier topographical poetry in forging its popular hybrid of scientific culture. The wealth of detailed literary descriptions of the Peak District shows how spectacular description, popular science, and local aesthetics all merge to form a variety of the "aesthetic materialism" outlined in the introduction to this book. The language and categories of the aesthetic participate in early geology, not as a rhetorical stand-in for "real" explanation, but as a form of knowledge that constitutes the objects of the science; early geology, in turn, explores and helps to define the aesthetic objects of Romanticism.

44. William Hamilton's *Campi Phlegraei*, a splendid folio volume with more than fifty color aquatints, paved the way for increasingly reproducible and affordable illustrated works of natural history, such as Friedrich Justin Bertuch's *Bilderbuch für Kinder*, the third volume of which begins with color plates of Vesuvius in eruption, modeled on Hamilton's. *The Volcano*, a "grand pantomime" first produced at Covent Garden in 1799, is worth noting again here. For examples in the other categories, see Richard Altick, *Shows of London;* and Barbara Stafford and Frances Terpak, *Devices of Wonder.* The witticism on Vesuvius is Roy Porter's (*Making of Geology*, 123).

# 4

## The Rock Record, Mineral Wealth, and the Substance of History

We must read the transactions of times past in the present state of natural bodies.

—JAMES HUTTON, *Theory of the Earth* (1795)

The bodies in question in Hutton's *Theory of the Earth* are rocks, which resist reading but transact a great volume of business. The rocky landforms of Romantic poetry—Mont Blanc, the Simplon Pass, Ben Nevis—also famously resist reading, generating images that articulate the otherness of the physical through the literal and metaphorical opacity of rock. This aesthetic response to the materiality of rocks and landforms is, however, inseparable from the emerging economic category of natural resources. Hutton's criterion of legibility is developed further by the more empirical geologists of the following generation, especially William Smith. Smith's economic geology preserves the paradox of a landscape both profitable and "romantic," a paradox expressed most vividly in the range of Shelley's descriptions. While Shelley describes Mont Blanc as an alien, "unearthly" landscape, the earth itself is domesticated as an "infinite mine" in his *Prometheus Unbound*. The model of an "infinite mine," with its latent natural history, generates what might be called a historiography of the earth.

John Clerk's famous engraving from Hutton's *Theory* gives a visual form to this historiography (Fig. 3.1). As a heuristic representation of the rock record, it distinguishes three basic phases of earth history. The engraving shows the side of a cliff exposing a set of vertically inclined strata underlying more recent, horizontal strata. If we understand these lower strata as having been uplifted and tilted from their original horizontal position, the image conveys some idea of the tremendous degree of subterranean heat and pressure by which, as Hutton argues, rocks are transformed and continents elevated. The deformation of these lower strata, especially to the right, evokes these immense plutonic forces, while the delicate and miniaturized landscape above suggests the contrasting "imbecility" of human capacities, as Hutton termed it. At the same time, this is recognizably the landscape of "improvement," and the image seems to figure the history of the earth as a narrative of improvement. William Smith was able to read more definitely, and to capitalize more fully on geological transactions, by focusing on the orderly sequence of horizontal strata. Through his system of fossil "characters," Smith claimed, the strata became "more intelligible and useful." When the history the earth has recorded in itself becomes legible, it is found to be a history of improvement, culminating in the human transformation of landscape. The tremendous age of the earth, of all the features of the earth's material so widely discussed during the Romantic period, becomes the province of an increasingly scientific geology, and the rock record today is still the basic paradigm of historical geology. In its early nineteenth-century form—particularly in the fossil archives of Smith and Shelley—the rock record represents the earth simultaneously as the substance and the text of history, generating a materiality located precisely between the two materialities recently competing for the objects of Romanticism, that of the letter and that of history.[1]

Toward the end of *Prometheus Unbound* (1819), an apocalyptic light illuminates "the secrets of the Earth's deep heart, / Infinite mine of adamant and gold," a fund of wealth embedded in a subterranean archive of prehistoric fauna and cultures (IV.279–80). The numerous scientific allusions in this passage have been traced to James Parkinson and Humphry Davy.[2] The "secondary strata" are, according to Davy, "monuments of the great changes

1. I allude to the deconstructive and new historical schools, respectively, of Romantic criticism in the last quarter-century. Rom Harré usefully distinguishes among three kinds of relationships in the study of literature and science: shared content, shared style, and shared "metatheory" ("What Is the *Zeitgeist?*" 3). While all three are at issue in this chapter, the paradigm of a rock record fits well into the third category.

2. *Shelley's Poetry and Prose*, 277. The reading "mines" (l. 280) is adopted by some other editors. Fraistat and Reiman (278n.) cite James Parkinson's *Organic Remains of a Former World* (1808–11), which Shelley owned, and Davy's influence is studied at length by Carl Grabo, *Newton among Poets*.

the globe has undergone. They exhibit indubitable evidences of a former order of things and of a great destruction and renovation of living beings. . . . The connection between their causes and effects is obscure but apparently within the reach of our faculties, and it is displayed in characters which can be deciphered only with difficulty, but which express sublime truths" (*Lectures*, 77–78). Shelley's light brings these phenomena entirely "within reach of our faculties," augmenting their sublimity at the same time. Similarly, Smith's famous map and his prose (1815–17) maintain the aesthetic provenance of the rock record while offering decipherable "characters" in the strata's fossils. Despite their differing idioms, both accounts expand the scope and utility of a rapidly accumulating body of geological knowledge, linking it to existing notions of social and economic progress. Both Smith and Shelley had investments on either side of the line dividing what we now call literature and science. While Shelley drew on Parkinson's account of fossils, which was partly based on Smith, and Smith himself wrote verse, their works are politically and generically too remote from each other to suggest direct influence.[3] Precisely because direct influence is unlikely, the striking parallels between Smith's and Shelley's fossil archives illuminate a significant cultural moment: the new, comprehensive geological map—whether visionary (Shelley) or practical (Smith)—reconfigures the established analogy between the "history of the earth" and the "history of nations" to produce a socially constituted materiality.

## Applied Geology and the Rock Record

William Smith, introduced briefly above, is the first English interpreter of a rock record still recognized as such by geologists, thanks to his major discovery: the importance of "guide" fossils for correlating rock strata. When a particular fossil organism occurs mainly in various outcroppings of the same rock type, it can be assumed that those outcroppings originally constituted one continuous stratum, deposited at a time when that organism thrived. Smith writes: "By the help of organized fossils alone, a science is established with characters on which all must agree. . . . The organized Fossils (which might be called the antiquities of Nature) . . . are so fixed in the

---

3. Dennis Dean discusses Smith's importance for the third volume of Parkinson's *Organic Remains* (*Gideon Mantell*, 23 and n.). Smith also cites Parkinson (*Stratigraphical System*, v), and Simon Winchester speculates on Smith's possible connections to Davy and John Whitehurst, whose geology is discussed at the end of this chapter (*Map That Changed the World*, 94, 123, 224). Simon J. Knell mentions Smith's own poetry in *The Culture of English Geology*, 37.

earth as not to be mistaken or misplaced; and may be as readily referred to in any part of the course of the stratum which contains them, as in the cabinets of the curious."[4] Smith joins the logic of orderly collection and the motives of improvement to earlier, more speculative notions of a rock record, such as Hutton's and Davy's. Smith's technique of stratigraphic correlation, traditionally emphasized by historians of geology, enabled him to produce his historic *Geological Map of England and Wales* (1815), recently in the limelight thanks to Simon Winchester's *Map That Changed the World*. I am concerned with Smith's largely neglected writings, but Winchester's emphasis on the map is convincing, and Smith made other important innovations in the graphic depiction of the rock record.

One of Smith's numerous tables provides a striking analogue to Shelley's description. The image (Fig. 4.1) represents the earth, like Shelley's drama, as both a stratigraphically organized mine of useful material and a subterranean archive of successions of prehistoric creatures. This table (published in 1817) is one of several remarkably comprehensive geological surveys of the island deriving from Smith's 1815 map, the first systematic representation of large-scale stratigraphic succession in our modern sense. The table is equally encyclopedic in scope, compressing into tabular form the comprehensively transparent rock record Smith had reconstructed from his fossil "characters." The second column names all the English strata in order of superposition, and its heading again emphasizes the logic of collection, which contributes to the narrative structure of the rock record. The first column lists the names of the fossil genera through which the strata become *intelligible*; the extreme righthand column lists the "products of the strata" through which they become *useful*. While Shelley does not trouble himself to specify what substances "make tolerable roads" (see the entry for "cornbrash" in Fig. 4.1), he is equally concerned to extrapolate a social and economic narrative through the transparency of the rock record.

Historical geology becomes a study of progress toward the "intelligible and useful," while nature's improvement of landforms is seen—on both sides of disciplinary boundaries less rigid than our own—as a model for the economic improvement of landscape and a basis for aesthetic categories. While the varying senses of "improvement" complicate this relationship, a fundamental connection between the knowable and the useful seems to underlie any notion—literal or metaphorical—of the rock record.[5] In Smith's version, the rock record emerges as a minutely structured narrative, animated

4. William Smith, *Strata Identified by Organized Fossils*, 1–2.
5. In Smith's rhetoric, the three strands of "improvement" identified by Raymond Williams—economic, aesthetic, and moral—are integrally linked and often difficult to unweave; *Country and the City*, 116.

by fossil "characters." The rigorous new distinction between the inorganic and these "organic remains," along with stratigraphic sequence itself, gives the rock record the shape of a narrative; this narrative seems to culminate in human civilization. Darwin was not the first to derive a notion of biological evolution from the history of the earth, but in the early nineteenth century this evolution remained teleological, as much for Shelley as for Smith.

The naturalist's act of reading embodies a hermeneutic sovereignty crowning the evolutionary narrative. Smith's act of reading is at once concrete, relying on his practice of applied geology, and a metaphorically suggestive literary project. He began his study of the strata "while employed in the underground surveys of collieries," and became convinced of their uniform orientation and succession while making his preliminary survey for the Somerset Coal Canal Company in 1796.[6] A manuscript note of that year indicates the relation between this discovery and Smith's pioneering insight into the distribution of fossils: nature, he observes, "has assigned to each Class [of fossils] its particular Stratum."[7] The historian David Allen stresses the excavations required by large-scale improvements such as mining, canal-building, and drainage—in all of which Smith was involved—as the greatest stimulus to the fledgling sciences of stratigraphy and paleontology. Allen points out that "for the first time" it became possible to "make a living . . . as an out-and-out consultant in [geology], providing the landed gentry with reports on the mineral and soil potential of their estates." The long delay between Smith's initial research and his authoritative publications of 1815–17 is partly due to the resulting sense that, as Allen puts it, "his data were of great commercial value."[8] Smith's stratigraphic system, based on the "guide" fossils, thus responds to economic impetus, but is also informed by the need for an accessible science and for moral justifications of utility. Its empirical rigor, motivated by economic interest, generates a uniquely complete and transparent archive whose clarity in turn vindicates utility, revealing an inexhaustible wealth that is equally moral, because it records a teleology of increasingly "organized" forms of life.

Smith's accessible new science—a study of "characters on which all must agree"—makes geological order identical to human ordering. He recalls becoming convinced of the earth's legibility during his early assignments as a

6. John Farey, "Mr. William Smith's Discoveries," 360–61.

7. Quoted in Joan M. Eyles, "William Smith," 146; see also 150. Stephen Jay Gould, in his review of Winchester's book, notes similar insights attained independently by Cuvier, Brogniart, and other pioneers of geological mapping ("Man Who Set the Clock," 53). See further Martin Rudwick, *Meaning of Fossils*, 139.

8. Allen, *Naturalist in Britain*, 51, 50. Simon J. Knell extensively analyzes the class issues surrounding geological collecting, often touching on these aspects of Smith's career; *Culture of English Geology*, esp. chaps. 1, 7, and 9.

# GEOLOGICAL TABLE of BRITISH ORGANIZED FOSSILS,

WHICH IDENTIFY THE COURSES AND CONTINUITY OF THE STRATA IN THEIR ORDER OF SUPERPOSITION;

AS ORIGINALLY DISCOVERED BY W. SMITH, *Civil Engineer;* WITH REFERENCE TO HIS

## GEOLOGICAL MAP of ENGLAND and WALES.

| ORGANIZED FOSSILS which identify the respective STRATA. | NAMES of STRATA on the Shelves of the GEOLOGICAL COLLECTION | COLOURS on the MAP of STRATA | NAMES in the Memoir and the PECULIARITIES of the STRATA. | PRODUCTS of the STRATA. |
|---|---|---|---|---|
| Echini, Rostellaria, Fusus, Cerithia, Nautili, Teredo, Crabs, Teeth, and Bones | London Clay | | London Clay forming Highgate, Harrow, Shooters, and other detached Hills | Septarium from which Parkers Roman Cement is made |
| Madrea, Turbo, Pectunculus, Cardita, Venus, Ostrea | Sand | | Clay or Brickearth with Interspersions of Sand and Gravel | No Building Stone in all this extensive District but Abundance of Materials which make the best Bricks and Tile in the Island |
| | Grey — Sand | | Sand a light Loam upon a sandy or absorbent Substratum | Potters Clay, Glad Grinders Sand, and Loam and Sands used for various Purposes |
| Flint, Alcyonia, Ostrea, Echini ... Plagiostoma; Terebratula, Teeth, Palates; Funnel-form, Alcyonia, Venus, Chama, Pecten, Terebratula, Echini | Chalk Upper / Lower | | Chalk | Flints the best Road Materials / Good Lime the Water Cements |
| Belemnites, Ammonites | Green Sand | | Green Sand parallel to the Chalk | Freestone and other sort Stone sometimes used for Building |
| Turrilita, Ammonites, Trigonia, Pecten, Wood | Brickearth — Sand | | Blue Marl | |
| Trochus, Nautilus, Ammonites in Millar; Ostrea in a bed; Bones | Portland Rock — Sand | | Purbeck Stone Kentish Rag and Limestone of the Vales or Fickering and Aylesbury | The finest Quarry and building Stone downward in the Series / Kimmeridge Coal |
| Venus, Modiola, Melania, Ostrea, Echini, and Spines | Oaktree Clay — Sand | | Iron Sand & Carstone which in Suzy and Bedfordshire contain Fullers Earth and in some Places Ochre and Glads Sand | Fullers Earth, Ochre, and Glads Sand / Some Lime used on these Sands in Sussex and Yorkshire |
| Belemnites, Ammonites, Ostrea | Coral Rag and Pialite — Sand | | | |
| Ammonites, Ostrea | Clunch Clay and Shale | | Dark blue Shale producing a strong Clay Soil chiefly in Pasture in North Wilts and Vale of Bedford | |
| | Kelloways Stone | | | |
| Modiola, Cardita, Ostrea, Avicula, Terebratula | Cornbrash | | Cornbrash A thin Rock of Limestone chiefly arable lying in Clay | Most valuable Resule |
| Produces, Teeth, and Bones, Wood | Sand & Sandstone | | | |
| Pear Encrinus, Terebratula, Ostrea | Forest Marble | | Forest Marble Rock thin Beds used for rough Paving and Slating | Coarse Marble, rough Paving, and Slate |
| Madrepora | Clay over the Upper Oolite | | | |
| Modiola, Cardia | Upper Oolite | | Great Oolite Rock which produces the Bath Freestone | The finest Building Stone in the Island for Gothic and other Architecture which require nice Workmanship |
| Madrepore, Trochi, Nautilus, Ammonites, Pecten | Fullers Earth & Rock | | | |
| Ammonites, Belemnites, &c in the under Oolite | Under Oolite | | Under Oolite of the Vicinity of Bath and the midland Counties | |
| | Sand | | | |

*Part on which Lime is rarely used as a Manure*

*Minerva Beds for Coal*

Plains — Coalhills — Clay Vale — Stonebrash Hills

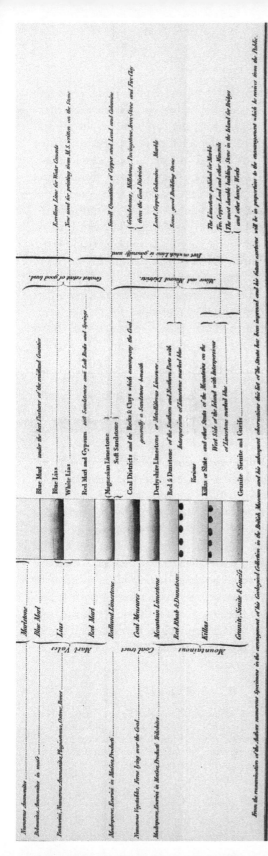

4.1 "Geological Table of British Organized Fossils," from William Smith, *Stratigraphical System of Organized Fossils* (1817). Courtesy of the Linda Hall Library of Science, Engineering, and Technology, Kansas City, Mo.

surveyor: "it was the nice distinction which those similar rocks required, which led me to the discovery of organic remains peculiar to each Stratum."[9] Smith claims that his system will bridge the gap between theory and practice because it is a science requiring only practical skills, operating on a self-evident and objective order. Fossils are arranged as perspicuously among the strata as in "the cabinets of the curious": to the trained eye, whether the landlord's or the artisan's, the rocks provide a transparent chronicle of the deposition of strata, and the guide fossils that Smith identifies permit an exact rendering both of mineral history and of earth history. He returns to the analogy of the cabinet in 1817, promising readers the ability to "search the quarries of different Strata . . . with as much certainty of finding the characteristic Fossils of the respective rocks, as if they were on the shelves of their cabinets." The transparency of the rock record is such that its intrinsic order can be confirmed by matching abstracted specimens with specimens still embedded in their respective strata. Any fossil specimen confirms the uniformity of stratigraphic succession: "the Geologist is thus enabled . . . to fix the locality of those previously found . . . and to find in all former cabinets and catalogues numerous proofs of accuracy in this mode of identifying the Strata."[10] Confirming these suggestions, Simon J. Knell argues that "the new science of geology found its focus . . . [in] fossils and their utility" largely thanks to Smith, who "invented the geological museum."[11]

The logic of orderly collection, once again, proves identical with the internal logic of stratified rock, and geological order seems fully revealed (Fig. 4.2). Stratigraphy, Smith writes, "must lead to accurate ideas of all the surface of the earth, if not to a complete knowledge of its internal structure, and the progress and periods of its formation; for nothing can be more strongly and distinctly marked than the line which separates the animal from the vegetable fossils, and the courses of numerous strata, which are designated by these and other characters, the most intelligible and useful."[12] Like Davy's sublime cosmological narrative, the intimation of perfect and comprehensive geological knowledge here legitimates the utility of the rock record. In Smith's version, such complete knowledge is obviously, rather than "apparently," "within the reach of our faculties," and in claiming the

9. Smith, *Strata Identified by Organized Fossils*, 2.
10. Smith, *Stratigraphical System*, v.
11. Knell, *Culture of English Geology*, 305, 74.
12. Smith, *Memoir*, 6. Cf. Arthur Aikin's *Proposals for a Mineralogical Survey* (1810): "of late years . . . it has been shown that the crust of the earth . . . instead of exhibiting those signs of ruin and confusion which a superficial examination might at first suggest, partakes in a very eminent degree of that order and regularity by which every other class of natural objects are characterized"; quoted in Hugh Torrens, "Arthur Aikin's Mineralogical Survey," 121.

entire planet as the territory of his stratigraphy, he simultaneously plants the flag of improvement on all the planet's mineral resources. As I suggested earlier, the idea of the globe as a geological entity owes as much to these economic concerns as to Alexander von Humboldt's discovery that the same strata succeed each other on widely scattered landmasses.

Smith's writings give voice to a fundamental shift in the cultural status of rocks. Hutton's *Theory of the Earth*, too, represented the deformation of strata and evolution of rock types as a legible text registering the motive forces of earth history. But Hutton appealed for evidence of these forces to older aesthetic categories of sublime power and alien physicality, and his reading of the rock record was derided—if partly on political grounds—as obscure and fantastic. Smith's theory reconfigures the strata as a text on which biological order inscribes itself: the rocks become a comprehensive archive of natural history. His recurring appeal to "characters" makes the analogy of text and strata more concrete. Smith's work epitomizes the new tendency to narrativize a scientific distinction between organic and inorganic substance. The intrinsic importance of rocks as a mysterious and alien territory decreases in proportion as they are found to be instrumental for a study of the more accessible organic past. Once the otherness of rock becomes institutionalized as a chemical distinction, it can be integrated into a hierarchically ordered cosmology. The aestheticized otherness of rocky landscapes remains an important topos in the era of Smith, Cuvier, Parkinson, and Buckland, but it appears side by side with an increasingly orderly rock record.[13] Smith's history is more strictly empirical, but also theologically more palatable (for the period) than Hutton's, and full of infectious wonder at the sheer quantity and distinctness of the facts shown by the fossil record. The definition of "intelligible and useful" strata provides a geological context for fossil organisms that ultimately makes intelligible the formation of the earth, as purely inorganic geology had not done before: "each layer . . . must be considered as a separate creation; or how could the earth be formed *stratum super stratum*, and each abundantly stored with a different race of animals and plants"?[14] Smith's religious rhetoric illustrates the symbiotic relationship, too often reduced to simple antagonism, between religious orthodoxy and scientific discovery.

13. Georges Cuvier (1769–1832) may be said to have founded the science of paleontology. Lord Byron and William Buckland, who became the first professor of geology at Oxford, were two of the more prominent Englishmen who acknowledged his influence. All three participated, in various ways, in an increasingly popular interest in paleontology and in catastrophism that is reflected in Shelley's passage on "the secrets of the Earth's deep heart" and its immediate source, Parkinson. On the chemical classification of rocks, see Rachel Laudan, *From Mineralogy to Geology*, chaps. 3–4; see also her chap. 7 on paleontology and stratigraphy.
14. Smith, *Stratigraphical System*, vii.

# SYNOPSIS OF GEOLOGICAL PHENOMENA.

| SOURCES OF EVIDENCE. | DEDUCTIONS. | RESULTS. | REMARKS. |
|---|---|---|---|
| Fishes in abundance .......... | indicate that ......... | Water prevailed. | |
| Petrified shells are* so filled ...... | as to evince minute solution ...... | Polarity of atoms. | |
| Crystals blunted prior to union in masses ...... | show aqueous action...... | 1. Crystalization of particles. | |
| Shells in rocks not crushed...... | therefore rocks hardened quickly ...... | 2. Aggregation, induration, and cementation of | |
| —— by hardening of matter ...... | were fixed. ...... | 3. mineralized masses in the *Stratification.* | |
| —— none in open joints and veins, ...... | therefore joints and veins opened subsequently to their fixation, ...... | | |
| Mineral veins obviously were cavities—are filled with spar and ore—spar, the principal filling. This therefore, traced through cavities of all sorts, in all the rocks, from chalk to mountain limestone, including cavities in fossil shells, and those caused thereby in blocks of stone, into which no gross matter could enter ...... | shows that all these, and thus by analogy, that ...... | Mineral Veins were filled by segregation. ...... | Where the mineral veins are cavernous, there, and there only, all the fine crystalline cabinet specimens of minerals occur. |
| The trade winds and sea currents ...... | are existing effects of the earth's centrifugal force. | | |
| Which force at the completion of the earth ...... | arrived at its maximum effect. | | |
| By the spheroidal figure of the earth ...... | water as well as land is 13 miles higher under the equator than at the poles, } General Action | This uplaying fluid action was therefore the origin of HILLS which with passes thro' the liquid matter of all strata being successively kept open was the origin of VALLEYS. | |
| Evidently such force was deflected by primitive rocks, and further as the matter formed, locally—the elevating force remaining greatest at greatest depths of water ...... | therefore the force which caused that general elevation under certain circumstances was adequate to the casual elevation of land, 3 or 4 miles higher than the ordinary level. ...... | | |
| By the remains of land animals mixed up with water-worn stones...... | We ascertain that the ...... | Earth was dry and inhabited. | |
| By the bouldered stones every where scattered over the earth's surface, ...... | There has been water in action. ...... | | |
| By the fossil shells in those boulders, identified with those in the stratified rocks, ...... | We ascertain the way of action ...... | THE DELUGE. | |
| By the height to which the boulders and sea-shells have been raised, ...... | We get the force of action and height of the water. | | |

# ILLUSTRATIVE EFFECTS OF THE DELUGE.

By alum-shale, organised fossils, those of coal, and mountain limestone, and boulders from all the rocks northward, in abundance, .......... } The effects of a great current from the N. are obvious on the Yorkshire coast. ..........

The first rush of water was by sea from the North.

By the same, .......... with the like effects, .......... Down the vale of York. from N.

By the absence of alum-shale fossils in the vale of Pickering, .......... } Filey cliff was not surmounted, which gives the height of .......... } First rush of water about 200 feet.

By bays being filled up, and low places inland, as at .......... } Staiths, Whitby, Scalby, and all Holderness, .......... } by westward uplaying.

By whinstone, porphyry, conglomerates, jasper, etc., etc. .......... on Suffield heights, etc. .......... From the N.

By Shapfellgranite, mountain limestone, etc. .......... } Cleveland hills, Lestingham,‡ Suffield hill, etc, to coast of Holderness, .......... } From the N.W.

By sea shells on the Lancashire coast.

By seas shells under 20 feet of gravel, ..........

By rounded chalk and flints, .......... } 1000 feet high in Snowden mountains. .......... on side of ditto, from north of Ireland .......... } With wonderful uplaying. from N.W.

By flints from the chalk hills, ..........

By bouldered chalk and flints, far in .......... } Vale of Taunton and below Bristol, on the hills near Bath, Cricklade common, .......... } Currents from S. and S.E.

By flints, .......... } Northamptonshire, Rutland and Huntingdon .......... Ashby de la Zouch, and vale of Trent, ..........

By lias fossils in gravel, of .......... Needwood Forest, .......... } From E. to W.

By the instances cited by Mr. Phillips of slate boulders on the side of Ingleborough. .......... } Moughton Fell, (see papers on Craven rocks,) 500 feet above the slate, .......... } Current deflected, with wonderful uplaying force.

By the absence of Cornish and Devonshire granite and schist boulders, and of Welsh schist and primitive boulders in the interior, } there seems to have been no general currents from W. and S.W.

W. SMITH
*Oxford, June 22, 1832.*

* In the original document this read "and ore," but these two words are crossed out and "are" inserted in ink.

‡ = Lastingham.

4.2 "Synopsis of Geological Phenomena," by William Smith (1832), a single folio printed in Oxford for the second meeting of the British Association and reproduced in Thomas Sheppard, *William Smith: His Maps and Memoirs* (1920). Courtesy of Widener Library, Harvard University.

The new geological context supplies natural history with a teleology as certain as stratigraphy: "there seems to have been one grand line of succession, a wonderful series of organization successively proceeding in the same train towards perfection." Because it preserves specimens perfectly and sequentially, Smith's archival rock record permits the reconstruction of earth history as such a teleological narrative. The archive holds "treasures of an ancient deep, which prove the antiquity and watery origin of the earth; for nothing," Smith argues, "can more plainly than the Zoophites evince the once fine fluidity of the stony matter in which they are enveloped." Zoophytes and shellfish must then be the most primitive animals, having existed before there was dry land at all, "as they have entered into the composition of a large proportion of the solid parts of the earth." The rock record here becomes the substance of history, as well as its text, because it embodies the successive creations it records. The teleology of increasingly perfect "organization" subsumes a progress from the liquid to the solid state. Fossil organisms seem the animating principle behind the geological evolution of a planet that eventually comes to sustain large land mammals like human beings. The strata themselves, rigorously distinguished as inorganic, become a neutral repository for vivid biological inscription, paradoxically permitting the virtual animation of specimens contained by them. If "many strata" are indeed composed chiefly of these remains, this fact underscores their function as a medium. They preserve fossil organisms "without violence" and display them "entombed . . . with all the form, character, and habits of life."[15]

Not only this teleology, but also the local and economic nature of Smith's geology, lead him to a quasi-Neptunist emphasis on marine deposition as the mechanism of rock-formation.[16] The theory of "watery origin" best accounts for the stratified sedimentary rocks that make up the majority of his geological map of the island, and particularly explains the perfect preservation of England's wealth of fossils and minerals. The most obvious link between these two is coal; Smith explains that "vegetable impressions particularly define in the collier's shaft the approach to coal." His historical interpretation emphasizes the aesthetic and moral profit afforded by the fossil record: "Thus endless gratification may be derived from mountains of ancient animated nature, wherein extinct plants and animals innumerable, with characters and habits distinctly preserved, have transmitted to eternity

15. Ibid., x, ix, x–xi.
16. I refer to the thesis of a "watery origin" for all the earth's material in its broad, culturally diffused form. On this point, see Knell, *Culture of English Geology*, 13–14, 18–19. Knell and others point out that Smith was charged with ignorance of the *theory* of Neptunism as articulated by Abraham Gottlob Werner (cf. Smith, *Stratigraphical System*, vi).

their own history, and the clearest and best evidence of the earth's formation." The recurring economic language in these elucidations of the rock record also evokes a gentility of taste organically linked to moral and fiscal improvement: "organized fossils are to the naturalist as coins to the antiquary; they are the antiquities of the earth; and very distinctly show its gradual regular formation, with the various changes of inhabitants in the watery element." Smith's rhetoric of superabundance, above all, calls attention to the economic subtext of this history of the earth. The passages just quoted conjure up an earth "abundantly stored" with "treasures of an ancient deep," "mountains" of "innumerable" fossils, an "immensity of animal and vegetable matter," and similar magnitudes. This wealth is metaphorical and literal at once, not only because of the connection between coal and vegetable matter, but because it is discovered under a utilitarian premise. Fossils, Smith insists, "are not the sports of nature . . . but they must . . . have their use."[17] With a stock of natural resources seeming to rise effortlessly—at home and in the colonies—to meet the needs of industrial expansion (one might cite the threefold increase in coal production between 1800 and 1830), all of nature falls under the rubric of inexhaustibility.

Smith's 1815 *Memoir* most explicitly reveals the economic context of his stratigraphic system of fossils. This text, which accompanies his geological map of England and Wales—the first of an entire nation—sets out to demonstrate that such a map is an object of "national concern": "The wealth of a country primarily consists in the industry of its inhabitants, and in its vegetable and mineral productions; the application of the latter of which to the purposes of manufacture, within memory, has principally enabled our happy island to attain her present pre-eminence among the nations of the earth." A geological map of the nation can evidently increase the efficiency with which these resources are exploited. Figure 4.3, a cross-section derived from the map, indicates the national scale of Smith's project. His *Memoir* goes on to detail the useful applications of minerals in every conceivable branch of the economy, producing the kind of catalogue later schematized in the table of fossils (Fig. 4.1 above). "Great benefits may accrue both to the landed and commercial interests of the country" from a knowledge of stratigraphy, as he urges in concluding this litany of economic opportunities. The regularity of stratigraphic succession, as proven by the distribution of fossils, guarantees the economic efficacy of the map, which must enable improvers accurately to predict geological conditions.[18] This reliable bot-

17. Smith, *Stratigraphical System*, x, ix–x, vii.
18. Smith, *Memoir*, 1–2, 4. Many other sources testify to this fusion of science and industry. The preface to the inaugural volume of the *Transactions of the Royal Geological Society of Cornwall* exhibits some striking parallels to Smith's vision. The epigraph to the volume states the eco-

tom line is certainly one reason why Smith recommends his master text to the "virtuoso," predicting that "his 'own house will be the best school of Natural History for the younger branches of the family . . . science will become more general. . . . No study, like that of organized Fossils, can be so well calculated for the healthful and rational amusement of youth."[19]

This pedagogy also intersects with the ethos of sensibility. The moral dimension of improvement equally informs Smith's recommendation of the rock record, read as a moral and religious text: "If these animals and vegetables had only to live and die, and mark respectively the sites of their existence in the mass of matter which now forms the earth, they have had their use, and will forever remain indefaceable monuments of that wonderful creative power."[20] The evolutionary narrative, with the act of reading that crowns it, provides an index of the naturalist's "virtuosic" sensibility. Literary texts attend increasingly to the telos of this evolution, the connection between natural and human history. The rock record in these cases is metaphorical in a different sense from the geologist's chronicle of literal geological events. While the geologist's rock record authorizes human agency, particularly the economic agency of prospecting and other forms of improvement, the "metaphorical" rock record speaks directly of human agency. In Wordsworth, for instance, it provides a natural model for the poetic practice of inscription, while for Novalis the history of the earth is its spirit, its identity with human agency. The scientific premises for the rock record are intrinsically of literary interest, especially the strong distinction between organic and inorganic matter. The earth's interior in *Prometheus Unbound* becomes an archive much like Smith's, in which the inorganic is illuminated by organic inscription, though in Shelley's version human cultures also participate in this inscription. At the same time, the older notion of "primitive rocks" persists as an impenetrable inorganic residue, again a figure with literary potential, against which human agency can be usefully set off.

It becomes more difficult to maintain these distinctions between "literary" and "geological" metaphors in light of the interpenetration of early earth science and the sphere of letters. This relationship is more plainly visible in German Romanticism, since there was an established state mining

---

nomic implications of geology succinctly: "A knowledge of our Subterranean wealth would be the means of furnishing greater opulence to the country, than the acquisition of the mines of Mexico and Peru" (unpaginated front matter). Arthur Aikin offers a similarly detailed list of economic applications of mineralogy (Torrens, "Arthur Aikin's Mineralogical Survey," 121). Concerning the increase in coal production, see Roy Porter, "Industrial Revolution and Rise of Geology," 321.

19. Smith, *Stratigraphical System*, v–vi.
20. Ibid., xi.

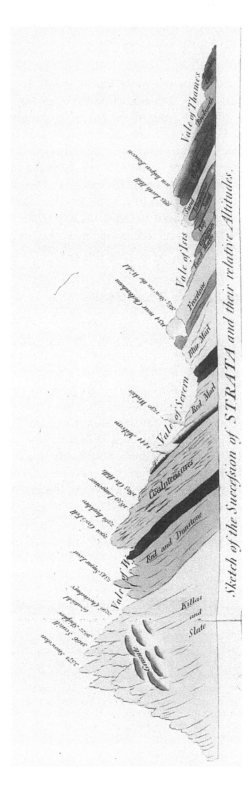

4.3 "Sketch of the Succession of Strata," from William Smith, *Geological Map of England and Wales* (1815). Courtesy of the Linda Hall Library of Science, Engineering, and Technology, Kansas City, Mo.

bureaucracy in Germany in which both Goethe and Novalis received training.[21] Shelley's poetry exhibits the more localized influence of geological reading, but also shows strong thematic affinities to earlier geologists such as John Whitehurst, whose account of apocalypse strikingly resembles the illuminated earth in *Prometheus Unbound*. Smith's exactly contemporary texts bring out the narrative sequence and the economic context that are fundamental to Shelley's account of the rock record as the substance of history. The pronounced differences between Shelley and Smith, on the other hand, reflect the gradual formation of disciplinary boundaries; it thus becomes clearer that the rock record is *both* a literary and a scientific metaphor, and that the aesthetic provides language necessary to imagine human control over resources, without being reducible to this economic function.

## The Secrets of the Earth's Deep Heart

Romanticism's illegible rocks have loomed large in the scholarship, at least partly because they pose such a hermeneutic temptation to critics. The blankness of the Simplon Pass in Wordsworth has prompted numerous influential readings, while the "naked countenance of earth" that Shelley sees in Mont Blanc is often clothed with signification in a critical gesture exemplified by Frances Ferguson's "What the Mountain Said."[22] But there are plenty of legible rocks in Romantic poetry, as in the "fossil scene" of Charlotte Smith's *Beachy Head* and the critique of applied geology in Wordsworth's *Excursion* or—in a different sense—Wordsworth's many "inscription" poems.[23] Paolo Rossi has shown that the practice of reading rocks as historical documents dates back to the late seventeenth century. Charlotte Smith and Wordsworth are in fact somewhat ambivalent readers of rocks because they find them too thoroughly inscribed with scientific theory and historical narrative. John Whitehurst's use of geology in 1778 to vindicate "the great antiquity of arts and civilization" shows how deeply rooted was the analogy between "the history of the earth and the history of nations" (in Rossi's phrase). This analogy, rather than the Wordsworthian atti-

---

21. See Theodore Ziolkowski, *German Romanticism and Its Institutions*, chap. 2; or Helmut Gold, *Erkenntnisse unter Tage*, for a more detailed study. I draw on both works for the consensus in Novalis criticism cited in what follows.
22. Patrick Vincent's "What the Mountain Should Have Said" typifies a new approach that makes the mountain legible by means available to Shelley's contemporaries, such as early theories of glaciation. See Alan Liu, *Wordsworth*, chap. 1, for one influential reading of the Wordsworth passage, incorporating much previous criticism.
23. See Charlotte Smith, *Beachy Head* (1807), ll. 368–419, in Smith's *Poems*.

tude of "Mont Blanc," moves Shelley to adopt the image of the archive for his representation of the liberated earth. *Prometheus Unbound* is thus affiliated not only with English applied geology (such as that of Whitehurst and William Smith) but also with the continental tradition identified by Rossi and its more recent exemplars in Germany.

German Romanticism provides some of the closest analogues for Shelley's liberated earth. Abraham Gottlob Werner formulated the eighteenth century's most systematic and influential model of the rock record, though his model of deposition was based on the density rather than the age of the strata. Werner's institutional forum, the Freiberg Academy of Mining, helps to account for his tremendous European influence—much greater than Smith's in both geology and poetry. Novalis was probably the most literary among the civil servants Werner trained for the state-run mining industry. Novalis' poetic goal of a liberated nature is deeply akin to Shelley's in *Prometheus Unbound*, but it also shows clearly the influence of Werner and economic geology. Novalis proposes, in *Die Lehrlinge zu Sais*, that "natural scientists and poets have always shown themselves, through a common language, as one people." A debate among the "apprentices" and their guests concludes by subsuming scientific knowledge of nature under poetic knowledge, as a humanizing form of domestication.[24] The Temple at Sais has generally been read as an allegory for the mining academy, a reading that makes the teacher or master of the temple a figure for Werner. Another such figure is the old miner in Novalis' novel, *Heinrich von Ofterdingen*.

In both works, Novalis strives toward a rejuvenated human relationship to nature, animated by a paradigm of utility drawn from mining: miners understand geology, in his view, because they have an authentic use for the earth's material. Alexander von Humboldt, who studied with Werner shortly before, articulates such a radical form of use-value in his analyses of mineral economy.[25] For Novalis, however, this utility inspires a reverence for nature, which in turn prompts the rocks to yield to miners a knowledge of the historical record that ultimately proves the identity of nature and spirit: "All divine things have a history, and should not Nature, the only whole with which human beings can identify themselves, be included in a

24. Novalis, *Schriften*, 1:84, 82: "Naturforscher und Dichter haben durch eine Sprache sich immer wie Ein Volk gezeigt." For Werner's theory, see his *Kurze Klassifikation der verschiedenen Gebirgsarten* (1786).

25. For Humboldt's analysis of mineral economy, see, e.g., *Political Essay on the Kingdom of New Spain*, 3:104–12. Novalis practiced briefly as a state inspector of mines, the profession for which he was trained at Freiberg. In *Die Lehrlinge zu Sais*, the Werner figure remarks that "where people are engaged in complex interaction and conflict with Nature, as in agriculture, navigation, animal husbandry, and in the mines . . . the development of this [understanding] seems to occur most easily and frequently" (*Schriften*, 1:108).

history just as human beings are, or have a spirit, which is the same thing?" Novalis' poetic domestication of nature, then, takes mining as its model and proposes to liberate nature by liberating the spirit in the rock record.[26] The rock record in Novalis provides both a tool for locating mineral resources, as in Smith and Werner, and the history that imbues nature with spirit, as demanded by the third traveler in *Lehrlinge*. Mining is poetic geology in the sense that it carries on the *Entwilderung* originating with poetry.[27] Beyond testifying to the spirit in nature, the rock record for Novalis is the privileged repository of history; in one fragment, he refers to the ordering activity of philosophy as "historical mineralogy,"[28] and in *Heinrich* (chap. 5) the presence of a studious hermit amidst the fossilized bones of a cave again links natural and human history.

The hermit's cavern with its grottoes full of bones provides one historical link between Novalis and Percy Shelley. In *Prometheus Unbound* IV, Shelley draws on James Parkinson's *Organic Remains of a Former World*, which stages a dramatic succession of fossil fauna in the famous cavern of Gaylenreuth (near Muggendorf)—also the probable basis for the Franconian cavern in Novalis' novel.[29] Shelley focuses less on the competition between poetry and science, perhaps because his scientific background is formed largely by such reading and lacks the practical bent of Novalis' scientific education. Nevertheless, Shelley's "scientific method," like Novalis', draws on the rock record to construct a vision of liberated nature; like Novalis, he is moved to such reflection by the narrative of zoological succession staged in the bone cave. Shelley's vision calls for a renovation of the globe, restoring the golden age of Edenic harmony that is one of the grand unifying themes traditionally seen as linking Romanticisms. But the resemblance between Shelley's and Novalis' mechanisms for achieving this renovation goes beyond this general thematic affinity. Panthea's vision of the earth in act IV literally enacts Novalis' protocol for domesticating the earth, to make its history legible through spirit.

Act IV celebrates the psychological liberation of humanity, now "sceptreless, free, uncircumscribed" (III.iv.194). As in *Queen Mab* VIII, where the

26. Novalis, *Die Lehrlinge zu Sais* (*Schriften*, 1:99): "Alles Göttliche hat eine Geschichte, und die Natur, dieses einzige Ganze, womit sich der Mensch vergleichen kann, sollte nicht so gut wie der Mensch in einer Geschichte begriffen sein oder, welches eins ist, einen Geist haben?" For the old miner in *Heinrich von Ofterdingen*, to extract metals *is* to liberate them (*Schriften*, 1:242; cf. the first interpolated lyric, "Bergmannslied"). The rape of nature, rather, as Novalis suggests in *Lehrlinge*, is enacted by theories that subjugate nature purely to human purposes, displacing reverence with instrumental reason. See further Dennis Mahoney, "Human History as Natural History."

27. Novalis, *Heinrich von Ofterdingen* (*Schriften*, 1:211).

28. Novalis, *Schriften*, 3:335.

29. Novalis, *Schriften*, 1:252–66.

revolution produces a universally habitable earth, Prometheus' liberation here extends to physical nature, and the description of that nature in act IV becomes a part of the hymn to universal liberation. Panthea and Ione, who function as quasi-choral figures throughout the drama, join the choruses of spirits in act IV in celebrating renewal as they witness the approach of the Earth and Moon, which will the share the majority of the dialogue. I am mainly concerned with Panthea's vision of the earth, especially the second half of her speech, which explicates the vision (270–318). The Spirit of the Earth is liberated within the globe, illuminating it from within. The earth appears as a sphere composed of thousands of spheres, all crystalline—solid, yet perfectly transparent, suffused with music and light (238–41).[30] The subsequent explanation of this radiant sphere in terms of recognizable features of the planet, as it was, relies on a notion of the literal and metaphorical opacity of rock (so prominent in Blake) that now dissolves into transparency. The source of light is a star on the Spirit's forehead, sending out shafts of light through and beyond the planet, like spokes on a wheel of which the planet itself would be the hub. These shafts of light

> as they pierce and pass
> Make bare the secrets of the Earth's deep heart,
> Infinite mine of adamant and gold,
> Valueless stones and unimagined gems,
> And caverns on chrystalline columns poised
> With vegetable silver overspread,
> Wells of unfathomed fire . . .
>
> (278–84)

In this form, the rock record locates mineral wealth more effectively than Smith or Werner could imagine; Shelley's program of liberation, again like Novalis', calls for natural resources to make themselves perfectly available. Novalis incorporates the exploitation of resources into a poetic program of domestication, but in Shelley's vision the struggle for subsistence (as idealized by the teacher in *Die Lehrlinge zu Sais*) is eliminated entirely. The domestication of nature here entails an absolute human sovereignty, to which resources willingly yield. The Earth, in its longest speech, describes human will as "compelling the elements . . . / . . . with a tyrant's gaze" (396–97) and "forcing life's wildest shores to own its sovereign sway" (411; cf. 418–20). The imagery suggests that the Earth is recalling the toppled tyranny of Jove, which Shelley indeed uses similar language to describe. This associa-

---

30. This is the image that Carl Grabo argues is based on Humphry Davy's model of atomic structure, which would strengthen the identification of the earth's material with materiality itself; *Newton among Poets*, 142.

tion undercuts the revolutionary difference between the ancien régime and Promethean "Love," which seems a less environmentally sensitive version of the miner's sexualized love for the earth in *Heinrich*. For Shelley, human liberation entails a complete liberation from economic necessity. The rock record answers this need perfectly by revealing a mineral wealth that is "infinite," "unfathomed," "valueless," and even "unimagined." The geology of the passage is thus an economic geology, a representation of the earth as a fund of resources. The poetry absorbs the environmental attitude of its time and makes it compatible with a poetic reverence for nature (in his preface Shelley credits the "divinest climate" and scenery of Italy with inspiring him to write the drama), producing an aesthetic vocabulary and a model of human agency equally amenable to emergent modern science. Shelley represents economic value by means of an economic sublime, which embellishes mineral wealth with an aesthetic value (as in the "chrystalline columns") anticipating the strata of human artifacts next illuminated by the Spirit (287–302):

> . . . the beams flash on
> And make appear the melancholy ruins
> Of cancelled cycles; anchors, beaks of ships,
> Planks turned to marble, quivers, helms, and spears
> And gorgon-headed targes, and the wheels
> Of scythed chariots, and the emblazonry
> Of trophies, standards, and armorial beasts
> Round which Death laughed, sepulchred emblems
> Of dead Destruction, ruin within ruin!
> The wrecks beside of many a city vast,
> Whose population which the Earth grew over
> Was mortal but not human; see, they lie
> Their monstrous works and uncouth skeletons,
> Their statues, homes, and fanes; prodigious shapes
> Huddled in grey annihilation, split,
> Jammed in the hard black deep . . .

The rock record in *Prometheus Unbound* is monumental. It incorporates human history more fully than any previous version: the transparent earth renders history as an open book. Shelley's imagery of the "ruins of time" evokes more than a century of cosmological speculation; "ruin within ruin," for example, recalls John Dennis' response to the Alps—"Ruins upon Ruins"—under the influence of Burnet and physicotheology.[31] The apocalyptic images of lapsed political power and lost civilizations recall the quasi-archaeological excursions of earlier geologists such as Whitehurst, while

31. Dennis, *Critical Works*, 2:381.

also amplifying the theme so definitively expressed in "Ozymandias." Once the perception of fossils as documents has blurred the boundary between nature and history, the "ruins of time" are of a piece with the "ruins of nature," both contributing to "the image of a world which is . . . the fruit of a slow but inexorable process of degradation and decay," as Paolo Rossi puts it. But the possibility of progress arises from the same seventeenth-century tradition, since fossils can also be taken to document a *human* history of unsuspected antiquity, implying a vaster, open-ended time-scale for "nations" as well as the earth.[32] William Smith's writings show how vigorously this positivism operates in the nineteenth-century context. Analogously, the ruins of *Prometheus Unbound* are a source of moral profit that complements the mineral wealth accompanying them; but rather than celebrating power, they entomb the tyranny and strife dispatched with Jupiter's fall. The fossilized instruments of war, for instance (290–95), appeal to geological stability to intimate an end to violence and all sublunary mutability.

The passage uses the techniques of natural history to represent the earth as the substance of a legible history. As in Smith's transparent rock record, the organisms (or artifacts) entombed in the strata compose a large part of their substance, and the mineral component becomes a medium or text on which the order of sentience is inscribed. Shelley's monumental episode particularly suggests the influence of James Parkinson. Parkinson believed that the earth was in large part composed of metamorphosed vegetable and animal substances. He devotes much space to the formation of coal and explains the nature of other minerals through a "Theory of the Petrifaction of Wood" (note Shelley's phrase "planks turned to marble"). Limestone, marble, "calcareous spar," and so forth are simply "the state in which [mollusk] remains were intended to exist in the present state of the globe." The proportion of fossil content, moreover, determines a hierarchy of aesthetic value: "In visiting the mansions of the rich, we shall in general find, that, in proportion to the wealth and consequence of the possessor, will the more solid parts of the building be composed of these remains of animated beings, which lived in a former world."[33] Shelley's account of animal remains, in the next portion of the passage, is the most visibly indebted to Parkinson and geologists such as Cuvier, Smith, and Buckland, who revived catastrophism and generated the fashion (roughly 1815–30) for deluges and dinosaurs. As John Wyatt puts it, "an aspect of Catastrophism which had special appeal to

---

32. Rossi, *Dark Abyss of Time*, 37 (cf. 4), 16–17.

33. Parkinson, *Organic Remains of a Former World*, 1: chap. 30; 2:3–4; 1:9. Parkinson argues that organic remains undergo "a process of bituminization, by which their conservation is secured, previously to their impregnation with earthy or metallic salts" (3:440). This theory reduces the entirety of geological process to an archival process for preserving the "organic remains" of vegetables (vol. 1), zoophytes (2), and animals (3).

artists and early paleontologists was the notion of a prehuman state where beast fought beast in a struggle, ending for the losing species in extinction."[34] Such a struggle informs Shelley's account of the ruined cities of prehistoric and even prehuman civilizations. The efficient cause of their destruction appears in the fossil skeletons of giant beasts that lie "over these" (302–18):

> . . . and over these
> The anatomies of unknown winged things,
> And fishes which were isles of living scale,
> And serpents, bony chains, twisted around
> The iron crags, or within heaps of dust
> To which the tortuous strength of their last pangs
> Had crushed the iron crags;—and over these
> The jagged alligator and the might
> Of earth-convulsing behemoth, which once
> Were monarch beasts, and on the slimy shores
> And weed-overgrown continents of Earth
> Increased and multiplied like summer worms
> On an abandoned corpse, till the blue globe
> Wrapt Deluge round it like a cloak, and they
> Yelled, gaspt and were abolished; or some God
> Whose throne was in a Comet, past, and cried—
> "Be not!"—and like my words they were no more.

The account of fossilized animal skeletons exhibits another kind of monumentality, equally relevant to science and poetry. This passage commemorates the dead as much as prehistoric events, and paleontology similarly produces epitaphs because it involves long-extinct species. Parkinson himself suggests that "many enormous chains of mountains are vast monuments in which these remains of former ages are entombed." Parkinson's "wood changed into marble" no doubt appeals to Shelley partly because marble is the right material for a monument, and other poets in the period also relate geological to cultural monumentality.[35] Shelley's account can offer a more

---

34. Wyatt, *Wordsworth and the Geologists*, 216.
35. Parkinson, *Organic Remains of a Former World*, 1:8. One related poem is James Montgomery's 1827 blank-verse narrative *Pelican Island* (45):

> Dust in the balance, atoms in the gale,
> Compared with these achievements in the deep,
> Were all the monuments of olden time.

According to its editor, Robert M. Hazen, this poem was quoted in "dozens of geological textbooks and treatises" (*Poetry of Geology*, 94). See also Erasmus Darwin, *Temple of Nature* IV.450; and William Drummond, *Giants' Causeway*, 50–51.

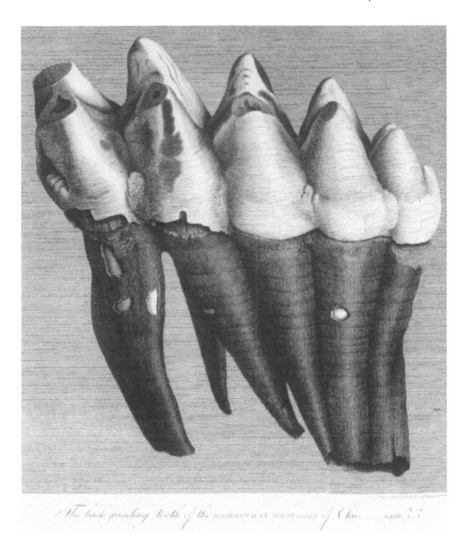

4.4 "The Back Grinding Tooth of the Mammoth or Mastodon of Ohio," from James Parkinson, *Organic Remains of a Former World* (1811), vol. 3, frontispiece. Engraving by S. Springsguth. Courtesy of the Linda Hall Library of Science, Engineering, and Technology, Kansas City, Mo.

spectacular concentration of remains, because it is not localized, but Parkinson's account of such "monarch beasts" is also inspired by their unparalleled size and destructive power (Fig. 4.4). The mastodon, he exclaims, is "one of the most stupendous animals known . . . whether we contemplate its original mode of existence, or the period at which it lived, our minds cannot but be filled with astonishment." The "inexhaustible accumulation" of such remains provides ample material for the Shelleyan allegory of tyranny ("monarch beasts") deposed, which takes its cue ("and over these") from the layering of these remains. The poem's sequence of events thus gains authority and necessity from the rock record. Parkinson points out, for example, that fossil crocodiles are uniformly found in strata lower than those containing warm-blooded quadrupeds, suggesting a large-scale succession that "explains" local folklore concerning races of "fairies" and "giants" whose conflicts are recorded in the strata. Parkinson cites a passage from Cuvier calling for the total annihilation Shelley requires for this passage, but disregards the implication of a sequence of catastrophic floods because of his investment in the literal truth of Genesis.[36] Since the religious rhetoric, as in Smith, accompanies economic incentives, it seems likely that Shelley's use of Parkinson is somewhat ironic.

The question of direct influence is in any event less interesting than the compatibility of this semipopular scientific discourse with Shelley's project. As Parkinson points out, a study of the stratigraphic relations among the substances in which fossils occur is necessary to reconstruct a historical record. Shelley loosely adapts this idea of a rock record in order to represent natural history as a discernible sequence of tyrannies, infestations of the earth finally arrested by a diluvial annihilation of such changes, strikingly represented as a process of organic decay. "Monarch beasts" are thus mere parasites upon the corpse of the earth until it is cataclysmically renovated by Prometheus' liberation (IV.313). Parkinson observes that the prehistoric times to which fossils can be dated are "so remote," and the evidence "so slight," that "the majority" of the theories he summarizes in the course of the work "rather resemble the fictions of poets, than the reasonings of philosophers."[37] Shelley's theory is no exception, and its breathtaking imagery and scope help to explain why Parkinson subtly offers the "poetic" quality as a selling point. But Shelley uses actualism just as impressively as catastrophism in the image of Prometheus chained to his rock in act I. A part of the curse that forms the centerpiece of this act involves a provocation of Jupiter (268–71):

> Let alternate frost and fire
> Eat into me, and be thine ire
> Lightning and cutting hail and legioned forms
> Of furies, driving by upon the wounding storms.

36. Parkinson, *Organic Remains of a Former World*, 3:352, 276–77; 1:4; 3:401, 449–50.
37. Ibid., 1:13–14. For economic incentives in Parkinson, see 3: xii.

These lines identify environmental upheaval as Prometheus' punishment; hence the necessity of abolishing it in the liberated world. Prometheus, meanwhile, becomes a living part of the rock record, receiving the impress of the forces gradually wasting the side of the mountain to which he is chained.

Panthea's concluding exclamation—"Be not!"—inverts the divine fiat, concentrating the destructive force accumulating throughout the passage into a single point. Here the apparatus of natural history serves the undoing of natural history, the abolition of change and decay, and the liquidation of history itself. The reference to the comet (IV.317) appeals to much earlier theories offering the approach of a comet as the efficient cause of the Deluge (an explanation also revived by Humphry Davy). Shelley acknowledges the disjunction between modern diluvialism and physicotheology by means of the "or" in line 316; he prefers the older theory not for the strength of its professed explanation, but because the passage requires a more literal divine intervention. This explanatory equivocation is one of the first signs of a literature self-consciously diverging from science; there are other examples in Romantic poetry of such pairings of alternatives that require, from a scientific point of view, mutually exclusive explanations.[38] The god's command "abolishes" the behemoths and, as a speech-act, terminates the speech of Panthea, the narrator here. Evolutionary succession—from demiurges to one prehistoric animal population to another—also terminates with this verbal Deluge; since the Earth's present-tense speech follows Panthea's narration, we know that human beings, with their contemporary fauna, represent the end of natural history. As Parkinson observes, "the last and highest work appear[s] to be man, whose remains have not yet been numbered among the subjects of the mineral kingdom."[39] Using geological metaphors to reveal it as an arrestable sequence of tyrannies, Shelley's vision liquidates human history as well. The illumination of the rock record completes that history by making available the resources at issue, along with psychological factors, in the struggles constituting history.

Shelley's passage incorporates several important functions of the rock record established by precursors as various as William Smith and Novalis. On the one hand, Shelley's illuminated rock record achieves the trans-

---

38. See, e.g., Wordsworth's "Ode: The Pass of Kirkstone" 11–12: "Left as if by earthquake strewn, / Or from the Flood escaped." Or "Mont Blanc": "Is this the scene / Where the old Earthquake-daemon taught her young / Ruin? . . . . or did a sea / Of fire envelope once this silent snow?" (71–74). The comet is a favored catastrophic device in Shelley; see "Epipsychidion" 368–73.

39. Parkinson, *Organic Remains of a Former World*, 3:455.

parency anticipated in Smith's claim that "encouragement only is wanting in this important branch of Natural History to unravel the mystery and simplify our knowledge of all the terrene part of the creation."[40] A wealth of positive knowledge is the least part of the profit yielded by this rock record. On the other hand, a renovative "Love," as the Earth puts it, "interpenetrates my granite mass" (IV.370). As in *Die Lehrlinge zu Sais*, a universal love lifts the veil of exteriority and elevates inanimate nature to the status of a humanized other: "Heaven, hast thou secrets? Man unveils me, I have none" (423). The earth appears, as in Novalis' work, as a rock record on the largest scale, domesticated through the legibility of this record, which provides evidence of its spiritual nature; or, as Shelley has it, domesticated along with nature, through the action of spirit. D. J. Hughes argues that "it is the very structure of things Shelley would dissolve through the action of mind upon matter."[41] The central image of such a recalcitrant materiality is of course the naked rock to which Prometheus is originally chained, "black, wintry, dead, unmeasured; without herb, / Insect, or beast, or shape or sound of life" (I.21–22). In act IV, Shelley's prophetic characters "measure" and overcome that resistance through the vision of a perfectly habitable world. As in the postrevolutionary world of *Queen Mab*, "all things are void of terror" (VIII.225), purged of deserts, polar wastes, boundless oceans, tempests—all hostile environments, but also, curiously, the poetic resources of the sublime.

The naturalist John Whitehurst (1713–88) is one earlier exponent of the geological sublime who seems to share very closely Shelley's vision of a monumental, dystopic earth history. Whitehurst, too, constructs apocalypse as a liquidation of natural resources. Whitehurst's catastrophe literally creates the "infinite mine" illuminated by Shelley's apocalyptic light: his "subterraneous convulsions" have the economic function of creating the stratigraphic discontinuities in which useful minerals are deposited, while inundations of "subterraneous fire" not only produce coal deposits but also envelop an astonishing variety of remains, many of them human artifacts. In combining this quasi-archaeological project with the economic one into a "subterraneous geography," Whitehurst establishes the rock record as the fundamental text of history. History thus guarantees the superabundance of resources; the substance of nature equals the substance of history, its domestication disrupted only by a catastrophe that ultimately restores the earth and reveals its plenitude. The parallel of a renovative Deluge—an ironic piece of theological orthodoxy on Shelley's part—accompanies the

40. Smith, *Stratigraphical System*, vii.
41. Hughes, "Potentiality in *Prometheus Unbound*," 611.

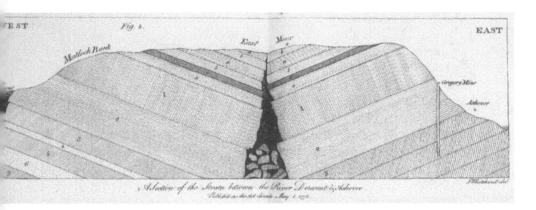

4.5   "Section of the Strata between the River Derwent and Ashover," from John Whitehurst, *Inquiry into the Original State and Formation of the Earth* (1778), fig. 4. Engraving from a drawing by Whitehurst. Courtesy of the Linda Hall Library of Science, Engineering, and Technology, Kansas City, Mo.

parallel narratives of successive depopulation in Whitehurst and Shelley, and both parallels inhabit the semantic field of "revolution," the politics of which is more central to *Prometheus Unbound*. G. M. Matthews, in a classic essay, has argued that Shelley deliberately interpolates obsolete and current science to give the geological features of the drama their political charge.[42]

Whitehurst takes the appearance of geological disorder as the point of departure for his theory of "subterraneous convulsions," inspired by the rugged countryside of Derbyshire, where he lived (see Fig. 4.5). Remarking on the ubiquity of "craggy rocks and mountains . . . steep, angular, and impending shores, [and] subterraneous caverns," he explains that "these romantic appearances" are the effect of "some tremendous convulsions, which have thus burst [the] *strata*, and thrown their fragments into all this confusion and disorder" (*Inquiry*, 61). At this point Whitehurst's theory resembles the physicotheology (especially Thomas Burnet's) of almost a century before. But its stronger empirical basis appears in the "illustration" of the theory, which examines evidence of environmental upheaval, including the recent catastrophes of Lisbon and Vesuvius, in order to establish, by analogy, the magnitude of the "great revolutions" occurring "anterior to history" (66). Whitehurst's scientific apparatus includes these spectacular anecdotes and his Derbyshire field observations, as well as a sound knowledge of contemporary physics and chemistry. He is probably the first English writer to appeal to the rock record for scientific testimony in the modern sense,

42. See especially the account of Demogorgon's ascent as both a literal and figurative eruption, evoked by a complex combination of archaic science (G. M. Matthews, "Volcano's Voice," 558–59), persistent social-natural analogies (564), and newer science (566–67).

chiefly in order to prove "the existence, force, and immensity of subterraneous fire" (115). In these respects, he anticipates both James Hutton's Plutonism and Smith's applied stratigraphy.

Like Shelley as well as these later geologists, Whitehurst allegorizes the stratigraphic paradigm of a rock record to derive historical evidence of "a period indeed much beyond the reach of any historical monument, or even of tradition itself" (*Inquiry*, 257). The rock record expands into a dimension that allows it to recover the human history of an antediluvian world, despite its disappearance from the "human record": "the history of that fatal catastrophe is faithfully recorded in the book of nature, and in language and characters equally intelligible to all nations, [and] therefore will not admit of a misinterpretation." Such notions of clarity and transparency are taken up by Smith and Shelley, in their differing idioms, but Whitehurst's earlier account makes clear that these idioms have common cultural sources. The *Inquiry* makes an apt bridge between the early nineteenth-century discourse and the philosophical tradition identified by Rossi, for whom recognition of the "dark abyss of time" coincided with reflection on the "boundless antiquity of 'nations.'" Rossi cites a number of late seventeenth-century thinkers who argue that "entire cultures" may have been swept away, just as fossil species have, and that fossils can serve to "document" or verify ancient myths.[43] These views are clearly present in Whitehurst, indicating how much the inherited narrative forms of geology contributed to its new industrial applications. A key piece of analogical evidence for Whitehurst's "revolutions" is that whole towns are erased by volcanic eruptions, as in the case of Calloa, Peru, in 1746. His direct evidence of even larger prehistoric cataclysms is necessarily more apocryphal. One detail agreeing with Shelley's account is the presence of ships buried at improbable depths, one of several hints that entire civilizations have been swept away, since it implies that "ships were in use at the time of that dreadful catastrophe" (134). Drawing on a wide array of evidence for lost civilizations on several continents—which culminates in a spectacular reading of the Giant's Causeway, a sublime site long associated with Atlantis (258–63)—Whitehurst concludes: "therefore to such fatal events, and to the conquests of civilized nations by savage barbarians, we may venture to ascribe the subversion of arts and sciences at sundry periods of time" (272).

The illustration of Whitehurst's theory is also an application of it, a subterraneous geography "leading mankind to the discovery of many things of great utility which lie concealed in the lower regions of the earth" (*Inquiry*, 178). Whitehurst's overarching Vulcanist-Plutonist thesis leads him not

43. Rossi, *Dark Abyss of Time*, x, 15–17.

only to explain the Giant's Causeway as the core of an extinct volcano, but also to make useful observations concerning the Derbyshire mines. He shows that mineral veins continue underneath interposing strata of "toadstone" (basalt and dolerite), while the mines below such strata are predictably dry (204–5). The theory makes this insight possible by imagining toadstone (in part correctly) as "actual lava, [which] flowed from a volcano whose tunnel, or shaft, did not approach the open air, but disgorged its fiery contents between the strata in all directions" (197). Whitehurst cites practical application as the primary motivation for his detailed and accurate sections (Fig. 4.5), innovative graphic depictions of the strata that are "the ultimate end of subterraneous geography" (210). Accurate stratigraphic representation, he observes, will enable miners and mine owners "to prosecute their mineral researches with more propriety and advantage to themselves and the public" (180). Whitehurst's evidence is itself "principally obtained . . . from experienced miners," though verified and augmented by original discoveries he made during his own "subterraneous visits" (181). He touts the sections, as Smith does his, for their capacity to help predict the location of coal and other commodities (204). The goal of an accurate representation of the strata, informed by economic interest, leads here, as in Smith, to the articulation of a rock record of great richness and complexity, both mineralogical and literary. The apocalypse or catastrophe on which the larger theory turns has the economic function of creating the spaces for mineral deposits. And it has the aesthetic function of astonishing readers and connoisseurs of natural history, not just by means of sublime depictions, but by establishing the historical legibility of sublime phenomena.

Geology's increasingly practical bent, its concern with physical geography and the improvement of land, resonates in surprising ways in Romantic images traditionally viewed as aestheticized representations of nature. Wordsworth, in a passage quoted earlier, uses the geological categories of primitive and secondary rocks to depict a record of improvement inscribed across the landscape by the totality of geological process: "Sublimity is the result of Nature's first great dealings with the superficies of the earth; but the general tendency of her subsequent operations is toward the production of beauty" (*Prose*, 2:181). If nature itself is an "improver," then the rock record is the monument of a self-domesticating landscape. Whitehurst, Smith, and Shelley all develop versions of this narrative, which is given visual form by the famous image with which I began, from Hutton's *Theory of the Earth*. Though no English poets had the professional geological training of Novalis or Goethe, broader cultural factors propelled them equally toward the rock record as a subject of poetic elaboration; as a scientific paradigm, the rock record achieved its currency through a geological mastery of literary technique, perhaps most obvious in Whitehurst. The discourse on the rock record works as a whole to condition early industrial perceptions of human agency vis-à-vis the natural world. Shelley's "infinite mine" provides evidence of the wide circulation of Romantic images of mines and mineral resources and, more crucially, of the aesthetic concept of economic agency that accompanies them. The sovereignty over nature that technology seems to afford corresponds to a human sovereignty established by scientific discovery, a sovereignty both hermeneutic and evolutionary. Geology and poetry alike, both still offering in their diverging ways to explain the materiality of nature, find the rock record a valuable instrument of the order required to balance the fascination of sublime disorder, of vast and alien geological forms and processes. It is startling to turn from Romantic rocks that *resist* reading, such as Shelley's "Mont Blanc" or the "huge stone" in Wordsworth's "Resolution and Independence," to the orderly archives the poetry shares with economic geology.

In *Prometheus Unbound*, this order takes the form of a liberated textuality, a revealed identity of textuality and materiality. Tilottama Rajan has argued that "the materiality of narrative and drama, of writing vision into the language of events" in *Prometheus Unbound*, "inevitably defers the re-visioning of history as the phenomenology of mind."[44] If we read Shelley's drama in the context of geology, it becomes apparent that the materiality of the text *is* the materiality of history. Shelley's instrument for the "re-visioning" or transfiguration of history is a phenomenology of the earth, identifying the substance with the text of history. This is not to say that the materiality of *Shelley's* text—its mediation of vision, or verbal and printed form—acts directly on the material conditions of history. Its "beautiful idealisms," rather, are anchored in a historically specific knowledge of the physical world as "intelligible and useful." "Love," says Shelley's Earth, "interpenetrates my granite mass," and the resulting narrative absorbs aesthetic and economic forms into one geological, material origin.

44. Rajan, *Supplement of Reading*, 302.

# 5

## *Aesthetic Objects and Cultural Practices in Erasmus Darwin's Geology*

Where science smiles, the Muses join the train.

—Anna Letitia Barbauld, "The Warrington Academy" (1770)

**T**he earliest scholarly book on Erasmus Darwin concludes with an assessment that sets the tone for twentieth-century scholarship on the author of *The Botanic Garden* (1791): "As the first English poet to interpret modern science, Erasmus Darwin deserves distinction."[1] In the nearly seven decades since James V. Logan published his study, scholars have explored Darwin's poetry, science, and professional life as a physician from a variety of perspectives that begin to reflect the prodigious diversity of his talents and interests. Maureen McNeil's 1987 study subjects Darwin's science and poetry to rigorous ideological critique, but returns to Logan's assessment in order to underscore the uniqueness of a career so productive in both fields. McNeil points out, however, that Darwin was far from being the first scientific poet, and constructs a genealogy of English scientific poetry to contextualize his writing.[2] I am also concerned to some extent with Darwin's predecessors, but in order to come to terms with Logan's assessment it seems equally important to decide in what sense Dar-

1. James V. Logan, *Poetry and Aesthetics of Erasmus Darwin*, 147.
2. McNeil, "Scientific Muse," 167; see also idem, *Under the Banner of Science*, 56.

win's science was "modern." The nexus of science and poetry in Darwin's writing has continued to fascinate readers since Logan's time because such a conjunction is not at all modern in the context of modern scientific disciplines. Darwin's poem, however, reveals much about the history of geology and of the disciplines in general. Darwin's geology situates ideas about the earth's material in the broader contexts of discipline-formation and popular science.

Darwin's "extravagant theories" about the formation of the earth and allegorical dramas of mineral deposition are embedded in a context of self-conscious reflection on the nature and status of natural philosophy as a discourse. Darwin occasionally uses the term "science," especially in *The Temple of Nature* (1802), in such as way as to suggest that he would have understood and embraced what Logan meant by "modern science." While it has since been pointed out that Darwin's evolutionary doctrine—which led Logan to refer to Darwin as a "modern prophet"—resembles Lamarck's more than his grandson Charles Darwin's, other aspects of his science do anticipate current views.[3] Although the content of Darwin's science was often modern in this sense, he did not himself refer to it as "science." With the few exceptions noted above, he uses the contemporary term "natural philosophy," and much of the scientific content of his poems belongs to the period's other major category of natural knowledge, natural history. Of the three branches of natural history—zoology, botany, geology—Darwin recognizes geology as being on the verge of discipline-formation. He invokes natural philosophy as the larger rubric that justifies *The Botanic Garden*'s distinctive synthesis of a huge cross-section of natural knowledge. Though "natural philosophy" traditionally referred to the sciences we now call physics and chemistry, Darwin saw the increasingly ambitious discourse of geology as a potential forum for integrating these laboratory sciences and the fieldwork associated with natural history. Geology offered the additional advantage of being connected with the aesthetics of landscape, a useful descriptive idiom for a poet interested in agricultural and industrial improvement. These are some of the important reasons why Darwin made geology the focus of *The Economy of Vegetation*, which became part I of the complete *Botanic Garden* in 1791. The independent success of *The Loves of the Plants* (later part II) in 1789 gave Darwin the license to place this poem

---

3. Desmond King-Hele refers to Darwin's version of the Big Bang as "scientific prophecy" and similarly praises his meteorology in *Erasmus Darwin: A Life of Unequalled Achievement*, 259, 269. Logan refers to Darwin as a "modern prophet" in the passage quoted above (147) and focuses on Darwin's evolutionary theory for his evidence that Darwin's is "the voice of modern science" (133, 135).

in a much more ambitious scientific context (as suggested by part I's 80,000 words of scientific endnotes).[4] There is much serious geology, along with other science, in the poem, but geology shortly took a very different direction that would have made serious contributions in verse impossible. The present chapter explores these formal and cultural senses in which Darwin's writing is foreign to "modern science."

The cultural currency of natural history in verse illustrates the importance of aesthetic forms for the transmission of scientific concepts. In Darwin's work, poetry becomes a dominant organ of scientific culture, a culture especially interested in geology because of its relevance to both natural history and natural philosophy. The polished couplets used by Darwin, William Drummond, and other poet-naturalists combine pleasure with instruction, science with entertainment, much like the other forms of popular science they synthesize: the collection and display of specimens, public lectures, picturesque-mineralogical tours, entrepreneurial applied science, and a growing tradition of scientific poetry. But although both Darwin and Drummond use spectacular images related to earthquakes and volcanism, Drummond's *Giants' Causeway* (1811) does not claim to contribute to natural knowledge. This contrast vividly illustrates the process of geological discipline-formation. It also opens onto broad historical issues surrounding the relationship of literature and science in the early industrial era, particularly the changing social dynamics of polite culture. Darwin's encyclopedic poem not only is more ambitious but also maps more explicitly the socioeconomic contours of his Derbyshire location.

## Poetry and Science

*The Botanic Garden* draws on the cultural visibility of volcanic eruptions for its elaborate geogony, which occupies much of *The Economy of Vegetation* (part I of *The Botanic Garden*), especially canto ii. Darwin's originary catastrophe is the eruption of a solar volcano, the image that Blake adapted for *The Book of Urizen* (*BG* I.ii.14–16):

4. The figure is King-Hele's (*Erasmus Darwin: A Life*, 257). The majority of these notes are geological (267). Darwin explains: "In the first Poem, or Economy of Vegetation, the Physiology of Plants is delivered; and the operation of the Elements, as far as they may be supposed to affect the growth of Vegetables. In the second Poem, or Loves of the Plants, the Sexual System of Linnaeus is explained, with the remarkable properties of many particular plants" (*BG*, p. v). There is still no critical edition of the complete 1791 version of *The Botanic Garden*. I have used a nineteenth-century reprint of the third London edition.

> From the deep craters of his realms of fire,
> The whirling Sun this ponderous planet hurl'd,
> And gave the astonish'd void another world.

A footnote affirms the plausibility of solar volcanoes, directing us to a substantial endnote on the subject, and goes on to explain that the "nucleus of the earth" is formed by this "solar lava," with the strata gradually accumulating over it (I.ii.17n.) In the longer note, Darwin cites Alexander Wilson's discovery that sunspots are "real cavities" to support his theory of seismic and volcanic activity on the sun. He concludes that all the planets have been "thrown out of the sun" in this manner (*BG*, p. 87). The verse moves quickly through an initial quiescent phase to an account of the first terrestrial volcanoes (lines 67–79), which rupture the globe more or less in Thomas Burnet's style, but also cause the moon to be ejected from it, as the earth was from the sun. Throughout the poem, the Goddess of Botany addresses Darwin's Rosicrucian figures—here, in the "earth canto," the Gnomes (77–79):

> Gnomes! how you gazed, when from her wounded side,
> Where now the South-Sea heaves its waste of tide,
> Rose on swift wheels the Moon's refulgent car.

A "Geological Recapitulation" concluding the endnotes shows how systematically Darwin's geogony puts these natural (or supernatural) catastrophes to work for the formation of the earth. According to this summary, "the whole terraqueous globe was burst by central fires; islands and continents were raised . . . and great valleys were sunk. During these central earthquakes, the moon was ejected from the earth" (p. 103). Though innovative in some respects, this is more an old-fashioned theory of the earth than it is modern geology. What is striking is its capacity for expression in verse; the verse and prose in these instances are continuous and compatible, uncolored by the skepticism with which Drummond, only twenty years later, approaches poetry's ability to represent scientific concepts. Darwin turns to these images again in the *Temple of Nature* (I.321–24):

> Next when imprison'd fires in central caves
> Burst the firm earth, and drank the headlong waves;
> And . . .
> Form'd lava-isles, and continents of shell.

Darwin frequently prefers such global narratives to local description, harnessing natural history to a systematic cosmology in verse that self-consciously balances poetic "fables" and obsolete science with progressive concepts of inorganic and organic development. In suggesting the "fermentation of morasses" (*BG*, p. 103), for example, as one cause of earthquakes, Darwin reclaims the improbable for its poetic usefulness, but also as an imaginative aid, rather than a hindrance, to scientific progress. This is one sense in which "Beauty beams amid tremendous fire" (I.i.156) in a volcanic eruption: the appealing imagery and spectacular style accompanying the description of environmental upheaval complement the scientific maneuverability afforded by catastrophes. The epic simile that follows this image of beauty—Venus in Vulcan's forge—illustrates another key element of Darwin's merging of poetry and science, what he calls the "ornament" of classical mythology, together with the pictorial aesthetic suggested by this term.

In the "Apology" prefixed to *The Botanic Garden*, Darwin defends this use of "ornament" and "extravagant" theories, arguing in effect that verse is a *natural* medium for scientific concepts. Though he sets out "to apologize for many of the subsequent conjectures," the poem's scientific content, as not being properly "philosophical," his "Apology" quickly assumes the tone of a manifesto: "Extravagant theories, however, in those parts of philosophy where our knowledge is yet imperfect, are not without their use; as they encourage the execution of laborious experiments, or the investigation of ingenious deductions, to confirm or refute them. And, since natural objects are allied to each other by many affinities, every kind of theoretic distribution of them adds to our knowledge, by developing some of their analogies" (*BG*, p. vii). Darwin's innovations, such as the theory of solar volcanoes, derive their extravagance in part from the literary qualities of his preferred geological sources. Darwin learned much of his field geology from John Whitehurst, who was a fellow member of the Birmingham Lunar Society, but Whitehurst's theory also appealed to Darwin on literary grounds (*BG* I.ii.17n., 36n.). Darwin also manages to cite Thomas Burnet without the embarrassment that had become *de rigueur* in any philosophical mention of the *Sacred Theory*. Even his most modern source (and regular correspondent), James Hutton, was rejected as too fanciful by numerous critics, though often for political reasons.[5] The materialism that made Hutton suspect in the French Revolutionary climate appealed to Darwin's republican

---

5. This is especially true of reviews of the two-volume edition of 1795, but a shorter version had been published in 1788, and by the time Darwin was writing "The Economy of Vegetation" in 1789–90 he had grasped the political potency of Hutton's theory in the milieu cre-

sympathies, while the history of geology has thoroughly vindicated his science. My point here is that Darwin recognized, through his chosen verbal medium, the imaginative power of conjecture. The verse form of *The Botanic Garden* necessarily places its science in a cultural context, and Darwin's choice of geological sources underscores the constitutive role of literary, political, and other cultural elements in natural history and philosophy—what Thomas Kuhn would call the "extra-scientific" elements of "pre-paradigm science."[6]

While Darwin defends "extravagant theories," in the "Apology," on philosophical (scientific) grounds, he defends classical mythology and poetry itself on literary grounds as properly aesthetic forms of natural knowledge. He remarks, first, that "many of the important operations of Nature were shadowed or allegorized in the heathen mythology"; the marriage of Venus and Vulcan, as an allegory of volcanism, is one of a great many such emblems adopted by him. The source of this knowledge, and of its literary form, Darwin believes, lies in ancient Egyptian "discoveries in philosophy and chemistry, before the invention of letters." The hieroglyphs expressing these discoveries were afterward "described and animated by the poets, and became . . . the deities . . . of Greece and Rome. Allusions to those fables were therefore thought proper ornaments to a philosophical poem." This narrative, subjoined to an account of the poem's Rosicrucian machinery, seems at first an innocuous explanation of its further eccentricities. But the emphasis on ornament is misleading: by invoking these classical precedents (including, implicitly, Lucretius), Darwin seeks to justify his entire project by naturalizing poetry and allegory as legitimate forms and concomitants of natural philosophy. Thus he adapts the seventeenth-century tradition of arguing for "the great antiquity of arts and civilization" (*Inquiry*, 272–73) to distinctively literary and commercial ends.

In his "Advertisement" to the first complete edition of *The Botanic Garden*, Darwin describes his project as follows: "The general design of these sheets is to enlist Imagination under the banner of Science; and to lead her votaries from the looser analogies, which dress out the imagery of poetry, to the stricter ones, which form the ratiocination of philosophy" (*BG*, p. v). The apparent subordination of poetry to science has to do with Darwin's goal of profitable entertainment. Poetic imagination is no less necessary to scientific reason here, however, than is implied in the "Apology." The poem itself begins with an address to readers that links the two discourses through

---

ated by the French Revolution. On Darwin's personal relationships with Hutton and Whitehurst, see King-Hele, *Erasmus Darwin and Romantic Poets*, 19, 10–11.

6. Kuhn, *Structure of Scientific Revolutions*, 16–17.

the faculties of taste and virtue: "Stay your rude steps! whose throbbing breasts infold / The legion-fiends of Glory, or of Gold!" (I.i.1–2). Ambition and avarice here figuratively disqualify aspiring "votaries" of poetry, and hence of science. By appealing to aesthetic sensibility as an index of moral feeling, Darwin allies scientific inquiry tacitly with the aesthetic sphere, as against the practical. His narrator then addresses the privileged reader for whom the scientific content, or allegories of the "operations of Nature," are reserved. In allegorical terms, the poem's Nymphs and Dryads (5–6) minister to the qualified "votary" by way of "soft vibration[s]": "But THOU! whose mind the well-attemper'd ray / Of Taste and Virtue lights with purer day; / . . . for thee" the poem provides access to a copious natural world enticingly delineated in the remainder of this address.[7] The successful pursuit of philosophy requires a poetic sensibility. If Darwin enlists poetry in the service of science as a prerequisite to the greater rigor of the latter, in the same gesture he also claims scientific subject matter for the aesthetic sphere.

Darwin carries on this project of "enlist[ing] Imagination under the banner of Science" in *The Temple of Nature: A Poem with Philosophical Notes* (published in 1802, the year of his death). The subtitle, identical with that of *The Botanic Garden*, nominally maintains a distinction between form and subject that is destabilized by this hybrid discourse. In the later poem, Darwin expresses this ambiguity by doubling his muse. Darwin invokes Urania to direct the progress of the poem, and in one place she is apostrophized as a higher muse to the poem's garden-variety poetic muse (I.167–75):

> PRIESTESS OF NATURE! . . .
> . . . . . . . . . . . . . . . . . . . .
> Oh grant the MUSE with pausing step to press
> Each sun-bright avenue . . .
> . . . . . . . . . . . . . . . . . . . .
> Led by thy hand.

Urania herself is both the scientific muse of astronomy and the "heavenly Muse" of Milton. While these doublings of the muse suggest division, Urania also personifies the identity of poetry and science. In her capacity of "celestial Guide," she consults with Darwin's muse on the proper subject mat-

---

7. *BG* I.i.9–24. Darwin borrowed much of this opening address (thirty-four of the first fifty-eight lines, including 5–11) from Anna Seward, an appropriation that has bewildered all his critics, beginning with Seward herself. (Charles Darwin referred to it as "highway robbery.") The address is based on a poem Seward wrote in reaction to the garden Darwin created at Lichfield, which she submitted to the *Gentleman's Magazine* (1783) under her name. See further King-Hele, *Erasmus Darwin: A Life*, 258; and Logan (who supplies the Charles Darwin quotation), *Poetry and Aesthetics of Erasmus Darwin*, 115, 109–10.

ter of poetry. Having dispatched the production and the reproduction of life (the subjects of cantos I and II, respectively), Urania and the Muse penetrate farther into the Temple of Nature, surveying the physical properties of the universe under such rubrics as geography, chemistry, and the physical laws of gravitation and heat (III.1–34).

The ambiguity of poetic decorum surrounding such subjects is significant because Darwin's project of "philosophical poetry" (unlike Wordsworth's) entails a relentless collapsing of generic and disciplinary boundaries. Darwin seeks to regulate this fusion by embracing first one hierarchy (the chief purpose of a poem is amusement, not instruction) and then another (the Muse's subordination to Urania), vacillating constantly between them. The human mind, as a subject for the poem, seems to take second place to this realm of phenomena, but by this point Urania is willing to indulge the Muse with an account of it (III.35–42):

> Now in sweet tones the inquiring Muse express'd
> Her ardent wish[:] . . .
> . . . . . . . . . . . . . . . . . .
> "Priestess of Nature! . . .
> . . . . . . . . . . . . . . . . . . . . . .
> Give to my ear the progress of the Mind."

Urania begins her account of this progress in the Hartleian physiological vein already familiar to Darwin's readers, as if to confirm the stability and high seriousness of his poetic idiom (49–56):

> The indulgent Beauty hears the grateful Muse,
> Smiles on her pupil, and her task renews
> . . . . . . . . . . . . . . . . . . . . . . . . . . . . . . . . .
> "First the new actions of the excited sense,
> Urged by appulses from without, commence."

The poem's natural history and its physiology begin with an account of spontaneous generation, in a similar vein (I.247–50):

> Hence without parent by spontaneous birth
> Rise the first specks of animated earth;
> From Nature's womb the plant or insect swims,
> And buds or breathes, with microscopic limbs.

Such subject matter clearly tests the limits of poetry, and a great deal spills over into the "philosophical notes" (Additional Note I; see also XIII on the "progress of the Mind")—notwithstanding Darwin's initial claim that the poem "does not pretend to instruct" but "simply to amuse" (Preface, *Temple of Nature*). This overflow of abstruse subject matter into prose remains in tension with Urania's insistence on rigorously philosophical verse.

Darwin's geology provides particular examples of a deep structural identity between scientific and aesthetic principles. One instance is the notion of granite as "primitive" rock, by this point a geological commonplace that also appears in Wordsworth's notion of a primitive sublime and secondary beautiful. In *The Botanic Garden*, an endnote on granite (Additional Note XXIV) accompanies an address to the Gnomes in which they are instructed to perform weathering on the most recalcitrant rock, to make the mountains habitable: "Climb the rude steeps, the granite-cliffs surround" (*BG* I.ii.523). The crucial word "rude" here is a commonplace from poetry describing such scenes, for which the stock phrase is "rude rocks." But in prompting the note, this phrasing also reveals the geological sense of "rude" as "primitive" or "original." In Darwin's particular inflection, "rude" and "primitive" designate granite's origin as solar lava, building further support for his theory of the earth as an eruption from the sun. By making it scientifically viable, he reclaims aesthetic vocabulary or shows it to be coextensive with natural philosophy. The *Temple of Nature*, reflecting its later date of composition, explicitly invokes the authority of "modern geology" for its account of the production of matter on this primitive foundation, according to which the sedimentary strata and all habitable land turn out to be "the recrements of organic bodies" (I.268n.).

Verse is integral to Darwin's synthesis of natural knowledge partly because of the very gaps between poetic images and philosophical explanations, which create the need for many of his substantive notes. Many of these serve to establish the plausibility of the images. Darwin explains "innocuous embers," for example, as a poetic conceit with a geological basis: the beauty of volcanoes corresponds to their beneficial function as escape valves for subterranean heat and pressure, implicitly linked to political as well as geological revolution. Darwin glosses another line in this passage, "Round her still centre tread the burning soil," with the assurance that "many philosophers" have proposed a molten core (*BG* I.i.139n.). Similarly, the apparently extravagant notion that "the whirling Sun this ponderous planet hurl'd" is made plausible by means of the elaborate theory of solar volcanoes, and even the apparently innocuous image of "vaporous air" is glossed with an account of how the hot gases adhering to this original planetary chunk condensed into the original ocean (I.ii.17n.). The subsequent "hurling" of the moon generates scientific correlatives for each of the moon's conventional poetic trappings: "Rose on swift wheels the Moon's refulgent car / . . . and roll'd round earth her airless realms of frost" (79–82). For an account of the "refulgent car," a footnote refers us to the lunar portion of the endnote on solar volcanoes, according to which the moon is literally driven out of the earth into the sky. Another note accounts for "realms of frost" by pointing out that water must either evaporate or freeze in the absence of an atmosphere (82n.). This continuity of image and explanation makes sense in light of Darwin's initial assertion that myths are direct allegories of scientific observations, and the insistence on plausibility suggests a defense of poetry as a medium for knowledge.

The aesthetic geology in Drummond's *Giants' Causeway*, by contrast, reflects the impact of twenty important years in the process of geological discipline-formation and cultivates a much stronger sense of disciplinary boundaries. Drummond's notes, confined to the end of the poem, are as likely to demystify literary representations of natural processes as to explain natural processes themselves. He is more conservative than previous poets and geologists alike, for example, on the central issue of earthquakes and volcanoes. One of his notes disavows the theory—toyed with in the verse—that Ireland is what remains of Atlantis.[8] Having made the conventional assertion that earthquakes and volcanoes defy description, this note resorts to dry philosophical prose, rather than verse or spectacular images, to offer the closest possible approximation. Yet the poem's chosen audience consists of philosophers. Drummond dedicates his work to a Dr. Bruce, "Principal of the Belfast Academy," and in one place directly addresses a learned society (8):

> Ye favoured few, whom nature's partial care,
> Leads through the realms of ocean, earth and air
> . . . . . . . . . . . . . . . . . . . . . . . . . . . . . . . . . . . . .
> But chief do thou Mac Donnell, taught to scan
> Each form and feature of the beauteous plan
> Declare . . .

The narrator asks Mac Donnell, a mineralogist, to declare whether the Giant's Causeway is a product of the ocean or of central volcanic fires, that is, to take sides in the continuing Neptunist/Plutonist debate. Darwin is more forthright about his own Plutonism, partly because of his equal qualifications as a philosopher and poet. Drummond, however, also demonstrates significant geological knowledge. Rather than true ignorance, his diffidence rhetorically signals a different conception of poetry, as primarily a moral discourse incommensurable with natural philosophy (a term he still uses) and the newly established science of geology.

Drummond's poem recognizes an aesthetic mode of explanation that cannot be reclaimed for science, unlike Darwin's "extravagant" theories and po-

---

8. "The supposition that Ireland is a fragment of the Atlantic isle, may answer the object of poetry, but the data are scarcely sufficient to justify its adoption in philosophy"; William Drummond, *Giants' Causeway*, 175. Compare Drummond's passage on "the black basaltic deluge" (83) and the corresponding note (181). James Montgomery's *Greenland* (1819), also in couplets, offers many parallels in its account of volcanic landscapes.

etic allegories. Following his address to the men of science, he turns his attention to "the sportive fancy of th'untutored swain, / To wonder prone, and slave to error's reign" (9). This "fancy" is the legend of Finn MacCool, the giant architect from whom the Giant's Causeway gets its name. Drummond does not consider this mythical explanation, as Darwin would, to be a scientifically viable allegory, but seems to feel that poetic decorum calls for him to begin with a myth. He apologizes for the "fond" notion that the Causeway might be a production of "human power" ("though reason spurn the thought"), but admonishes his philosophical audience: "Nor let the sage, in lettered pride severe, / The simple legend with impatience hear." Drummond then spins out this legend, elevated through his formal poetic idiom and augmented with numerous substantive changes, over the next 150 lines (9–16). The first task of poetry, then, is to render an elegant and up-to-date version of myth, which places it in dialogue with other branches of literary culture, including the scientific. We can read this paradigm in succession with the two offered by Darwin: poetry can be substantially enlisted under the banner of science (1791); "philosophical" poetry aims chiefly to amuse, and less to instruct (1802); and finally, the literary is capable of an elegant dialogue with the scientific, exalting scientific practice without really participating in it (1811). The actual claim of poetry on scientific subject matter does not diminish in the movement through these rhetorical devices. These, rather, reflect the gradual formation of disciplinary boundaries as a matter of social and intellectual convention.

In addition to staging current geological debate over the origin of the Causeway, Drummond's poem asks whether it might be a work of art, and not a "production of nature" at all. This dilemma over the Causeway's unusual prismatic basalt columns had influenced printed accounts since the seventeenth century, making it a familiar one. Preceding the address to naturalists, Drummond's initial address to readers draws more heavily on such familiar tropes, integrating the new scientific dilemma with the aesthetic and cultural issues surrounding the Causeway. His scientific disavowal of the Finn MacCool legend is thus canceled by an appeal to aesthetic discourse and cultural nationalism; the implausibility of art as an original cause of the formation is no obstacle to the powerful construction of these rocks as a monument of "awful wildness" (6–7):

> O thou whose soul the muses' lore inspires,
> Whose bosom science warms, or genius fires,
> If nature charm thee in her wildest forms,
> Throned on the cliff 'midst cataracts and storms
> Or with surpassing harmony arrayed,
> In pillared mole, or towering colonnade,

> Seek Dalriada's wild romantic shore
> . . . . . . . . . . . . . . . . . . . . . . . . . . . . .
> Here broken, shattered, in confusion dread,
> Towers, bridges, arches, gods and temples spread
> Stupendous wrecks, where awful wildness reigns!

Somewhat like Darwin, Drummond addresses the "votaries" of science and of poetry together. Unlike Darwin, he separates them by means of the "or" in the second line, but implicitly joins them again under the rubric of "genius," a characteristic move for Drummond. Scientific appeal—in the form of "pillared mole[s]" and "wonders of volcanic fires"—is embedded within the conventional appeal of the "wild romantic" landscape, with its aestheticized otherness and completely internalized architectural metaphors (this allows the strange equation between artifice—"temples," "bastions"—and "awful wildness"). Drummond's sublime set-piece integrates all forms of cultural production (including science) within the claim that the Giant's Causeway competes with the monuments of antiquity:

> Let folly's sons to lands far distant roam
> . . . . . . . . . . . . . . . . . . . . . . . . . . . . . .
> In scenes like *these* did Collins first behold
> Pale Fear . . .
> And mighty Shakespeare breathed heaven's pure ethereal fire.
> (6–7, emphasis added)

In Drummond's descriptions, geology serves more to confirm landscape as the province of imagination than to convey natural knowledge. At the close of a quasi-scientific episode investigating the local customs and economy, as well as the landforms, of the Causeway, Drummond's narrator exclaims (50):

> Through limestone vallies, and basaltic caves;
>                          . . . let me go
> to view the wonders of the world below;
> What roots of rock thick-woven, and entwined,
> Those giant steeps to earth's fixed centre bind.

There follows an inventory much like Shelley's aestheticized rock record in *Prometheus Unbound*, featuring "mines of gold," shipwrecks, "vast cemet'ries," and "wrecks of worlds unknown." In Drummond, as in Darwin, the relation of poetry and science is constantly shifting; previously presented as the obsequious handmaid of science, here the verse uses geological terms and issues to gesture at a metaphysics that subsumes them both: "Amazing world! how vain the thoughts of man, / Thy depths, thy terrors, and thy wealth to scan!" Poetic imagination fuses scientific, aesthetic, and economic categories in filling the place left vacant by the mind's inability to "scan" the earth's interior methodically. While Drummond maintains, in his demystifying endnotes, that scientific knowledge must be distinguished from the "merely poetic" (189), he praises modern geology above all because it partakes of the aesthetic sensibility the poem is trying to cultivate: "The modern geologist, who pursues his investigations, not in the narrow bounds of the closet, or laboratory, but . . . among the precipices of the mountain, sees that nature [everywhere] displays . . . beauty, order, and design" (192). Drummond, despite his more insistent division between poetry and science, constitutes geology as aesthetic practice; but unlike Darwin, he relies on the natural theology reestablished in the climate of anti-Jacobinism, especially by William Paley, in order to do so.

This inspired and yet diligent contemplation of actual landscapes best embodies aesthetic geology. But the poem's language also achieves an unusually complete synthesis of aesthetics and geology, especially given the context of geology's increasing divergence from letters in 1811 (the first year of the *Transactions of the Geological Society*). On the one hand, the wholly geological portion of the poem (book III) assimilates scientific theory increasingly to artistic practice. After sketching the scenario implied by the Vulcanist explanation of the Causeway, Drummond turns to the Neptunist account: "Now see how other hands this mole design" (84). In characterizing the Neptunist theory as a "design," he brings out its distinct set of artistic implications; it gives ampler scope to sacred history and provides a formidable sublime object in the Deluge (86–87). On the other hand, this head-to-head contest between the two schools leads the poem directly to confront the scientific content of the debate. The Neptunist episode is considerably briefer, and requires more untenable positions, which Drummond then has to disavow in the notes. These improbabilities and the conflict of the two theories lead to a new episode in which the narrator, tormented by indecision, states the fundamental scientific cruxes presented by the Causeway: "Druids of science, to the muse disclose / From what vast source th'overwhelming Deluge rose" (89). The more local and immediate questions following from this old chestnut concern the specific cause of the characteristic dislocations of the Causeway, with its vertically inclined strata, and the cause of superposition "adverse to attraction's laws," that is, denser basalt over lighter sandstone (laterite) strata. The poem fully internalizes and dramatizes the ethos of the scientific debate: "O that the light of some celestial ray / Would touch my soul and clear these doubts away!" A deus ex machina, in the person of a "hoary sage," promptly appears to resolve these excruciating doubts. The sage proves to be a dyed-in-the-wool Plutonist, specifically a Huttonian; and here, at the end of the poem, the consistent pretense of an impartial and subaltern presentation of competing theories, with all due poetic ornament, drops away.

Drummond's aesthetic geology entails a dual relationship between poetry and science. The Huttonian conclusion of the poem brings out the synthetic aspect of the relationship, which traverses disciplinary boundaries even as it maintains them. The earlier Vulcanist account begins with a long poetic digression on fire (73–75), using poetic commonplaces—"the noblest element that Nature wields"—to establish the cultural authority of this hypothesis. The more sophisticated Plutonist thesis, which connects volcanic manifestations to a molten core and continental uplift, provides the more spectacular and original imagery with which the poem concludes. The Huttonian sage resolves the narrator's dilemma with a geogony in miniature, teaching "that central fires up-heaved the earth / From ocean's depths, and gave these wonders birth" (90). The sage's lengthy speech capitalizes on such features of Hutton's geology as the uniformitarian cycle of wasting and renovation ("perpetual circle of harmonious strife") and sheer vertical scale ("arched o'er hell's gulf their strong foundation spreads"), at the same time vividly sketching the prehistoric scenarios called for by the theory.[9] While Drummond takes pains, in the notes, to present counterarguments and to register his reservations about Hutton, here, too, he shows greater confidence as a geologist than before, and the verse really seems to decide the issue. The impassioned appeal for resolution, followed by the appearance of the Huttonian sage, has the effect of claiming for the verse the power to resolve a real scientific dilemma, precisely because of the objective distance created by the initial distinction between poetry and science. The impartial treatment the poem initially claims to give to the competing theories also results from this distance, and only strengthens the preference expressed at the end. Drummond's conclusion seems to imply that more viable theories are more susceptible of poetic treatment, while the poem as a whole offers a cultural context for such theories. Disciplinary boundaries come into play, however, as they do not in Darwin, when we consider this cultural context: for Drummond, geology's associations with landscape aesthetics and natural theology—as *distinguished* from its scientific content—make it especially suitable for poetry.

Drummond's synthesis of poetry and science, then, is ultimately consistent with stronger disciplinary boundaries. The invocation to Lucretius that begins book III resituates poetry as the handmaid of science. Lucretius (whom Drummond had translated) is the prototype of the poet–natural philosopher, and in Drummond's account an important precedent for po-

---

9. Dennis Dean convincingly identifies the sage as John Playfair, who popularized Hutton's theory ("Geology and English Literature," 143). Drummond's poetic emphasis on fire makes more sense in the context of phlogistic chemistry, which informs Hutton's *Dissertation upon the Philosophy of Light, Heat, and Fire* (1794).

etry's assimilation of the prestige of science (71). Drummond generalizes from Lucretius' case to a eulogy of "science" (here more closely identified with "natural philosophy" than it would have been twenty years before):

> Thrice happy he, whose truth-illumined soul
> With Science wanders through the boundless whole;
> No angry fiends of night her skies deform.
> . . . . . . . . . . . . . . . . . . . . . . . . . . . . . . . . .
> And scanning Nature's universal laws,
> Mounts from the second to th'eternal cause.

"Science" in this view not only liberates from superstition but also provides an immediate pathway from efficient to final causes. Poetry's task is partly to praise it and partly to record, in antiquarian fashion, the debunked superstitions, such as the Finn MacCool legend. This view is consistent with Drummond's self-presentation in his preface as an "amateur . . . who has not advanced beyond the threshold of the Temple of Geological Science" and must "beg indulgence for presuming to touch this subject" (xxvii). Whereas Darwin synthesizes material from almost every branch of science into bold, original theory, Drummond focuses on one reified discipline. Though Drummond lacks much of Darwin's expertise, the key difference is rhetorical. Drummond professes himself a novice, a discursive position made possible by the erection of disciplinary boundaries. Under the ethos of eighteenth-century natural history, no educated author would have laid claim to such naiveté, but Drummond can plausibly suggest that it makes him an impartial student of competing views. This diffidence, however, does not extend to his *poetic* persona, who has no qualms about ranging himself repeatedly among "heaven-rapt bards" and the like. When inspired by the god, as we have seen, Drummond is *not* impartial; and this contradiction makes his aesthetic geology a vivid record of the factitious nature of the disciplinary boundaries his poetry helps to construct.

## Science and Culture

The contrast between Darwin and Drummond sheds light on a local process of discipline-formation, but also opens onto a long critical tradition of bracketing georgic and scientific poetry. This tradition begins with Aristotle's contention that "Homer and Empedocles have nothing in common except their meter."[10] John Aikin's *Essay on the Application of Natural History to Poetry* (1777) is one of many eighteenth-century works informed by this

---

10. Aristotle, *Poetics* (chap. 1), 541.

tradition. Aikin rejects past attempts at scientific poetry, but with the revisionist aim of allowing "modern" poets to make legitimate use of natural history. He argues that "descriptive poetry has degenerated into a kind of phraseology" because of a "scarcity of original observations of nature . . . only to be rectified by accurate and attentive observations, conducted upon somewhat of a scientific plan." "Scientific" in 1777 does not refer to modern scientific disciplines such as geology, but does serve Aikin's attempt to distinguish natural history, as empirical, from natural philosophy. Aikin holds up James Thomson as a model of his method: "No poet before Thomson had thought of *studying* in fields and woods."[11] Aikin gives numerous examples from *The Seasons* to support his claim, but he neglects evidence of the poet's equal investment in natural philosophy. Thomson made his name with an elegy on Newton, extended in Newtonian passages in *The Seasons* (e.g., "Summer" 1706–57). These passages not only exalt natural philosophy but also embrace the "fabulous ideas" and obsolete science that in Aikin's view mar earlier scientific poetry.[12] He thus rejects an entire tradition that actually includes Thomson and several of his imitators, such as Henry Brooke, who attempted to popularize natural philosophy by means of involved allegory and archaic terminology. This rejected tradition is very much alive in *The Botanic Garden*, with its "extravagant theories" and mythological "ornaments."[13]

Darwin's scientific poetry succeeded, though it was hardly what Aikin envisioned. But Darwin, too, challenges the Aristotelian tradition of excluding science from poetry. Their differing literary solutions respond to the same set of historical problems, and both illuminate what was at stake in changing the relationship between poetry and science. Aikin's skewed account of Thomson suggests that the classical distinction between poetry and science became inadequate to explain changes in the structure of natural knowledge. Changes in the relationship between poetry and science reflect the progress of discipline-formation and the development of a popular natural history facilitating this process, as the case of geology illustrates so admirably. If Brooke's poetry fell short in its popularizing aim, it did so because a popular scientific culture, of the sort that Aikin and Darwin helped

11. Aikin, *Essay*, 5, 10, 67. Aikin is quite polemical on "the insipidity of modern poetry" (1–2).

12. Ibid., 26, 31.

13. Although Brooke's *Universal Beauty* (1735) claims to offer an "easy" introduction to natural philosophy, its complex theories require frequent glosses. A number of attributes, including its popularizing aim, make this poem a forerunner of *The Botanic Garden*: the invocation of Urania, the involved allegory, the universal scope (Brooke professes to survey "most branches of natural philosophy" in the poem), the complex economy of natural processes, and archaic components such as alchemy.

to create, did not yet exist.[14] By emphasizing natural history, Aikin substitutes a social distinction for an epistemological one: poetry and science are not incommensurable ways of knowing, but discourses with historically specific forms. He argues that scientific content was unsuited to verse in the past mainly because "false opinions were then received" in natural history, which in turn inhibited poets from making their own observations.[15] Ironically, then, Aikin's defense of "scientific" poetry helps to explain why the vogue of *The Botanic Garden* was relatively short. Darwin drew on the tradition of minor Miltonic poets such as Henry Brooke and Mark Akenside, whom critics increasingly rejected, with Aikin, as mere versifiers of "false opinions." Nineteenth-century poets such as John Clare continued to pursue natural history somewhat as Aikin envisioned, but neither Aikin nor Darwin succeeded in creating a lasting synthesis of literary and natural knowledge.

The attempt to establish such relationships, however, reveals a great deal about the cultural environment of early industrial Britain. Aikin's *Essay* sheds light on Darwin's project partly because both emerged from a new provincial culture that favored applied science and merged natural history with landscape aesthetics. In a 1991 essay, John Guillory locates a new class consciousness in the nexus of English poetry and English landscape. Such a consciousness creates a social matrix uniting the critical debate about poetry and science with other cultural components of Darwin's aesthetic geology. Guillory analyzes Anna Barbauld's "The Warrington Academy" (1770), a poem prefixed to an anthology produced for classroom use in the Dissenting institution named in its title (whose faculty included Barbauld's and Aikin's father, John Aikin senior). The passage of the poem (also known as "The Invitation") that supplies the epigraph to this chapter also includes a brief sketch of the objects of natural history, as part of Barbauld's account of possible careers for the academy's graduates. Barbauld integrates natural history with literature and the professions, much as Aikin and Darwin do, to map the cultural terrain of a new social class. Guillory makes English landscape the center of this culture, arguing that it used topographical poetry to substitute associations with vernacular literature for the classical commonplaces originally informing this genre. The vernacular anthology including the poem, according to Guillory, exemplifies the new pedagogy which "institut[ed] a standard of grammatical speech as a credential of gentility, and

14. The popularity of natural history was just beginning to burgeon in the 1730s; the seminal events in this history include the publication of Buffon's *Histoire naturelle* (1749–1803); the accession of the scientifically minded George III in 1760; and the founding of the Linnaean Society in London in 1788, with other societies following in its wake. See David Allen, *Naturalist in Britain*, chap. 2, "The Rise to Fashion."

15. Aikin, *Essay*, 31.

discovered in the form of anthologised vernacular works the means of directing this speech toward a political objective." The Warrington Academy curriculum creates a vernacular canon in order to substitute useful knowledge and acquired wealth for classical languages and the aristocratic privilege they signified. An autochthonous middle-class culture thus succeeds the gross distinction between propertied literacy and illiterate poverty.[16] The poem itself presents natural history as another such form of useful knowledge, anticipating the development of scientific professions and the new economic connotations of landscape as an aesthetic form.

Darwin's geology lies very close to this intersection between Dissenting culture and the newly acquired wealth of the early Industrial Revolution. James Pilkington, who studied at the Warrington Academy under Aikin and the mineralogist J. R. Forster, wrote to Darwin with a question about the hot springs at Matlock and Buxton as he was preparing to publish *A View of the Present State of Derbyshire* (1789). Darwin gave him permission to include his twenty-page reply in the book and refers readers to it in *The Botanic Garden* (I.iii.166n.). Darwin's intense interest in geology originated in the fervor surrounding applied science in the Lunar Society of Birmingham, and he shared his interest with both naturalists (Whitehurst) and manufacturers (Wedgwood) in the group. Under Whitehurst's guidance, for example, he supplied Wedgwood with mineral specimens for his experiments in search of the best materials for mass-produced pottery, and his suggestions in mechanics and other areas benefited Boulton and Watt as well.[17] Maureen McNeil makes the compelling point that "Darwin's poetry presented certain features of industry: raw materials, the mechanical inventor or factory owner, machinery, and products. What is missing is any sense of the labor process."[18] Margaret Jacob's analysis of Darwin's circle shows that mystification of the labor process is only one aspect of the crucial ideological function of science in early industrial culture. Historians, she says, have replicated the period assumption that reduced industrialization to "the application of surplus capital to raw materials." Along with the Dissenting ideology of individual achievement, the "availability of scientific culture"—epitomized by Darwin's poetry—was crucial to these economic decisions, which Jacob recasts as "scientific culture at work."[19] The scientific culture

16. Guillory, "English Common Place," 17, 15, 18.
17. See King-Hele, *Erasmus Darwin: A Life*, 77–79. There are also two relevant letters to Wedgwood (*Letters of Erasmus Darwin*, 40, 43). The culture of Dissent is most clearly influential with James Watt, who came from a radical Calvinist background.
18. McNeil, *Under the Banner of Science*, 17.
19. Jacob, *Scientific Culture*, 187–88. The most important feature of this culture, for Jacob, is that "a distinction between the 'scientist' and all others simply did not exist" (133), as Darwin's circle helps to illustrate (128–29). Similarly, she argues, there was no distinction between

surrounding rocks is built on a dialectic that involves the aesthetic and economic senses of "raw materials." The economic sense is especially concrete in Darwin's geological poetry because of its social and geographic setting.

This social history is integral to the culture of landscape and also provides a way of synthesizing the intellectual history originally charted by Marjorie Nicolson and M. H. Abrams. Nicolson's narrative begins well before the Romantic period, tracing "the great interest shown by . . . poets in scientific theory," an interest catalyzed by Newton's death in 1727. Abrams' discussion in *The Mirror and the Lamp* focuses on the Romantic resistance to "scientific theory" that Nicolson also sketches at the end of *Newton Demands the Muse*.[20] The year 1802 is a landmark for this discussion because of the competition between the Poet and the Man of Science introduced in Wordsworth's revision of the *Preface* to *Lyrical Ballads*, with Keats' anti-Newtonianism—in *Lamia* (1819), for example—marking a terminal point in both discussions. The 1770s texts by Aikin and Barbauld, as illuminated by Guillory, reflect essential changes taking place between the 1730s, when "Newton gave color back to poetry," and the early nineteenth-century moment when he was seen as unweaving the rainbow.[21] Aikin's essay shows that science does not merely supply poetry with material by linear transmission, but also interacts with it. That interaction is part of the rising middle-class culture of landscape animating Barbauld's poem, a broader culture succeeding the aristocratic " 'grotto fad' that peppered English estates with artificial retreats" and capitalizing on the way in which, as Nicolson points out, "the new geology merged with old literary conventions" in such settings.[22] Increasing social mobility meant increasing physical mobility in the landscape, which thus became the most important territory shared, and eventually contested, between poetry and natural history. Nature in the form of landscape provides the scene for the movement from books to observation, from "secondary" to "primary" pleasures of the imagination, that is the foundation of the culture shared by literature and the emerging modern sciences.

---

"pure" and "applied" science (4). The first version of this book gives more nuanced accounts of this issue and of the politics of Darwin's circle; see *Cultural Meaning*, 137, 164–68.

20. Nicolson, *Mountain Gloom and Mountain Glory*, 339n.; idem, *Newton Demands the Muse*, 164–75.

21. Nicolson argues particularly for the importance of the *Opticks* as an impetus to poetry: "It is no exaggeration to say that Newton gave color back to poetry . . . [by] produc[ing] a new scientific grasp of a richer world of objective phenomena peculiarly sympathetic to the poets"; *Newton Demands the Muse*, 22–23.

22. Nicolson, *Mountain Gloom and Mountain Glory*, 342. The term "geology" is somewhat misleading in this context, since it did not come into use until the 1790s and had no real existence, during the period Nicolson discusses, independent of physicotheology. But this slippage only illustrates the importance of the broader cultural shift by which landscape became the province of practitioners from a variety of class backgrounds.

The nineteenth- and twentieth-century accounts of the Romantic attitude toward science surveyed by Abrams display common failings as means to a historical understanding of the late eighteenth- and early nineteenth-century culture that unites poetry, aesthetics, and natural history.[23] Abrams' history registers the increasing influence of disciplinary boundaries. In the half-century since Abrams wrote, it has become clearer that such intellectual divisions fundamentally distort our perceptions of the period's culture, and it now seems that this distortion serves to project a mythical origin for those very divisions onto Romantic thought. The view of science as an abstract discourse of "classification and generalization" is much less a product of that culture than of subsequent developments, however accurate it may be as a description of Victorian and modern science.[24] Likewise, scientific deism was thoroughly established by the late eighteenth century, and offered nothing like the theological challenge of Victorian uniformitarian and evolutionary thought. Even James Hutton's science is a long way from being secular. The *Anti-Jacobin* railed at the elder Darwin on both political and religious grounds, but this sort of pressure tended to bring intellectuals from diverse fields together, as in the Birmingham Lunar Society or Thomas Beddoes' circle in Bristol (which included Davy and Coleridge). The views that Abrams cites, and to some extent shares, tend above all to occlude the historical importance of poetry in *explaining* the natural world. Darwin's revitalization of myth and pagan machinery makes this general epistemological claim of poetry impressively (if garishly) visible. For a fuller historical understanding of the period's conceptions of the relation between science and literature, we must recognize that it lacked a cultural basis for divisions of knowledge that became fundamental later in the nineteenth century. The shared culture of "letters" and a relatively unified religious culture, along with the class and institutional parameters delineated by Guillory, Jacob, and others, are some of the factors that distinguish the cultural landscape of Romantic science, particularly natural history, the protean complex of literary and material practices against which geology was beginning to assert its independence.

It would be a gross exaggeration, of course, to say that the period ac-

23. According to the most influential of these views, "poetry is true in that it corresponds to concrete experience and integral objects, from which science abstracts qualities for purposes of classification and generalization"; M. H. Abrams, *Mirror and the Lamp*, 315. Abrams' notion of the poem as "heterocosm" (297) also perpetuates the emphasis on antiempiricism that still informs received ideas of Romanticism and obscures the deeper affinities that originally subsisted between Romantic poetry and science.

24. Such generalizations are never wholly accurate, but it is significant that, as commonplaces, they are historically conditioned by a later parochialization of the sciences and of intellectual life.

knowledged no division between literature and science, especially natural philosophy. Darwin's synthetic project—one of the many encyclopedic projects, in verse and prose, that are typical of the period—is possible only as the result of a perceived gap that must be bridged; hence the elaborate "Apology" in *The Botanic Garden*. It is a project, however, that registers the facticiousness of divisions of knowledge and presumes a confidence in the epistemological function of poetry, in its ability to provide an *organon* of the modes of knowledge. This may be Darwin's greatest affinity with the canonical Romantic poets: he offers an alternative formal model for the integration of knowledge, and his model directly confronts the incipient specialization of scientific language and concepts. Darwin and most of his readers agreed that verse was a viable and important forum for the transmission of scientific concepts, and even skeptical readers such as Coleridge and Wordsworth used *The Botanic Garden* as a resource for scientific ideas. The poem's popularity testifies to an economic and cultural demand linked to the fact that intellectual professionalization was at an early stage; literature was somewhat more established than science as a profession, but both were also avocations increasingly diffused among the middle class.

Darwin's poetry provides an index of numerous cultural changes, loosely tied to changes in class structure, that shaped the history of science, including changes in public discourse and postsecondary education. To the extent that portions of what we now call "science" performed the function of entertainment, Darwin's project helped to consolidate a transition from aristocratic to middle-class audiences as the focal point of this entertainment. Though Darwin's main subjects of botany and geology are branches of natural history, he clearly privileges natural philosophy (essentially Newtonian science plus chemistry), echoing such popularizing treatises as William Nicholson's *Introduction to Natural Philosophy* (1782). The pedagogical emphasis on natural philosophy reflects weaknesses in the system of formal scientific education: the Newtonian revolution was primarily a product of the attention paid to mathematics in Britain's major universities, which also offered lectures in natural history; but specialized scientific courses developed only in response to the popularization, professionalization, and commercialization of science, which also generated educational venues outside the university.[25] The career of Humphry Davy in one such venue, the Royal Institution, illustrates more fully than Darwin's "how the leadership of the nation might be transferred to another social class."[26] Darwin's appeal to natu-

25. Even chemistry, according to Jan Golinski, was a marginal presence in the universities until the 1780s, "and even then the improvement was not sustained for long"; *Science as Public Culture*, 53.

26. Morris Berman, *Social Change and Scientific Organization*, xxii.

ral philosophy in order to authorize and systematize the discourses of botany and geology anticipates Davy's more permanent transformation of chemistry into an independent discipline of natural philosophy. *The Economy of Vegetation* demonstrates Darwin's ambition through its broad range of subjects, including cosmological issues traditionally classed with natural philosophy. Though Darwin rhetorically privileges natural philosophy, natural history—strongly associated with fieldwork and hence more accessible to a generalist public—is more important to his project as a distinctively literary form of natural knowledge. *The Botanic Garden* reveals geology as a major vehicle for the commerce of literature and natural history, the publishing industry and overlapping readerships that also sustain Romanticism and landscape aesthetics.

This discursive context illuminates the literary, scientific, and social senses in which Darwin's mythological allusions and other machinery provide "proper ornaments to [his] philosophical poem" (as promised in the "Apology"). The propriety of these ornaments derives from their utility, their explanatory power, as well as their cultural content. In some cases the mythological illustrations are simple and elegant, as in Darwin's glossing of the "hieroglyph" of Venus rising from the sea on a shell as an allegory for the initial rising of land out of the sea from the accumulation of seashells (in the form of limestone) (*BG* I.ii.47 and n.). In a more complex instance, Darwin uses the seduction of Venus by Mars to illustrate the explosiveness of saltpeter ("nitre"), but then goes on lavishly to describe Vulcan's strategy for exposing the miscreants (143–182). The sequence here—from the "treacherous courtship" of "Azotic Gas [and] the virgin Air" to the "wanton arms intwined" of Mars and Venus—is the reverse of what Darwin proposes in the "Advertisement": to lead the "votaries" of Imagination from these "looser analogies to the stricter ones" of Science. Such instances seem to continue the project Tim Fulford identifies in *The Loves of the Plants*: "the Ovidian sexuality" of the poem "was not 'proper' but seductively playful, especially since Darwin presented his verse as an amusement for ladies in their dressing rooms," and led to its condemnation as dangerously licentious.[27] While this condemnation intensified in the 1790s cli-

27. Fulford, "Coleridge, Darwin, Linnaeus," 127. The "courtship" of Azotic gas (nitrogen) and "vital" (or "virgin") air (oxygen) is literally "treacherous" in the sense that it produces a hot red cloud of nitrogen dioxide. Darwin's analogies here are looser than usual in both senses: he moves quite abruptly from naturally occurring nitre or saltpeter (143–46) to the nitric acid (147–50) produced artificially in the laboratory by Henry Cavendish (as explained by the note), who performed the reaction allegorized as "courtship." But since Cavendish sparked the gases over a sodium hydroxide solution, his experiment would have generated saltpeter (potassium nitrate) as well. In either case, the product signifies Vulcan's inflamed reaction to the indiscretion of Venus (oxygen) and Mars (nitrogen). Thanks to Professor John E. Adams for help with the chemistry.

mate of reaction, Darwin may already be responding to charges of frivolity in *The Economy of Vegetation* (which followed, belatedly, as part I in 1791). Such charges would add a social motivation for his move to geology in this poem: gender and sexuality are social issues, and geology, as we shall see, expands the poem's scope to more "weighty" social issues such as political economy.

In *The Economy of Vegetation*, Darwin finds additional ways to capitalize on the poem's allegorical structure, which allows cultural content to seep into the fissures between apparently incongruous scientific content and literary execution as in the seduction fable mentioned above. Darwin's copious notes provide graphic and verbal expression of this heterogeneity, the gap between disciplines ultimately subsumed by cultural content and poetic form. The footnotes (not to speak of the endnotes) seem to crowd out the verse, often filling more than half the page and entirely absent on only a few pages. Their more technical scientific content, too, seems at times to supplant the imaginative explanations of the verse, and in these instances one wonders if the votaries of poetry are to be led to science through a vertical movement down the page, rather than horizontal movement through the poem. The supervening unity of form, that of "a poem with philosophical notes," provides the solution to this dichotomy. It might be argued, of course, that what was acceptable in Henry Brooke's copiously annotated poem is no longer acceptable in the age of Wordsworth and Blake, or that Darwin has taken the convention of footnotes too far, qualifying at most as a "versifier" (Abrams) and not as a poet. Ironically, such Aristotelian strictures exclude the cultural content that unites the notes with the verse and makes *The Botanic Garden* a literary work. The verse is just as likely as the notes to bring news from the laboratory to a wider public and celebrate the beauties of ancient sculpture or the progress of manufacturing in the same heroic/documentary vein. A glance at the argument prefixed to this canto will confirm this range.

## Aesthetic Objects and Aesthetic Practice in Darwin's Scientific Poetry

The following extract from the "Argument" to one canto of the first part of *The Botanic Garden* introduces a bewildering range of topics, suggesting that the poem is a textual form of the cabinet:

> Argument of the Second Canto.
> . . . [VI.]2. Coal; Pyrite; Naphtha; Jet; Amber. Dr. Franklin's discovery of disarming the Tempest of its lightning. Liberty of America; of Ireland; of France,

349. VII. Ancient central subterraneous fires. Production of Tin, Copper, Zink, Lead, Mercury, Platina, Gold, and Silver. Destruction of Mexico. Slavery of Africa, 395. Destruction of the armies of Cambyses, 431.

Analysis of some of these apparent digressions from the main purpose of the canto, the sketch of geogony, makes it possible to reconstruct the unity as well as the breadth of scientific culture. The ekphrastic digressions on classical sculpture and on pottery, for example, introduce familiar subjects of polite knowledge that are also relevant to specimen culture and applied science. Darwin seems to retreat even further from scientific ambition in the preface to *The Temple of Nature*, which "does not pretend to instruct by deep researches of reasoning. . . . its aim is simply to amuse by bringing distinctly to the imagination the beautiful and sublime images of the operations of Nature." But the poem's more ambitious scope suggests that Darwin is simply disguising his appropriation of aesthetic discourse for the diffusion of science, an investment nevertheless signaled by the phrase "the operations of Nature." *The Loves of the Plants* invokes this "philosophical" register markedly less than the two later texts, partly because it aims at popularizing one particular system (the Linnaean), while Darwin propounds his own theories in the later texts. In 1789 he informed his readers that the publication of *The Economy of Vegetation* would be delayed "for the purpose of repeating some experiments on vegetation."[28] The project of amusing readers, stressed particularly in the front matter of *The Loves of the Plants*, is realized in the principle of "poetic exhibition" that governs both parts of *The Botanic Garden*: Darwin exhibits both natural and artistic "specimens" throughout, and presents cultural uses of the earth's materials in a variety of historical and political contexts.

Both form and content bear out the profession that *The Loves of the Plants* (1789) is a "trivial amusement," and its strategies of amusing help to identify the means by which *The Economy of Vegetation* (1791) and *The Temple of Nature* (1802) continue to function as entertainment. Specifically, the poem carries out its intention that we "contemplate" its contents "as diverse little pictures, suspended over the chimney of a Lady's dressing-room, *connected only by a slight festoon of ribbons*" (*BG*, p. vii). Fulford's essay shows how richly this scenario itself repays close reading.[29] The metaphor of pictures supersedes the one of shadows in the camera obscura, arguably moving from the private space of the parlor game into the more public one of the gallery or cabinet. While the exhibition seems confined at first to the most

---

28. "Advertisement," in Erasmus Darwin, *Loves of the Plants*, unpaginated front matter.
29. Fulford, "Coleridge, Darwin, Linnaeus." See further Ann Shteir, *Cultivating Women, Cultivating Science*, esp. 26–27; and Londa Schiebinger, "Private Life of Plants."

private space of all (the dressing room), two pieces of the poem's synoptic apparatus markedly expand the scope of this metaphor. These are the index, presented as a "Catalogue of the Poetic Exhibition" (*BG*, pp. 197–98), and the three "Interludes," dialogues between the poet and the bookseller. The *ut pictura poesis* principle is the common thread uniting these dialogues, in which the Poet maintains that the chief task of poetic language is to stimulate the sense of sight. But he qualifies the Bookseller's distinction between the notes as the site of "sense" (151) and the verse as that of mere amusement by explaining that "the Muses are young ladies, and we expect to see them dressed"; this poetic "sense" remains distinguished from "ratiocination" by its suggestive embellishment of "nature" (162). The poetry's capacity for "amusement" thus derives partly from its visual emphasis, underscored by recurring analogies to contemporary painting. But because Darwin makes this capacity contingent on some degree of accuracy—the garb of the Muse—it remains compatible with his form of scientific explanation.

*The Economy of Vegetation* considerably expands the poem's scientific apparatus and investment but maintains the form of the "poetic exhibition." There are a few actual paintings and objets d'art among the fanciful scenes of the 1789 poem, and in the 1791 text such objects—the Portland Vase, Wedgwood's cameos—become more prominent: the later volume includes plates depicting these objects and devotes some of its more copious notes to them. This treatment is an important tactic of aesthetic geology, situating itself deliberately in the realm of polite knowledge while explaining the structure and function of the earth's material (here marble and clay). The presence of these artifacts in the canto devoted to geology, along with specimens of ancient sculpture and useful minerals, a rhetoric of collection, and accounts of relevant social circumstances, reveals Darwin's project as a compendium of polite knowledge. His natural philosophy obtains a wide cultural currency by being situated within this range of subjects and by virtue of his capacious poetic idiom, effectively summarized by Jonathan Wordsworth: "Darwin's readers were to enjoy botany through poetry, and enjoy the poetry through nostalgia for the lost world of beaux and belles at the court of Queen Anne."[30]

The double sense of "specimens" becomes especially clear in the account of marble. The poem first presents the various "lime-stone rocks" in situ, according to the manner of their formation (*BG* I.ii.93–100):

30. Darwin, *Loves of the Plants*, second page (unnumbered) of Jonathan Wordsworth's introduction. Darwin's style is encyclopedic in the sense that it incorporates elements from Thomson as well as Pope, Richard Payne Knight as well as Lucretius, Richard Jago as well as Henry Brooke.

> *Gnomes!* you then bade dissolving *Shells* distil
> From the loose summits of each shatter'd hill
> . . . . . . . . . . . . . . . . . . . . . . . . . . . . . . . . . . .
> [Till] in white beds congealing rocks beneath
> Court the nice chissel, and desire to breathe.

This passage alludes to the trope of "breathing marble" from ekphrastic poetry about sculpture (marble being metamorphosed limestone), and it motivates the subsequent display of "worked" marble rather than strictly geological specimens: "Hence wearied HERCULES in marble rears / His languid limbs, and rests a thousand years" (101–2). Darwin proceeds to list specimens of sculpture belonging to the neoclassical canon, such as the *Apollo Belvedere*, but concludes the digression by citing modern sculpture in the nationalist framework prescribed by neoclassicism in its late phase. The claim of aesthetic permanence is typical for this discourse: General Wade's sculptured tomb, like the *Apollo*, will "conquer time," and the statues of Mrs. Damer will "enchant another age" (109–14). This account of monuments politicizes the durability of marble, a geological fact, but monumentality is also a geological category for Darwin. His theory of mountains as "monuments of past delight" (*TN* IV.450) doubles the association between sculpture and permanence and seems to prove that a knowledge of geology is compatible with a knowledge of sculpture.

The intimate connection between marble and sculpture (as in the name "Elgin Marbles") provides a model for presenting other minerals as aesthetic objects. The naturally crystalline form of salt, for example, endows an extensive salt mine in Poland with Gothic grandeur: "Thus, cavern'd round in CRACOW's mighty mines, / With crystal walls a gorgeous city shines" (125–26). Another product of the "chissel nice," a statue depicting—unsurprisingly—Lot's wife, also adorns this underground city (135). The notes characteristically hasten to make this scenario plausible, citing a travel narrative for the geography and Hutton's *Theory of the Earth* for the geology (119n.). The note's authenticating function here seems akin to Aikin's demand for accurate natural history, but the sources and additional details it provides also bind natural history more firmly to cultural history. As Darwin moves through the catalogue of useful minerals, he gives each one its proper narrative; nitre, for example, occurs with the story of Venus' infidelity, as we have seen. These narrative digressions give social form to the allegory that is the basic method of all Darwin's poetry. Less significant minerals, such as the semi-precious stones, are more briefly sketched through personification allegory: "playful Agates weave their colour'd threads; / . . . And changeful Opals roll their lucid eyes" (224–26). *The Temple of Nature*, with its Loves and Graces, further amplifies this Spenserian strain, and allegorizes geogony itself as architecture in the figure of the Temple, "unwrought by mortal toil": "deep in earth the ribbed vaults descend" (*TN* I.67–70).

This notion of minerals themselves as works of art becomes most explicit in the account of clay, in which ceramic arts appear as extensions of the processes forming their materials. This aesthetic paradigm depends on the more active participation of a major set of allegorical figures, the Gnomes (inspired, like the couplets, by Pope). In the case of clay, the Gnomes not only are "the guards and guides of nature's chemic toil" (*BG* I.ii.272) but also work on rocks to produce the useful mineral (297–300):

> *Gnomes!* as you now dissect with hammers fine
> The granite-rock, the nodul'd flint calcine;
> Grind with strong arm, the circling chertz betwixt,
> Your pure Ka-o-lins and Pe-tun-tses mixt.

The Gnomes suggestively combine the agency of geological process, that of the field geologist with his "dissecting hammer," and that of "CHINA's sons, with early art elate" (281). Darwin inserts the invocation halfway into an account of pottery (having first defined clay as "obedient to the whirling wheel"). This account begins with the "huge dragons" and "cobaltic blues" of the "early art" of Chinese porcelain. Both "kaolin" and "petuntse" are Chinese terms for porcelain clay.[31] The history continues with an "Etrurian" or Etruscan phase, based on a false attribution of the Greek vases collected by Sir William Hamilton. The narrative sequence suggests that the Gnomes intervene at this point further to refine the materials in question (as suggested by the wonderful onomatopoeia for the induration of chert at 296). Darwin's history concludes triumphantly that as a result, "A new Etruria decks Britannia's isle" (304).

Several interlocking factors, Darwin implies, contribute to the new golden age of ceramic art in England, signaled by the Gnomes, who "pleased on WEDGWOOD ray [their] partial smile" (*BG* I.ii.303). Clay belongs, like coal and most of the other minerals catalogued in this canto, to a class of strata defined by their status as commodities. Hence its geology is best understood through the history of its uses, partly expressed as the extension by human art of that of the Gnomes. For the same reason, credit for our knowledge of clay belongs partly to the "dissecting hammer" of applied geology. Darwin's teleology carries the imperialist suggestion that British

---

31. Lydia Liu relates these terms to English and European literary history in "Robinson Crusoe's Earthenware Pot."

clay and British rocks in general are more highly evolved than those of China. The Gnomes "smile" with partiality on Britain and on Wedgwood in particular, favorably influencing every detail of Wedgwood's products: "Charm'd by your touch . . . / The bold Cameo speaks, the soft Intaglio thinks" (307–10). Here, to adapt Darwin's earlier terminology, the votary of ceramics is led to an interest in the geological properties of clay by an elaborate identification of the one with the other, in which the substance is both formed and processed by one "art," and the aesthetic practice of admiring Wedgwood pieces is extended to the minerals themselves. At the same time, as Maureen McNeil points out, Darwin's personifications entirely efface the human labor involved in this production process.[32] Plates depicting two of Wedgwood's cameos and the Portland Vase, which inspired them, accompany a long ekphrastic episode devoted to these objects (311–48), which thus resembles an exhibition catalogue (and to some extent a commercial one). The episode concludes with a claim to aesthetic permanence ("nor Time shall mar"; 347) that echoes the earlier account of marble; clay is also Britain's marble in the sense that it is a native resource. Within the nationalist, industrial context of Etruria, however, Darwin chooses to gloss Wedgwood's cameo of the African slave, "from Britain's sons imploring to be free" (I.ii.315), bearing out Margaret Jacob's suggestion that intellectuals in Darwin's circle "could also be republican even, or perhaps more especially, when [they were] busy being entrepreneurial or industrial."[33] The fierce backlash against Darwin would also seem to vindicate his politics as progressive; but the same ideology of progress, in this historical moment, dictates his effacement of labor and the resulting equation of raw materials (clay) with aestheticized commodities (cameos).

Darwin's account of mineral art relies on the expanded role that "specimens" play in the period's visual culture. The artistic specimens that operate verbally and visually in Darwin's poem take their place alongside the geological specimens that play an active role in its composition. The notes refer repeatedly to specimens in Darwin's "possession," and in one instance he writes of a nodule of iron ore "which now lies before me" (*BG* I.ii.183n.). Martin Rudwick observes that in this period, "mineralogy was first and fore-

---

32. McNeil, *Under the Banner of Science*, 20; see 16–30 on Darwin's vision of Etruria.

33. Jacob, *Cultural Meaning*, 164. McNeil objects to this commercial use of the antislavery cause (*Under the Banner of Science*, 28) and points out that Darwin tends to "project the social problems of industrializing Britain onto nature" ("Scientific Muse," 196–97). The agency of Darwin's Gnomes, on the other hand, seems designed to generate a superabundance of resources, as suggested in his apostrophes to various "products of the strata" (e.g., *BG* I.ii.183–201; cf. 540–42).

most a matter of mineral specimens: specimens collected, sorted, named and classified."[34] Natural history collections were still assembled in cabinets and museums, where they mingled comfortably with sculpture and other artifacts, as in Darwin's poem. This mode of collection, however, was also the basis of fieldwork and hence of applied geology; geological collections increasingly carried the economic importance I discussed earlier. Mary Anning's successful trade in fossils at Lyme supported the growth of her scientific expertise, which gained her the epithet "hand-maid of geological science." Darwin describes one of Wedgwood's cameos as having been made deliberately of clay from Botany Bay and sent there to inspire its colonists. Clay is one of many products collected in Botany Bay by naturalists that proved to be valuable resources and greatly expanded the role of natural history as an instrument of imperial economy. The applied antiquarianism of Wedgwood's imitation of "Etruscan" vases also resembles the applied geology of William Smith, who represents himself as collecting "the antiquities of the earth." By installing the Portland Vase (1810) and Smith's fossil collection (1816) in the British Museum, Parliament implicitly acknowledges mining and manufacturing's contributions to the greater glory of the nation.

While art and commerce depend on geology, geology—in the form of collections, popularizing lectures, and even theoretical texts—also relies on specimens for their aesthetic interest. Illustrated volumes of natural history became increasingly affordable in the nineteenth century. Important examples include the twelve-volume *Bertuch's Bilderbuch* and Gideon Mantell's *Wonders of Geology* (1838), which boasted a frontispiece by the painter John Martin. While much else had changed in geology over forty years, Mantell was still, as Darwin and James Hutton were, a passionate collector of specimens. The French naturalist Faujas de Saint-Fond, in his accounts of meetings with British geologists, devotes much space to accounts of their collections; he found common ground in discussing these where he differed on matters of theory from older geologists like Hutton and Whitehurst.[35] These collections have a wide range of uses, but they are equally indispensable for every level of geological explanation. Specimens function as a familiar medium of information in Darwin's poem, in short, because many of his readers would have seen such collections, if they were not collectors themselves. The "Section of the Earth" reproduced in Darwin's notes (Fig. 5.1) also relies on visual appeal; later geological illustrations, such as Smith's

---

34. Rudwick, "Minerals, Strata, and Fossils," 266.
35. Barthélemy Faujas de Saint-Fond, *Journey through England and Scotland*, 1:20–22 and passim; cf. 88 on the British Museum.

sketch of the strata across England (Fig. 4.3), rely similarly on graphic possibilities of compression, though Darwin was, characteristically, the only
one to attempt a section of the entire globe in that period.

While the aesthetics of collection helps to make Darwin's poetic geology
accessible and entertaining, the cultural content of such rocks as marble and
the utility of clay and other minerals also combine to justify the study of
natural history as a lens on political economy. The emphasis on utility in-
creases steadily through the accounts of clay and coal to those of the metals.
A long apostrophe to steel, echoing Richard Jago's *Edge-hill* (1767), celebrates utility in lieu of mythological associations: "Hail, adamantine *Steel!*
magnetic Lord! / King of the prow, the plowshare, and the sword!" (*BG*
I.ii.201–2).[36] The political implications of this utility come to the fore in the
accounts of coal and especially silver and gold, which generate the longest
narrative digression. Darwin introduces coal and related minerals with a
brief set of allegorical personifications, from "sable Coal" on "his massy
couch" to "bright amber" on his "electric throne" (349–53). The mechanism of association here seems quite loose, linking the electricity of amber
to "immortal Franklin's" famous experiment with the kite, which in turn becomes a metaphor for Franklin's role in the American Revolution: "The pa-
triot-flame with quick contagion ran, / Hill lighted hill, and man electrised
man" (367–68). The succeeding account of the French Revolution seems de
rigueur in this context, at least given the level of liberal support in 1791, and
the return to geology at this point shows that the chain of association can
lead from culture back to science as well: Darwin's controversial suggestion
that occasional violent upheaval contributes to the health of the system is
more safely articulated in terms of fissures allowing the escape of fecund
steam from "central subterraneous fires" (398n.; cf. I.i.152n.). Gold and silver present a cultural content more directly linked to their economic importance, the history of New Spain, and, by extension, bloodthirsty imperialism
in general. Darwin moves here from Spanish "Avarice, shrouded in Religion's robe" (I.ii.415) to British complicity in the African slave trade, and
from there to Cambyses' invasion of Egypt, as the crowning parable of ac-
quisitive tyranny.[37]

---

36. Jago's longer and more openly nationalist account moves from the landscape around
Birmingham to a step-by-step account of steel production, which incorporates prospecting,
mining, smelting, and forging into a grand heroic narrative; *Edge-hill*, 72–75.

37. The "blood-nursed Tyrant" and his marauding troops, already decimated by famine and
sandstorms, are duly swallowed up by an earthquake: "Wave over wave the driving desert
swims, / Bursts o'er their heads, inhumes their struggling limbs" (I.ii.489–90). Drummond
makes a similar connection between political and geological revolution (*Giants' Causeway*, 179).

Darwin's poem is not merely a compendium of polite knowledge or a textual version of the cabinet. He is outspoken in politicizing the traffic in resources, though this traffic in its legitimate forms, such as Wedgwood's socially conscious industry, remains for Darwin a valorized application of natural history. As Maureen McNeil points out, it is optimistic, if not cynical, for Darwin to suggest that industry is enlightened simply because it uses wage rather than slave labor. But the digression on Revolution (which also sends Liberty, seditiously, to Ireland) provides a clear instance of the politics that made the poem seem dangerous as the 1790s progressed. It seems more likely than not that the ensuing controversy, which resulted in parodies (such as the *Anti-Jacobin's* "Loves of the Triangles") and other attacks, would have contributed to the poem's popularity. But the controversial content, at the same time, moves it to the margin of polite knowledge. The tendency toward applied science in the geological canto and throughout the poem—as in its account of the cotton plant—sheds light on a different set of political implications in the poem's articulation of scientific culture. The poem's cultural currency, and hence its transmission of scientific concepts, depend on the increasing democratization of intellectual life: the popularity of natural history, of collecting and collections; the wide dissemination of neoclassical principles in the visual arts, and the appeal of the visual itself; the established appeal of the couplet form; and the recognizable nature of the particular aesthetic objects and practices the poem invokes. Wedgwood's china, if not mass-produced in our late industrial sense, still bridges the divide between beauty and utility, and relocates aesthetic objects in the domestic sphere, as does the initial camera obscura metaphor.

Popular entertainment, by attempting to engage the culture already shared among a wide readership, makes a suitable mode for a "philosophical poem" aiming to unite the forms of knowledge and explain the natural through the social world. This solution makes "trivial amusement" not only compatible with but indispensable to Darwin's vision of natural philosophy. It would be going too far to attribute the emergence of modern disciplines such as chemistry and geology, in the two decades following *The Botanic Garden*, to Darwin's success; but the poem marks a crucial moment in the history of both poetry and science because it engages and galvanizes the shared culture that is essential to the understanding of both. The poem's form and cultural content both encourage and depend on a relation between natural history and aesthetics that operates in the public as well as the domestic spheres and unites public display, economic incentive, and private amusement.

## Darwin's Geology

Darwin's incorporation of aesthetic objects makes his poem resemble a textual cabinet, but this poetic exhibition also draws on textual forms of natural history. He takes his botanical system from Linnaeus while drawing botanical particulars from travel narratives—such as, in one famous instance, a Dutch naval surgeon's account of a Javanese plant translated for the *London Magazine*.[38] Likewise, the canto on geology moves freely among such sources as Helen Maria Williams' *Letters from France*, Hutton's *Theory of the Earth*, and the *Philosophical Transactions*. While the diverse geological materials also serve, to some extent, for the concrete elaboration of a scientific system, Darwin's geology is more thoroughly his own than his botany, and his range of geological references more fully transforms this science as it was then understood. The geology is remarkable for its compression and for its cavalier assurance that a complete account of geogony can be subsumed under the project of explaining "the operation of the Elements, as far as they may be supposed to affect the growth of Vegetables" (*BG*, p. v). The geology's subordinate position relative to botany, its reliance on the agency of the Gnomes, and the compression of the canto all point to Pope, whose influence on the poem conditions its use of the more recent materials, both popular and scientific.

Darwin uses the Rosicrucian machinery of *The Rape of the Lock* to represent the workings of the four elements in the service of plant growth, and this move can be related in two ways to Pope's own account of this machinery. The original motive for poetic machinery, according to Pope's satirical preface, was this: "the ancient Poets are in one respect like modern Ladies; Let an Action be never so trivial in it self, they always make it appear of the utmost Importance."[39] The use of "trivial" here almost certainly informs Darwin's "trivial amusement" in the 1789 Proem. Pope emphasizes the frivolity of "modern Ladies" by comparing them to classical poets, rendering their trivial incidents still more trivial, and their bathos more bathetic than that of the poets. Darwin's adoption of the Gnomes for the "operation" of the earth thus would seem to stress the subservience of geological process to plant growth and organic life, itself a "trivial" subject. King-Hele is not the

38. *BG* II.iii.219–58 and 238n. The Dutch account is translated in full in the endnotes. Many critics have also noted that this account of the upas tree turns up in Blake's "A Poison Tree" and his later Tree of Mystery.

39. Pope, *Poems*, 217. The machinery in both poems consists of Sylphs, Gnomes, Nymphs, and Salamanders, standing for Air, Earth, Water, and Fire, respectively. Pope's influence on the sexual politics of *The Botanic Garden* is also worth considering.

only critic to dismiss the Gnomes and their colleagues, too readily, as "distracting, useless, and best ignored."[40] I would argue that Darwin ironizes Pope's conception inasmuch as he uses the Gnomes to deal in a very serious way with large issues of geogony and cosmology—which is, if anything, an instance of litotes rather than bathos. While there are many famous (and well-parodied) instances of the latter in Darwin, of the most banal physical processes absurdly rendered in full Augustan dress, it is important to remember that he also holds the Rosicrucian spirits to have originated as "hieroglyphic figures representing the elements" (*BG*, p. vii)).[41] Darwin does not simply recommend this machinery for the "Lady's dressing-room," then, as Pope does to Arabella Fermor, on the strength of its supposedly "novelistic" or fantastic character. Darwin also seems more invested in the idea of a female readership.

The primary function of Darwin's Gnomes seems opposite to that of Pope's: they serve to give complex natural processes a frivolous poetic appeal, rather than endowing frivolous incidents with a mock-heroic grandeur. The Gnomes also serve the more serious function of making it possible to visualize geological agency, an exigency underlined by Hutton's elaborate provisos about the human mind's inadequacy to conceive the magnitude of subterranean heat and pressure. Once again, the device that Darwin uses self-effacingly to render his subject matter appealing to ladies—and gentlemen—also serves a scientific purpose. Darwin's adoption of Pope's closed couplet accounts for the remarkable compression of the verse; but this compression also tends to increase the volume of footnotes, since it often forces science to be presented (in the verse) through the barest of allusions:

> *Gnomes!* how you shriek'd, when through the troubled air
> Roar'd the fierce din of elemental war;
> When rose the continents, and sunk the main,
> And earth's huge sphere, exploding, burst in twain.
> *Gnomes!* how you gazed, when from her wounded side,
> Where now the South-Sea heaves its waste of tide,
> Rose on swift wheels the Moon's refulgent car.
> (I.ii.73–79)

The two closed couplets beginning this sequence feature the antithesis ("shriek'd"/"roar'd") and inversion (line 75) characteristic of Pope, while

40. King-Hele, *Erasmus Darwin: A Life*, 258.

41. Darwin's success is clearer by contrast with the wanton doggerel of one of his imitators, John Scafe. Scafe's "Granitogony" (from one of his three didactic geological mock-epics) allegorically stages the loves of the minerals to explain the "birth" of granite and other rocks (reprinted in Robert M. Hazen, ed., *Poetry of Geology*, 19–20).

the open couplets that follow drive the narrative to its climax. The Gnomes merely bear witness to a primal catastrophe associated in this short space with three levels of geological agency: the literal "explosion" of the earth; the resulting metaphorical "wound," with its personification of the earth; and the apparently innocuous allegorical device of the moon's "car." The metaphor of elemental war is common in earlier poetry, particularly in accounts of catastrophes, such as David Mallet's account of volcanism in *The Excursion* (1728). It also serves Darwin in moments when the Gnomes are more active, as when they are instructed to perform erosion: "Climb the rude steeps, the granite cliffs surround, / Pierce with steel points, with wooden wedges wound" (523–24). Darwin supports this military image with an allusion to Hannibal's crossing the Alps and organizes the Gnomes into "squadrons" and "hosts" in other passages.

The "Geological Recapitulation" in prose, at the end of the poem, summarizes the forms of geological agency and process in a more literal scientific language (*BG*, pp. 103–4), but also offers visual confirmation of the canto's mythical geography in its "Section of the Earth." This section (Fig. 5.1) offers a striking adaptation of the system in place for representing the strata of an outcrop or a mine, magnified to a planetary scale. The image harks back to Burnet's figures of a cracked sphere or "Mundane Egg," and may well have influenced Blake's cosmic maps (e.g., *Milton* 32). The planet's protuberances in any case have a decidedly organic cast, drawing on the recurring image of the wound. The rocks and minerals are distributed in the image just as they are organized in the verse: all the metals, for example, are grouped together at the bottom. The scientific figure, in other words, corroborates the visual emphasis that Darwin claims for the verse in the "Interludes" of *The Loves of the Plants*. The section also expresses graphically the tremendous compression of the verse. The "Recapitulation," on the other hand, abstracts the chronology of the verse, beginning with the planet's projection from the sun and proceeding in fourteen steps to the most recent geological events. Here Darwin uses the passive voice throughout; "the whole terraqueous globe," for example, "was burst by central fires" (p. 212). This sequential abstract makes clear that the essential contribution of the verse account is to visualize these processes and to attach them to concrete agencies as yet inaccessible to "rigorous" science. Here the section stands in for some of these functions.

Two implications of Darwin's geogony become clearer in the prose apparatus. First, his system is cosmological in scope, unlike Hutton's or Whitehurst's. Second, the "Recapitulation" and notes amplify the geogony's function in explaining the origin of matter itself, in keeping with Darwin's overarching materialism. The long note propounding the theory of solar

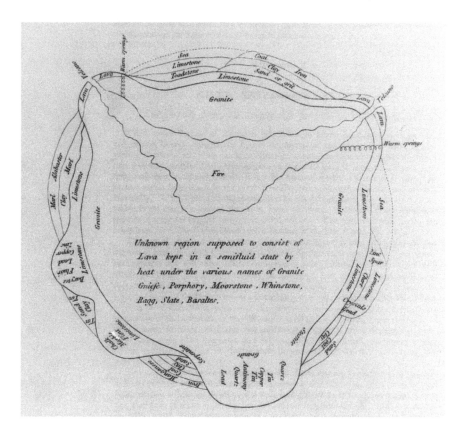

5.1 "Section of the Earth," from Erasmus Darwin, *The Botanic Garden* (1791). Engraving by T. Conder. Courtesy of the Linda Hall Library of Science, Engineering, and Technology, Kansas City, Mo.

volcanoes advances the notion of an originary chaos, based on Herschel's hypothesis that the solar system is moving around "some other centre . . . which may be of opake matter, corresponding with the very ancient and general idea of chaos" (*BG*, p. 170). The matter constituting the "original nucleus of the earth" is of another sort, flammable matter from the sun's surface, generating the earth's central heat in its molten form and visible in its solid form as "granite, porphyry, basalt, and stones of a similar structure" (p. 212). Darwin invokes the legacy of Buffon on this subject, proposing to correct the Frenchman's theory but also advancing a claim to maintain natural history on the comprehensive scale so influentially established in the *Histoire naturelle* (p. 171). Within the verse, Darwin's use of Jove's earthly loves, as an allegory for the union of oxygen with solid elements to form all the common mineral compounds, also operates on this grand scale (I.ii.229–70). This process of combination brings the earth's original matter, or "lava projected from the sun," into circulation in a general economy that also involves organic matter (35n.).

The "circulation" of matter is of the utmost importance in Darwin, supporting a kind of hylozoism:

> *Gnomes!* with nice eye the slow solution watch,
> With fostering hand the parting atoms catch,
> Join in new forms, combine with life and sense,
> And guide and guard the transmigrating Ens.
>                                    (571–74)

Darwin explains "the transmigrating Ens" in a note: "The perpetual circulation of matter, in the growth and dissolution of vegetable and animal bodies, seems to have given Pythagoras his idea of the metempsycosis [*sic*], or transmigration of spirit."

Darwin comes back to geology and the circulation of matter in *The Temple of Nature*, in order to develop this ambitious idea outside the context of botany. The theory comes to fruition in Darwin's exclamation that "mountains . . . / are MIGHTY MONUMENTS OF PAST DELIGHT!" (IV.449–50) because they are formed from the shells of living organisms that renovate the pool of matter by giving "the Bliss of Being to the vital Ens" (446). In the earlier poem, the Venus and Adonis myth provides the classical analogue for this economy (and also concludes the canto). Darwin explicates the two characters as "hieroglyphic figures representing the decomposition and resuscitation of matter; a sublime and interesting subject, which seems to have given origin to the doctrine of transmigration" (*BG* I.ii.576n.). The implicit substitution of matter for spirit in both allusions to the transmigration of souls is a consequence of Darwin's attempt to elevate natural history to the status of natural philosophy. The identification of the earth's original substance makes it possible to trace the history of the earth, including its organisms, as a continual material evolution; to establish this paradigm, Darwin must unify the three branches of natural history into one science, which he attempts most fully in *The Temple of Nature*. In both poems, he adopts the common assumption that rocks are "primordial substance," explored in chapter 3 above. This is a crucial premise for Whitehurst, while Hutton disputes it (in an uncanny allegiance with Blake)—a difference that helps to explain why Darwin draws on Whitehurst rather than Hutton for the initial stages of his geogony.

Whitehurst's treatise represents one of many textual genres synthesized—

along with visual and other material—within Darwin's aesthetic geology and his verse as a whole. All these genres—Augustan satire, natural history and travel narrative, theories of the earth, and descriptive poetry—contribute to the poem's cultural currency, as do its links to collection and visual culture. Though Darwin's use of Pope's machinery and tone as a backdrop for the poem may be primarily an appeal to nostalgia, the couplet was still the dominant English verse form in the 1790s, as Wordsworth was famously to complain at the end of the decade. All these factors contribute to the poem's efficacy in the transmission of scientific concepts. The reception of *The Botanic Garden* depends on a scientific culture in which the couplet form and the collection of specimens are cognate forms for the organization of knowledge. The poem is, perhaps, the last monument of a form of aesthetic experience that accommodated both science and poetry and judged both on one standard of taste. In *The Giants' Causeway*, the topographical genre subsumes aesthetic geology, indicating a major shift in the divisions of knowledge and the nature of aesthetic experience. After Darwin, the values associated with Romanticism increasingly drive aesthetic experience in the direction of "nature," both as a domain of physical experience and as a prototype for literary rhetoric and forms. Or rather, the idea of nature itself becomes identified with the external world as a result of the cultural and economic factors that generate Romantic aesthetics as well as the modern scientific disciplines.

**6**

# Wonders of the Peak

There seemed to be an embargo on every subject. At last she
recollected that she had been travelling, and they talked of
Matlock and Dove Dale with great perseverance.

—JANE AUSTEN, *Pride and Prejudice* (1813)

**G**eology plays a major role in *The Botanic Garden* not
only because, as Darwin suggests, a knowledge of botany is incomplete
without an understanding of soil and minerals, but also because the poem
was written in a region celebrated for its geology. Vibrant intellectual com-
munities in the environs of Birmingham and Derby also contributed to the
scientific richness of Darwin's thought. The Birmingham Lunar Society, especially, brought him into contact with applied science and industry as well
as local geology. He undertook extensive geological tours of the Peak District with geologist and fellow Lunar Society member John Whitehurst,
whose influence helps to account for the practical and local emphasis of
Darwin's poetry. His theory of metals resembles Whitehurst's and is based
on some of the same observations of Derbyshire mines and outcrops.[1] The

1. Both, for instance, cite the example of a local barium ore to illustrate the deposition of
metals by steam traveling up through fissures from the center of the earth (*BG* I.ii.398n.). Darwin also derives the image of "wooden wedges," as the Gnomes' instrument for performing
erosion, from a local practice for extracting millstones (524n.).

descriptions of Derbyshire landscapes embedded in Darwin's poetry also draw on descriptive poetry for their aesthetic and economic geology. The catastrophe that shatters the original globe, based partly on Whitehurst, takes shape in the commonplace tropes of topographical poetry: "Now rocks on rocks, in savage grandeur roll'd / Steep above steep, the blasted plains infold" (*Temple of Nature* I.46–47). Even such generic topographical motifs point to the importance of Darwin's location, because the Peak District was the subject of an especially large number of topographical poems. Beginning with Thomas Hobbes' *De Mirabilibus Pecci. Being the Wonders of the Peak in Darby-shire* (1636; first English edition 1678), these poems helped to create the categories through which Derbyshire, and consequently rugged English landscapes in general, were understood. After Darwin, their geological content became increasingly divorced from their aesthetics. But for well over a century the popular and accessible Peak District provided the most important local setting for the geological discourse I have called "aesthetic geology."

In his Derbyshire episodes, Darwin takes advantage of topographical poetry's increasingly scientific associations. *The Botanic Garden* draws on the entire tradition of scientific poetry extending back to James Thomson and the classical prototype of Lucretius, as well as the more recent georgic trend of incorporating applied science into local poems, as in Richard Jago's *Edgehill* (1767). By highlighting the Derbyshire landscape in his poem, Darwin also participates in a fashionable literary culture that grew up around his particular landscape. The Peak District (mostly in Derbyshire, but extending into Staffordshire and Cheshire) is the English place most strongly associated with "convulsions of nature" because of its striking geology: deep river gorges, pinnacles and other oddly weathered blocks of ancient limestone, and highly visible deformations and dislocations of the strata. These landforms and the aesthetic ideas associated with them helped to generate many kinds of writing on landscape and fueled the pursuit of specimen collection and the scenic tour. The texts and practices together register the cultural impact of aesthetic geology. The Peak District was one of the first geologically important sites to stimulate broader interest in rocks and landforms, and it did so in the context of its importance as an English landscape. The resulting profusion of verse and prose testifies to the complexity of the local landscape as a category with political, scientific, and aesthetic dimensions. The regional genealogy of Darwin's poem bears out this multiplicity. His precursors include Whitehurst's *Inquiry* (1778); a spectacular London pantomime, *The Wonders of Derbyshire* (1779); Joseph Wright of Derby's "glowing paintings" of the Derwent Valley (1780s); John Sargent's Rosicrucian verse drama, *The Mine* (1785); and James Pilkington's *View of the Present State of Derbyshire* (1789), which included a preview of *The Economy of Vegetation.*[2]

The importance of geological features in the Derbyshire landscape makes it an apt region for illustrating the aesthetic materialism associated with rocks and landforms. The *style* produced by literary descriptions of rocks and landforms in all registers is an essential vehicle of the scientific *content* of earth science, which erects a new disciplinary framework around the content and eventually abandons the style to local poems such as *The Giants' Causeway*. The list above provides just a hint of the range of Derbyshire materials that contributed to produce this style in the eighteenth century. The rocks of Derbyshire provide aesthetic resources for the vision of a representative landscape. Their "primitive" character authenticates a vision of nature that incorporates wildness and productivity, the distinctively English and archetypally global. The fusion of these categories—which now seems paradoxical—was crucial to industrialization and empire-building at a time when global resources seemed inexhaustible and English productivity seemed natural. The specific character and setting of Derbyshire's rocks allowed them to serve as evidence of planetary processes in Whitehurst, as models for landscape design in Whately, as images of wildness in poetry, as a topic of polite conversation in Austen, and, above all, as the basis of a tourist economy that attracted writers and created markets for their publications. In Richard Hamblyn's phrase, the Wonders of the Peak provided "the first structured tourist itinerary in Britain."[3]

The "primitive" appearance of the Peak's well-traveled and productive landscapes provided access to the fundamental properties of matter, most explicitly of interest to the naturalists' accounts.[4] In the aesthetic register, this appearance prompts expressions such as "savage grandeur," which designate the focal point of interest to poetry, the tour, and the art of landscape. The material basis of this consensus is evident in the economic interest common to all these accounts; Darwin, following Wright of Derby, makes Richard Arkwright's cotton mills a prominent feature of the landscape around Matlock (*BG* II.ii.85–87; cf. iv.175–77). But the architectural metaphors of Darwin's description ("Masson . . . / with misshapen turrets crests the Peak") also testify to an interest in the form and structure of mat-

---

2. *The Botanic Garden* mentions most of these figures explicitly. Stephen Daniels offers a concise account of their affiliations via Rosicrucianism and freemasonry in his chapter on Wright; *Fields of Vision*, 57–61.

3. Hamblyn, "Landscape and Contours of Knowledge," 16. Most of Hamblyn's numerous case studies come from the Peak District, and I have learned a great deal from his work.

4. Whitehurst and Moreton Gilks, for example, draw connections between the theory of matter and the accretion of calcium carbonate in the hot springs of Matlock Bath. Geologically interested travelers also benefited from unusually good roads in the region.

ter, in the geological paradox of form through deformation. Beginning with such topographical motifs, this chapter examines a variety of landscape-driven genres with particular reference to the Peak District, concluding with an account of the particular places favored by this literature, with their highly particularized geological features: the famous Peak Cavern near Castleton, Matlock Bath, and especially Dovedale. The Peak District as a whole represented wilderness for eighteenth-century England, as reflected in these authors' aestheticized estrangement from landforms they regarded as primeval and alien. In our current environmental crisis, we tend to suspect the idea of wilderness of occluding the political economy of land use.[5] While much eighteenth-century "improvement" aimed at such a mystification, the striking juxtaposition in the Peak District of early industry and barren, unclaimed cliffs prompted aesthetic responses that registered the impact of development on these landscapes. These early industrial landscapes are so instructive because they depended on, and in some cases maintained, a sustainable level of development at a time when no one could anticipate the degree of environmental devastation that has led to our polarized ideas of wilderness and development.

## Topographical Motifs

The Peak District becomes an exemplary English landscape because it provides a local basis for abstract concepts such as the sublime and picturesque and gives a specific shape to these conventions. Stock phrases such as "savage grandeur" and "rocks on rocks" have their literary genealogies, but they seem to cluster especially around the Peak District, and their aptness for describing this particular type of landscape seems to be the main reason they circulate so widely, becoming "commonplaces." Many critical accounts of the sublime and the picturesque in eighteenth-century England have emphasized the taste for Claude Lorrain and Salvator Rosa as a crucial factor in forming these categories. These painters became popular partly because their aesthetic values could be readily relocated to the English landscape. Anna Seward, for example, adopts Salvator for the Peak District in a letter that actually denies sublimity to Italian poetry and then moves to English poetry and landscape: "The first scenic objects that met my infant glance, and impressed me with their lonely and romantic grandeur, were the mountains, the rocks, and the vales of Derbyshire. Nursed in their bosom . . . poetic descriptions and penciled resemblances please me best when they take the Salvatorial style." Though "the Salvatorial" may seem indispensable as

---

5. See, for example, William Cronon, "The Trouble with Wilderness."

an aesthetic category, Seward goes on to insist that pastoral and descriptive poetry flourish "much better in our temperate climate, than on the banks of the Tyber."[6]

Seward's catalogue of the sublime, including Ossian as well as Salvator and the Peak District, at first seems incongruous and indiscriminate because of her efforts to articulate a revisionist British canon. In "Eyam" (1788), a poem from about the same time, Seward develops the childhood reminiscences in a spirit very similar to Wordsworth's in the early books of *The Prelude*, as when he addresses another Derwent (the major stream of the Peak District also bears this name). In Seward's poem, the Derwent "lave[s] / the soft, romantic vallies, high o'erpeer'd / By hills and rocks, in savage grandeur rear'd."[7] In the later eighteenth and early nineteenth centuries, such landscapes and the stock phrases associated with them become increasingly important in narratives of nativity. In Seward's account, the Peak is important as her native place, but also as a native *English* place, another concern picked up by Wordsworth and amplified especially in his *Guide to the Lakes*. The more widely known Peak landscapes initially provide the best match for Salvator's Italian landscapes and a challenge to aristocratic taste. The Peak is an indispensable model for landscapists such as Whately or Uvedale Price, who want to relocate the picturesque landscape to Britain, and in Seward this preference seems deliberately opposed to the aristocratic canon of the Grand Tour. Seward's letter illuminates the canonical status the Peak landscapes have attained by the later eighteenth century. Many travelers to the region expected that their taste would be formed or confirmed in the manner she describes.

Awareness of the landscape's economic functions often accompanies admiration for the Peak's "savage grandeur." Much poetry on the Peak shares Seward's lexicon and sensibility. In *The Botanic Garden* (on which Seward involuntarily collaborated), the confrontation with "savage grandeur," at one of its most popular sites, requires that such a sensibility come to terms with the presence of Richard Arkwright's cotton mills. Seward's Eyam is at some distance from Cromford Mills, but the mills are just downstream from scenic Matlock Gorge, and the situation loosely parallels Wordsworth's omission of a busy ironworks (among other things) in "Tintern Abbey." Wordsworth may well have been registering tacit disagreement with an established school of thought that regarded factories as capable of enhancing the aesthetic character of a scene. Whately, for example, describes a "truly

---

6. Seward, *Letters*, 3:131–35 (April 2, 1792). On the connections between painting and the taste for scenery see, for example, Samuel H. Monk, *Sublime*, chaps. 9–10.

7. Anna Seward, "Eyam," in *Poetical Works*, 3:1–4.

great and awful" cascade on the Wye, a few miles north of the abbey, as "more interesting and important" because of the loud and gloomy-looking iron forge situated there.[8] A "Late Religious Lady," writing in Seward's vein, seems to disagree more openly with such an aesthetic. Addressing the "craggy steeps" of Matlock, this poet bids "Farewel, where commerce now assumes a place, / where art infringes on wild nature's face."[9] Thomas Lowndes, more like Darwin, registers no contradiction between the "rich romantic vale" with its "misshapen rocks" and "Cromford's stately mills":

> Here Nature kindly o'er an Arkwright show'rs
> To swell our commerce, the mechanic powers
>     . . . teaching us still more,
> The great Mechanic of this World t'adore.[10]

While the earlier anonymous poem describes the relationship in terms of "infringement," seemingly a pre-echo of twentieth-century environmental sensibilities, in most of this local writing—as in Wright's "Arkwright's Cotton Mills by Night" (1782–83)—there is no contradiction between wildness and economic development.[11] The interest of these landscapes, rather than being divided between aesthetics and economics, is material in an integrated, historically specific sense.

Many topographical narratives also give attention to the economics of scenery. According to the guidebook of Sarah Murray Aust, enjoyment of Peak District scenery requires negotiations with an established tourist economy. She urges her readers to leave the guide behind in order to see "the most beautiful part" of Dovedale, warns them disgustedly about the tips expected by the servants at Chatsworth, and conscientiously evaluates eating and lodging establishments—all in the service of a highly sophisticated landscape aesthetics.[12] John Farey's *General View of the Agriculture and Minerals of Derbyshire* provides lists of "highly curious and interesting phenomena, of which Travellers may in future avail themselves: it is to such Valleys

8. *OMG*, 109. See also Marjorie Levinson, *Wordsworth's Great Period Poems*, 30–31.

9. "A Late Religious Lady's Farewell to Matlock Bath."

10. Lowndes, "An Account of Matlock Bath," in *Tracts in Prose and Verse*, 155–58. No date of composition is given, but the poem is likely to predate by many years the 1825 publication of *Tracts*.

11. Making a somewhat similar comparison, Heather Frey pits Ann Radcliffe's "proto-ecofeminist" *Observations during a Tour to the Lakes* (1795) against Wordsworth's less environmentally sensitive *Guide to the Lakes*; see "Defining the Self, Defiling the Countryside." While some dissent against the ideology of improvement can be observed before 1800, I would argue that there is insufficient environmental or economic basis in eighteenth- or even early nineteenth-century England for a sharp binary opposition between wildness and development.

12. Sarah Murray Aust, *Companion and Useful Guide*, 1:6–7.

also, that Mineralogists and Geologists must principally resort." The "great object" of this work is professedly "to state and shew the economic purposes, to which the various Mineral Products of these Districts are or may be applied." Farey duly notes the "highly picturesque and beautiful" scenery, along with such improvements as a paper mill operating at Matlock Bath, but corrects the spectacular descriptions offered by earlier geology: "The Valleys, contrary to Mr. Whitehurst's too confident expressions on the subject, have not the Strata in their bottoms broken up, and deep 'horrid chasms.' "[13] Farey's text is less a guidebook than a survey, sponsored by the Board of Agriculture, and similar works were soon undertaken under the auspices of the Geological Survey; but he inherits his approach from earlier picturesque-mineralogical tours. William Maton, for example, writing some years earlier on the West of England, observes the correlation of "savage grandeur" and mineral wealth with less geological fastidiousness: "Cornwall, a county of quite a primeval aspect with regard to the stratification of substances, contains an inexhaustible store of metal in its bowels. The bold mountains of Dartmoor and Mendip also are not without their metallic treasures, and here too nature appears in her rudest and wildest form. . . . Of sublime as well as decorated scenery the most striking specimens will be found."[14] As in Jago's *Edge-hill* and other topographical poems, sublime scenery can act as an economic incentive. The barren, primitive landscape in particular becomes a sign of mineral wealth.

Topographical poetry disseminates the architectural metaphors through which these sites are often understood. Aesthetics and geology merge in the architectural category of ruin, both in such creations as the "grotto" and in Miltonic descriptive poetry, where the "Ruins of Nature" retain a theological dimension. John Dyer, in *The Ruins of Rome* (1740), depicts Roman monuments as

> tow'ring aloft,—upon the glittering plain,
> Like broken rocks, a vast circumference!
> Rent palaces, crush'd columns, rifled moles,
> Fanes rolled on fanes, and tombs on bury'd tombs![15]

Though Dyer inverts the comparison—the ruins of time mimic the ruins of nature—his syntax and diction (echoing the stock phrase "rocks on rocks") testify to the influence of "ruin" as a descriptive category for the natural

13. Farey, *General View*, 1:72–73, viii, 472, 490.
14. Maton, *Observations*, 1:ix.
15. Dyer, *Ruins of Rome*, 19–20.

world. Dyer's archaeological use of these conventions strengthens the association underlying one of a growing number of poems on Dovedale in 1756:

> There, threatening mute amazement's fixed eye,
> Yon huge cathedral rock in ruin frowns,
> Whose rifted brow, of Gothic structure high,
> The blasted yew with shaggy foliage crowns.[16]

The poet, Anthony Champion, figures the stones themselves as a temple, associating the uninitiated with unbelievers: "Lo! where sequester'd deep from vulgar sight / Dove's wizard waters wild and broken roll" (3–4). "Vulgar" has its social connotations here as well as religious ones. As it will for Anna Seward, a generation later, the taste for this scenery counts as cultural capital, and the Ossianic sensibility invoked by Seward is already in place in Champion's appeal to the "Druids, eldest bards," to attend the Dove's "wizard waters." While specific Peak District landforms call forth an especially high proportion of architectural motifs (cf. Darwin's "misshapen turrets"), there are also substantive cultural reasons why the classicizing meditations of earlier poetry become naturalized in this region. The Wonders of the Peak, unlike most English places, have their own literary tradition extending back to the seventeenth century, and this tradition in turn is bound up with the economic history reflected in another set of topographical motifs: the Peak is a developed "wilderness," with spas, mining, and manufacturing industries flourishing relatively early in the eighteenth century.[17] Dovedale epitomizes the sort of wonder sought out by a class of tourists who had the means and the vigor to travel farther than Bath, to whom the Peak District offered fashionable resorts as well as a national and local brand of sublimity. The physical character of these landscapes, like their social character, conforms to established protocols for the sublime, but also shapes those protocols in new and specific ways. Champion's poem captures much of this local aesthetic, with its allusions to the sublime and the Gothic as well as tourism and local geology.

---

16. Anthony Champion, "To a Friend. The Scene Dovedale. 1756," 49. Champion, like Darwin, is the author of a long "philosophical poem," and some degree of interest in natural history is evident in such images as "Time's dumb file" (weathering) and "crystal tears" (stalactites).

17. Michael Drayton's *Polyolbion* (1613) contains probably the first literary mention of the Peak District, and its tradition is established in poems by Thomas Hobbes (*De Mirabilibus Pecci*, 1636, 1678) and Charles Cotton (*Wonders of the Peake*, 1681). On the early date of industrialization in the district, see Andy Wood, *Politics of Social Conflict*, esp. chap. 14.

## The Peak as Representative Landscape

The images of ruin in Champion's poem are less naturalistic than his images of stalactites and erosion, but they carry strong associations with geology: geology generally, according to Robert Aubin, became "popular with
topographical poets, who easily transferred their Day of Judgment melodramatics to a topic more in accord with the march of mind."[18] Geology itself,
of course, retained the diluvial melodramatics, if not the Armageddon, of
Burnet, well into the nineteenth century. John Whitehurst, like Champion
and the other poets, relies on the general category of ruin, which becomes a
literal component of his natural history. Aubin again provides a useful connection, linking the "the interest in natural curiosities like Wookey Hole
and the Peak Caves" to the "grotto fad" in gardens.[19] Whitehurst writes (*Inquiry*, 63):

> The mountains in Derbyshire, and the moorlands of Staffordshire appear to
> be so many heaps of ruins ... for, in the neighborhood of Ecton, Wetton,
> Dovedale ... the *strata* lie in the utmost confusion and disorder. They are
> broken, dislocated, and thrown into every possible direction, and their inte-
> rior parts are no less rude and romantic; for they universally abound with sub-
> terraneous caverns; and, in short, with every possible mark of violence.

The analogy to grottoes may play a role here, but the placement of "for"
near the end of the excerpt draws attention to the causal aspect of Whitehurst's explanation. The caverns constitute *evidence* that the landscape is indeed "romantic." The romantic quality of these local landscapes, in turn,
provides the central evidence for Whitehurst's philosophical romance of
"subterraneous convulsions": they are a proof, he later claims, of the effects
of these convulsions, and prove "likewise that mountains are not primary
productions of nature, but of a very distant period of time from the creation
of the world" (189). In basing his theory of planetary processes on these
local phenomena, Whitehurst shows his limitations as a local geologist, as
modern geologists have been eager to point out. Farey, a generation later,
already strives to correct Whitehurst's excesses of style, archly noting that
"the surface of Derbyshire is much diversified, yet far less enormously so,

18. Aubin, *Topographical Poetry in XVIII-Century England*, 109.
19. Idem, "Grottoes, Geology, and the Gothic Revival," 410.

than many writers have represented it."[20] Whitehurst's fascination with the diversity of this landscape, however, may be vindicated by observing that this fascination was widely shared among his contemporaries, most of them neither geologists nor Derbyshire natives. The long quotation above provides an authoritative explanation for the general appeal of these landscapes, one facet of which was precisely that they seemed to provide a glimpse of the "original state of the earth."

One is tempted to speculate that this geological fascination was among the commonplaces exchanged by Elizabeth Bennet and Fitzwilliam Darcy during their moment of embarrassment at Pemberley. Austen herself may have been immune to this fascination, but the Peak District is nonetheless essential to the plot of *Pride and Prejudice*. Austen ironically acknowledges the existence of an ample travel literature on the region chosen by Elizabeth's aunt and uncle as the destination of their tour. She disavows the object that might well have been expected by readers of Radcliffe, Scott, or tours in verse and prose, stating that the landscapes in question are "sufficiently known": "It is not the object of this work to give a description of Derbyshire. . . . A small part of Derbyshire is all the present concern."[21] Pemberley (Darcy's estate) is the "small part" of Derbyshire that sets the scene for the novel's climax, the reconciliation that paves the way for Elizabeth's engagement. Though the Lake District had been the original choice of destination for their holiday, Elizabeth's aunt in particular has her reasons for seeking out the Peak: "The town where she had formerly passed some years of her life, and where they were now to spend a few days, was probably as great an object of her curiosity as all the celebrated beauties of Matlock, Chatsworth, Dovedale, or the Peak." Austen's tone here and an earlier reference to this period suggest that Mrs. Gardiner herself may have been wooed in the fictitious village of Lambton. Derbyshire certainly provides an apt setting for the sudden intensification of Elizabeth's courtship. Her outdoor activities there accentuate her unfashionable tan, which readers are invited to perceive, with Darcy, as a glow of health that signals her eligibility.[22]

20. Farey, *General View*, 1:4. In addition to rejecting Whitehurst's epithet "horrid," Farey challenges Whitehurst's science, rejecting his notion of a volcanic origin for toadstone (274–76; cf. 473n.). Portions of the long Whitehurst quotation above have appeared in previous chapters.

21. Austen, *Complete Novels*, 375. Austen had traveled in Derbyshire in 1806. *Jane Eyre*'s journey is similarly based on Charlotte Brontë's visit to the Peak.

22. Ibid., 375, 392. I am indebted here to Amy Mae King's brilliant work on the use of language from natural history to describe the social and sexual maturation of the girl. See her *Bloom*, esp. chap. 3. See also, in Austen, Elizabeth's return from her muddy walk early in the novel, 251. On her aunt's youthful residence in Derbyshire, see 317. The chronology here and

Pemberley's grounds also share the Peak's proverbial power to surprise: "Its banks were neither formal nor falsely adorned. Elizabeth was delighted. She had never seen a place for which nature had done more, or where natural beauty had been so little counteracted by an awkward taste. . . . at that moment she felt that to be mistress of Pemberley might be something!"[23] Pemberley's extensive woods signal its natural "character" by combining timber value with the surprising effect of picturesque vistas suddenly appearing through the trees.

Pemberley's power to surprise also testifies to the cultural significance of the Peak landscapes. Elizabeth inwardly resists the Gardiners' plan of going to Derbyshire, because she knows Pemberley is there, and she objects explicitly to visiting Pemberley itself. She is also disappointed with the whole idea of Derbyshire on aesthetic grounds because "she had set her heart on seeing the Lakes." A trip to the Lakes had been the original plan, and Austen marks Elizabeth's "ready and grateful" acceptance of the Gardiners' initial invitation with strong dramatic irony: "what delight! . . . Adieu to disappointment and spleen. What are men to rocks and mountains? Oh! what hours of transport we shall spend!" Forced to resign herself to the change of destination, Elizabeth decides that she "may enter his county with impunity, and rob it of a few petrified spars without his perceiving [her]."[24] While Darcy's presence certainly helps to dampen her enthusiasm for "rocks and mountains," the example of specimen-hunting provides a clue to cultural factors underlying this disappointment. The collection of mineral and fossil specimens is a long-established practice of tourists in the Peak, while the geology of the Lake District is less favorable to it; the practice seems trivial to Elizabeth here. The wonders of the Peak themselves—Matlock and Dovedale are mentioned twice in this episode—do not seem to make an impression on her. Perhaps these places are too fashionable—the Lakes were becoming popular by 1813, but they still carried stronger touches of the wild and the exotic. This fashionable character and Elizabeth's disappointment, however, underscore the Peak's importance to Austen's plot. First, the reduced expectations with which she enters on the trip make the unsuspected and unassuming pleasures of Pemberley all the more striking. Second, the very fact that Matlock and Dovedale are topics of polite conversation makes it possible for Elizabeth to converse with Darcy in a socially

---

the tone of all the previous references to Lambton make it at least a reasonable inference that she met her future husband during her residence there.

23. Austen, *Complete Novels*, 376. By remarking that Pemberley's discreet improvements have made its grounds "in character with the general air of the scene" (381), Elizabeth signals her debt to picturesque theory.

24. Ibid., 375, 374, 324, 375.

acceptable way. Thus, they speak "with great perseverance" of the places that ultimately provide the catalyst for their engagement.

Wordsworth's *Guide to the Lakes*, first published three years before Austen's novel, contains lengthy discussions of disappointment concerning landscapes. In Austen, the Peak appears as an ersatz Lake District. Wordsworth fears that his mountains will be approached as ersatz Alps. His discussion of disappointment brings out the Englishness of Pemberley and of Austen's pedagogical narrative. The Peak disappoints Elizabeth because she had expected a "transport" associated with the remoteness and the larger scale of the Lake District. Wordsworth treats hypothetical cases in which visitors bring similarly disproportionate expectations to the Lake District itself: "A stranger to mountain imagery naturally looks out for sublimity in every object that admits of it; and is almost always disappointed" (*Guide*, 99). Both districts function as representative national landscapes. The Peak, because of its physical and cultural peculiarities, represents both decorous wildness and economic solidity for Austen and her many predecessors, while the Lake District represents a somewhat different set of values, especially in Wordsworth. The topos of disappointment, however, links the two accounts as preludes to the recognition of a quietly compelling, authentic English nature that rewards minute discrimination.

Wordsworth offers two strategies for realizing the nationalist project most explicitly stated near the end of the *Guide*: "my object is to reconcile a Briton to the scenery of his own country" (106). Both strategies might be termed practical or applied aesthetics, and the first is psychological: "It is upon the *mind* which a traveller brings along with him," Wordsworth insists, "that his acquisitions, whether of pleasure or profit, must principally depend" (98). Elizabeth's mind, by contrast (both to Wordsworth's norm and to the earlier Peak District narratives), is too exclusively attuned to a rarefied aesthetic pleasure—until she is confronted with the bounty of Pemberley. Her initial disappointment proceeds from an absence of the "disposition to be pleased" that Wordsworth here calls "the best guide to which in matters of taste we can entrust ourselves." The discussion that follows can equally be applied to Elizabeth's case: "Nothing is more injurious to genuine feeling than hastily and ungraciously depreciating the face of one country by comparing it with that of another." There follows a sustained comparison between the Alps and the Lake District, which is itself curiously self-deprecating, pitting "gigantic torrents" against "brilliancy of water," "irresistible violence" against "such impetuosity as [English streams] possess," and "ravages of the elements" against "unimpressive" but "grateful" "stability and permanence" (98–99). By emphasizing that Elizabeth anticipates the violent sort of "transport," Austen captures precisely the "injurious" type of expectation Wordsworth seeks to prevent. His *Guide* strives to "correct" more enthusiastic narratives of Lakeland scenery. Wordsworth also offers bodily tactics toward this end: "It is not likely that a mountain will be ascended without disappointment, if a wide range of prospect be the object," unless it is ascended either at sunset or at sunrise, when the vapors open to reveal the "whole region" (97–98). These lessons are staged in a somewhat different way in *The Prelude*, in the contrast between the famous disappointment of the Simplon Pass and the tranquil sublime of Snowdon, ascended strategically before sunrise. The *Guide* seems to confirm that one cannot expect or "look out" for sublimity.

The crowning touch of this tactical discussion, however, guarantees a certain measure of sublimity through physical calculations. Wordsworth offers this empirical guarantee under the guise of an antiempiricist argument:

> As far as sublimity is dependent upon absolute bulk and height . . . it is obvious that there can be no rivalship. But a short residence among the British mountains will furnish abundant proof that after a certain point of elevation, viz. that which allows of compact and fleecy clouds settling upon, or sweeping over, the summits, the sense of sublimity depends more upon the form and relation of objects to each other than upon their actual magnitude; and that an elevation of 3,000 feet is sufficient. (102)

The ostensibly objective standard here is obviously both national and local: the highest peak in the Lake District—and in England—happens to be just over 3,100 feet. There is some merit in the meteorological basis for Wordsworth's figure, but the chief attraction of these peaks seems to be that they are native, both to the country and to Wordsworth himself, just as the Peak District was for Anna Seward. The earlier objection to nonnative plants (which always "remind us that they owe their existence to our hands") illuminates these values: "surely there is not a single spot that would not have, if well managed, sufficient dignity to support itself, unaided by the productions of other climates" (83). By harmonizing with the local landscape, the Pemberley plantations exhibit precisely this native "dignity," counteracting Elizabeth's initial appetite for "transport."

These parallels, however, belie an essential local difference, one that demonstrates the aesthetic importance of the local. The slate peaks in the Lake District—Skiddaw, Saddleback—are typically smooth and rounded; the volcanic central peaks are more rugged, but still more uniform than those of the Peak District.[25] A partly eroded fringe of younger limestone

25. There are no "petrified spars" to collect in the central Lake District—an absence suggesting higher seriousness to Elizabeth Bennet—because of the high temperatures involved in

surrounds these peaks. The ancient, denuded limestone plateau of the White Peak, by contrast, is penetrated by igneous intrusions and broken by faults that make the underlying strata visible, accounting for the higher incidence of minerals and mines. In addition, fast-moving streams such as the Derwent supply economically vital water power, less available in the Lake District. The Peak District presents a national disorder, as against the order suggested by Wordsworth's emphasis on proportion, transparency, and tranquility, and also against the national order insisted on by historical critics as the master trope of all literary landscapes. The "primitive" or disorderly karst topography of the Peak landscapes and their water power accompany increasing economic productivity. In *Pride and Prejudice*, Pemberley's grounds offer a stylized forum for this healthy disorder: the woods are wild and "luxuriant," but their occasional openings offer startling glimpses of orderly abundance; Elizabeth's wholesome tan stands out against the artificial accomplishments of other young women; and the primitive, in short, becomes the authentically natural. Pemberley might have been located in the Lake District, where there are a few old family seats (such as Rydal), but it might also have been located much nearer to London. Derbyshire offers both a long history of development—the gentry are not mostly, as Wordsworth says of the Lake District, "alien improvers"—and a significant residue of wildness. Many Peak District narratives connect these representative working landscapes explicitly to their geology. For most of the writers covered so far, the primitive is not compromised by mining and manufacturing, but literally complements them, since the dislocation of the strata has created the discontinuities where lead and other minerals are deposited. The nationally representative character depends, more fully than that of Wordsworth's landscapes, on a relation between wildness and cultivation, on a productive disorder.

## The Local and the National

John Guillory's political reading of topographical poetry helps to illuminate the Peak District texts, but it can be usefully complicated by restoring landscape, as a set of real physical conditions, to the history of the genre. Guillory argues that the acceleration of the rural enclosure process in the eighteenth century fundamentally changed topographical poetry: "The imposition of the grid of property upon the land . . . makes possible the *descrip-*

---

forming both the metasedimentary and volcanic peaks. Both sets of peaks were "slate" in the geological parlance of that time.

*tion* of landscape as a perspective upon the social order itself. Such a perspective represents the interests of property, but represents these interests as
*universal*, and therefore as capable of expression by recourse to the rhetorical commonplace." Guillory contextualizes, through a notion of emergent
middle-class culture, the dynamic relationship between aesthetics and rhetoric within topographical poetry in the later eighteenth century. By pointing out how this relationship is repeatedly reconfigured in response to social
conditions, Guillory diverges from Dr. Johnson's account of the genre
through a set of fixed conventions. The specificity of Guillory's historical
moments remains somewhat ambiguous, however, because he follows John
Barrell in linking topographical poetry across the century to "national
order" and pictorialism. Descriptions of prospects, as Guillory puts it, "depict the view as a representative image of the national order," an order that
operates by "depopulating [the landscape] of many of its residents."[26] Guillory draws three broad assumptions concerning the social space treated by
descriptive poetry from Barrell's influential analysis of estate-poems: landscape is strictly pictorial, representations of "land" stand for "the nation,"
and landscape as both aesthetic and political category is organized around
property divisions. Guillory lucidly traces a shift in rhetorical emphasis in
topographical poems from Denham to Barbauld, but his narrative presumes
the continued stability of these assumptions. Changes in the *use* of landscapes weaken these assumptions: poets increasingly see the land as foreground, as locally specific and often *dis*orderly, as is particularly the case
with the Peak. Far from invoking the landscape itself as a national and pictorial "totality" serving to regulate this chaos, much of the topographical literature on the Peak displays an aesthetic materialism, a fascination with
physical details and a landscape that resists organization.

Local poetry registers social change, not only by means of its rhetorical
repertoire but by approaching landscape itself in new ways, which is clearer
when we observe how many of these poems venture beyond the confines of
the park. Later eighteenth-century descriptions complicate the Guillory-
Barrell paradigm by presenting working landscapes (Barrell himself ultimately presents John Clare's departure from landscape conventions in this
light). In discussing Jago's *Edge-hill* (1767), Barrell notes that such "hill
poems" are always about the prospect from the summit and never about the
hills themselves.[27] It is true that Jago, even more than James Thomson, devotes much visually oriented description to the parks and estates found in

26. Guillory, "English Common Place," 6, 3, 4.
27. John Barrell, *Idea of Landscape*, 21. In the following discussion I refer particularly to
Jago's *Edge-hill*, 72–75.

the prospect, duly footnoting each with the name of the owner. Jago's poem, however, is also very much interested in what Barrell calls "terrain," as opposed to "landscape": he includes lavish accounts of minerals and metal production, and other aspects of land use. In terms of Guillory's genealogy, this interest would be a throwback to the earlier classicizing mode, with Jago appealing to Virgil's *Georgics* as a sort of antipastoral authority. Many of the eighteenth century's poems on "unpoetic" subjects are, like Jago's, locodescriptive. An interest in locally specific minerals or landforms, in productive landscapes, is compatible with the cultural legitimation of acquired wealth that Guillory describes; as a consequence, landscape becomes a less exclusively pictorial category. In accounts of the Peak District, with its unique combination of industry and scenery, the chaos of rude rocks and raw materials is held consciously in tension with the pictorial national order inherited from the estate poem.

Though landscape functions broadly as a domesticating category, it operates on a continuum between wildness and cultivation and across a range of social and geographic contexts. The way in which a topographical poem carries out this domestication varies with the position of its locale along this continuum. The cultural content of the genre, as tied to gentlemen's estates ("the nation as a landscape or park"), differs essentially from the content produced by its application to sites like Dovedale or the Giant's Causeway. In a temperate climate, wildness, as an absence of cultivation, depends chiefly on geology: rugged topography, bad soil, or both inhibit most forms of development. If rocks are seen as matter in its simplest form, and matter as the sphere of human ordering, then rockiness marks the resistance without which ordering is inconceivable. The category of wild landscape, like the sculpted landscape of the park or garden, is initially relocated from Italy, especially the Alps, with their more spectacular geology. But while the "Salvatorial" strain remains a factor in the reconstructed landscape—the later picturesque of Price and Richard Payne Knight, for example—increasing attention to wild British landscapes leads them to be distinguished from these prototypes. Jago's "native British ore," in a landscape assertively re-populated with labor, industry, and other unaesthetic phenomena, is an early instance of this tendency (the wilderness here is the "rude unsightly form" of the heath around Birmingham, with its secret freight of ore). The resistance to Alpine prototypes continues to inform such nineteenth-century projects as Wordsworth's Lake District pedagogy and William Drummond's agenda of native sublimity in *The Giants' Causeway*.

The accessibility of "wild" landscapes in Derbyshire makes the Peak an early, influential stimulus to changing literary descriptions of landscape. From the early eighteenth century, the Peak District literature displays the importance of the wild, the local, and the practical, sometimes working with and sometimes against the interest in social and national order. These descriptions provide prototypes for accounts of more remote landscapes—the Lake District, County Antrim, the Hebrides—that proliferate later in the century. The accounts of places like Dovedale and Matlock High Tor— partly because they are fashionable and make viable subjects for publication—help to establish the aesthetic importance of the local and of physical detail, which causes the focus of descriptive poetry to shift from the horizon to the foreground in the course of the century. On the one hand, these sites are unimproved, and break the pattern Guillory identifies by being dissociated from the context of land ownership; the most spectacular sites functioned as de facto public parkland, much like the Cumbrian peaks, until this function was formalized by the state in the twentieth century. On the other hand, they exist side by side with the cotton mills, the baths, and the lead mines, so the poetry is often required to place wildness in some relation to productivity. These conditions distinguish the Peak as a topographical locale, as does its long tradition of local poetry.

The Peak District, as a literary site, has a complex role in cultural history. It is a prominent domestic example of wildness, with all the paradoxes this entails. Peak landscapes are "domestic" in the sense that their character is unique, narrowly local, but also because they are English, or economically productive, or through some relation of these factors. Pemberley signifies in a national context as a healthy deviation from an artificial national order. In the Lowndes poem quoted earlier, the unique coexistence of "misshapen crags" and Arkwright's mill generates a higher species of devotion, and Matlock becomes a sort of deist Eden. Dovedale, in Anthony Champion's 1756 poem, generates a purely individual aesthetic response that psychologically mimics the site's idiosyncratic wildness. In Whitehurst, on the other hand, the local "ruins" open out onto planetary processes; while Whately, working on an equally large scale, makes scenes like Dovedale represent pure nature. All these representations bring fundamental questions to bear on the contested notion of wild landscapes—economic questions, such as how these relate to property, religious and scientific questions about their origin, aesthetic questions about their physical appearance, and social questions about productivity and landscape.[28]

---

28. William Drummond explains how landscapes "give rise" to poetry and how the genre, in turn, influences the perception of landscape (*Clontarf*, xii). In his history of the topographical genre, Robert Aubin shows that "poets were gradually encouraged to pass from a generalized account of an ideal and selected nature to something more exact and detailed" (*Topographical Poetry in XVIII-Century England*, 63; see further 96–97). See also George Dyer's "On the Use of Topography in Poetry," which recommends the "romantic" pursuit of "grotesque, disrupted rocks" (*Poetics*, 2:112–13; cf. 119).

## Wonders of the Peak

The physical character of individual landscapes gives rise to categories of the wild, the local, and the practical, which function in the topographical genre(s) alongside the categories of the social, the national, and the pictorial proposed by Guillory. The literary description of places must account for foreground as well as background, fragments as well as totalities; and the former two factors seem especially important in descriptions of the Peak District sites. The fractured, "primitive," aberrant character of these sites allows them to serve as aesthetic resources for a wide array of literary projects; such diverse figures as the "Late Religious Lady," Austen, and Whitehurst exemplify this range, which accurately represents the breadth of literary culture in the mid to late eighteenth century. The literary context for descriptions in verse includes geology, travel narrative, and garden design, as the case of Derbyshire makes particularly clear. This complex literary context mirrors the complexity of landscape as a category. For these reasons, I conclude my illustrations of aesthetic geology by reading a set of literary descriptions of rocky Peak District landscapes, grouped around three of the proverbial "Wonders of the Peak": the Devil's (or Peak) Cave, Matlock Bath, and especially Dovedale.[29] The descriptions of these places employ a shared set of social and aesthetic categories: wildness, ruin, sublimity; the native, the local, and the practical; property, productivity, and the nation; natural history and the earth; and finally materiality itself. While the accounts in question negotiate these large categories in various ways, collectively they establish the aesthetic importance of physical and local conditions by modeling a physically situated aesthetic experience. Landscapes are at once social and geological, and the fascination of the Peak survives partly because its remarkable geological features remain constant amidst literary and cultural change.

One of the Peak's Wonders appeared earlier as the centerpiece of my Derbyshire discussion in the Blake chapter (see especially *Jerusalem* 25): the Peak Cavern just south of Castleton, formerly known as the Devil's Arse, the Devil's Bottom, and the Devil's Cave, the latter being the most common in print by the late eighteenth century. Unlike many of the "show caves" now attracting tourists in the area, this cavern is natural and not a disused mine.

29. Of these three "wonders," only the Devil's Cave (originally the "Devil's Arse") appears in the list of seven "Wonders of the Peak" established in Thomas Hobbes' *Wonders of the Peak* and embellished in Cotton's poem of the same name. The remaining six are Chatsworth, Mam Tor, Eldon Hole, Tydes-well, St. Anne's Well, and Poole's Hole. The Devil's Cave, Matlock, and Dovedale are probably the three most widely discussed sites in the later eighteenth and early nineteenth centuries.

The experience of the cavern, however, differs from that of more accessible natural sites in that it must be seen with the help of a guide. Three representative accounts from the last three decades of the eighteenth century all devote considerable attention to the guide and his manner of packaging the experience. The earliest of these is a letter published in the *Gentleman's Magazine* in 1772 by a practical traveler, skeptical of aesthetic illusions, who nonetheless documents the cave's success as an attraction. Sarah Murray Aust's account is professional, in the sense that she includes it in her *Companion and Useful Guide to the Beauties of Scotland;* here the experience is systematically aestheticized, creating a revealing contrast to the traveler's matter-of-fact descriptions. The third account, by the French naturalist Barthélemy Faujas de Saint-Fond, presents conflicting responses of scientific precision and rationalized wonder. All three accounts advise the reader in detail on how to navigate safely within the cavern, making it clear that its dangers are not just a matter of rhetorical convention. The pragmatic strand of each narrative, however, draws on its own distinct symbolic register. These differences also inform the handling of other pragmatic concerns, such as the poor people living in the mouth of the cave—accounts of whom apparently attracted Blake's interest—and the machinations of the guide and his assistants.

The narratives of the *Gentleman's* correspondent, Murray Aust, and Faujas de Saint-Fond all confront the challenge of integrating accounts of human labor into aesthetic discourse concerning a natural wonder. Two human spectacles in particular draw the attention of all three: the thread- or ropemakers living in the mouth of the cave (see Fig. 6.1) and the singing children dispatched by the guide to create a Gothic illusion in the cavern. The letter-writer, James Ferguson, seems little moved by the cavern's multiple chambers themselves, but reports being "agreeably surprized by a melodious singing" issuing from an opening called "The Chancel," sixty feet off the ground in one of the interior caves. The surprise seems no less agreeable for his having spied several boys with candles climbing the wall before he entered the cavern.[30] Sarah Murray Aust, on the other hand, embraces this aesthetic illusion as eagerly as the vastness of the cavern and the guide's expert organization of the experience. She aims at maximizing aesthetic illusion rather than demystifying it. Her recommendations for visitors to the cave, accordingly, display a surprising range: "if females, dry shoes, stockings, and petticoats will be requisite," but also (regardless of sex) "snuff and tobacco, which will be grateful offerings to the old witch-looking beings,

---

30. James Ferguson, "Mr. Ferguson's Description," 518. "The roof of this place," he writes, is "all of solid rock, and looks dreadful over head, because it has nothing but the natural side-walls to support it."

6.1 "Entrance to the Peak Cavern," from Thomas Moule, *The Landscape Album: or, Great Britain Illustrated* (1832). Engraving by E. Finden from a drawing by William Westall. Courtesy of Special Collections, Art and Architecture Library, Washington University, St. Louis, Mo.

spinning in the dark mouth of the cave." The "beings" meant here are clearly the "poor people" who carry on a "a pack thread manufactory" in the mouth of the cave, in Ferguson's rather different vocabulary. Although Murray Aust points out that the cost of a tour will depend on "the number of guides, men, women, and singing children engaged," she relates her experience of "The Chancel" in purely Gothic terms, comparing the cavern to Westminster Abbey.[31]

Faujas de Saint-Fond's 1784 account, though assertively scientific, shares Murray Aust's willing suspension of disbelief vis-à-vis the guide's aesthetic illusions. The French geologist, in fact, has more in common with the guidebook-author than with the amateur "philosopher" (Ferguson) who insists on precise measurements of the cave and reports that he has braved the most dangerous passages unaffected. Murray Aust and Faujas de Saint-Fond also shared the same itinerary, having taken in the Wonders of the Peak on

31. Murray Aust, *Companion and Useful Guide*, 1:8–9, 10, 12. This report is largely based on her initial visit in 1790. Some sources refer to Aust by her maiden name, Murray, because she published the first edition of her *Companion* in 1799, before her marriage.

the road between London and the Scottish Highlands. While Murray Aust's account of Staffa, in the Hebrides, includes historical as well as geological reflections, Faujas de Saint-Fond traveled there specifically to test his theory that columnar basalt was volcanic in origin, and his account of the Peak District, too, includes a long technical chapter on the classification of rocks. Despite these substantive differences, the sentiments of the scientific and the picturesque traveler often coincide as much as their itineraries. Faujas de Saint-Fond shows a keen interest in social conditions, which sometimes mirror the "dismal" landscapes around Castleton, as illustrated by the limekiln workers, who "scoop out their dwellings among heaps of cinders and lime-refuse." Compared to these workers, at least, the pack-thread manufacturers living in the mouth of the Devil's Cave seem comfortable; by 1784 this industry seems to have expanded to rope, lace, and tape, and many of the workers live in two large houses built entirely inside the cavern. The "pretty young girls" who work in the lace workshop, however, fail to convince the Frenchman that a certain stalactite in the cavern resembles a "leg of pork": "the more we examined this pretended 'leg' the more did we find it to resemble an object which young girls are scarcely allowed to examine, and still less to allow others to examine." While maintaining the truth of his own associations, the account of this cave finds Faujas de Saint-Fond subscribing increasingly to the aesthetic illusions it presents. He shares Murray Aust's idea of the underground stream as the river Styx: "it is impossible . . . not to see in this scene a picture of the passage of the dead in the fatal bark."[32]

Scientific incredulity struggles in this episode against the aesthetic character of the place and the talents of the guide. While noting that the cavern is "vulgarly known by the name of 'The Devil's Bottom,'" Faujas de Saint-Fond also relates with due gravity that "this grotto, regarded from all time as the chief of the seven wonders of Derbyshire, has been celebrated by several poets." Though he chastises the poets for their inaccuracy, he concedes that "it excites real astonishment to find such extensive natural excavations in the centre of the hardest rock, and one is lost in conjecturing what has become of the materials which once must have occupied those vast empty spaces." A geologist entering a large limestone cave gradually abandons his scientific agenda for "real astonishment" at a phenomenon that seems "as perfectly made as if it had been a work of art." Faujas de Saint-Fond seems to give himself up to the experience after losing his hat while trying to iden-

32. Barthélemy Faujas de Saint-Fond, *Journey through England and Scotland*, 2:288–89, 320, 323. Faujas comments extensively on social conditions in this area (see also 272, 277–78, 282). Wood remarks that "to the gentry outsider, the Peak Country and its plebeian people were . . . defined within the interlocking of geological and social difference" (*Politics of Social Conflict*, 7).

tify the composition of the exceedingly low roof over the "Styx." He visits the same Gothic and classical themes as Murray Aust, expanding on them considerably, and cheerfully suspends his disbelief concerning the guide and the "little tricks of his trade." During the musical episode, for example, he charts the elaborate distraction orchestrated by the guide to prevent the party from seeing the singers as they ascend. The singers, appearing in this account as white-robed "chantresses" singing an "air in some words of Shakespeare's," make a "very vivid" and "very pleasant impression"—even though it is clear that the guide is "playing off his grand machinery for our entertainment."[33] The ultimate effect, as in the other accounts, is tied to the depth and dimensions of the cavern, but the cultural predisposition to see a "natural curiosity" as a work of art is also magnified by the various human contexts that impinge on the scene and make it a spectacle.

The renowned landscape painter and scenic designer Philippe-Jacques de Loutherbourg capitalized on the geological spectacles he witnessed on a sketching tour of the Peak by translating them to the London stage. Advertisements for the 1779 pantomime *The Wonders of Derbyshire* announced: "All the Scenery, Machinery, &c. designed by De Loutherbourg and executed under his direction. Nothing under full price will be taken." The pantomime's extraordinarily long and profitable run prompted "A Derbyshire Man" to publish, under that pseudonym, a guide for playgoers who had "expressed a desire for a short Account of those celebrated views to refer to during the Representation" (Fig. 6.2).[34] The guide, which includes portions of the Sheridan libretto, comments extensively on such scenes as the "prodigious cave" at Castleton, noting that the cave-dwellers "live by rope-spinning, and shewing the place."[35] Numerous reviews praised the "sublime stile of the paintings" and scenery, but at least one of them challenged the accuracy of the representations: "we suppose Mr. de Loutherbourg was not at Matlock or Dovedale; for we imagine they could not have escaped his pencil." The same critic, evidently priding himself on his connoisseurship as a tourist, endorses de Loutherbourg's staging of the Devil's Cave, but com-

---

33. Faujas de Saint-Fond, *Journey through England and Scotland*, 2:315–16, 324, 329, 326.

34. Charles Beecher Hogan, ed., *London Stage*, part V, 1:227; "Advertisement," in *Account of the Wonders of Derbyshire*, unpaginated. The *London Evening Post* exulted that the "noble, picturesque scenery [was] perhaps the finest that was ever exhibited on the English stage" (quoted in Ralph G. Allen, "Wonders of Derbyshire," 64).

35. *Account of the Wonders*, 14. A surviving design of the "cave scene" shows that de Loutherburg made the rope-spinners' huts a prominent feature of the scene (Stephen Daniels, *Joseph Wright*, 66). Though the rope-spinning industry died out before World War II, the present guides have demonstrated that the entrepreneurial spirit of the cave lives on by reinstituting the name "Devil's Arse" after a lapse of more than 200 years. In 2001 the cave director reported that ticket sales had risen by 30 percent within a few months of the name change. See http://www.showcaves.com/english/gb/showcaves/Peak.html.

# AN

# ACCOUNT

## OF THE

# WONDERS

## OF

# DERBYSHIRE,

### AS INTRODUCED IN THE

# *Pantomime Entertainment*

### AT THE

## THEATRE-ROYAL, DRURY-LANE.

LONDON:

Printed by G. BIGG, Denmark-Court ;
AND SOLD BY
W. RANDALL, (No. 13) Catherine-Street, Strand.

1779.

6.2   Title page of *An Account of the Wonders of Derbyshire* (1779). Reproduced by permission of the Huntington Library, San Marino, Calif.

ments that "the Caverns themselves cannot be represented. . . . Castleton is among the few Wonder's [*sic*] which may be described and painted, but after all exceed the Traveller's expectation."[36] The *Account* of the "Derbyshire Man" singles out the first three geological scenes of the pantomime for special mention:

> Matlock, Dovedale, and the Lead-Mines, although not ranked among the Wonders of Derbyshire, are places so remarkable for beauty and public resort, that it would have been arraigning the taste of Mr. de Loutherbourgh, had he not conducted the Dramatis Personae through such delightful scenes. . . . the mines . . . are famous, not only for their romantic situation, and the riches they produce, but likewise for those petrefactions, and mineral substances, from which are made the beautiful vases, urns, &c. &c. we so often see and admire.[37]

The concluding reference to mineral products locates a more concrete form (besides entertainment) in which the geology of the Peak shaped metropolitan culture. Richard Hamblyn has reconstructed the region's souvenir and specimen industry to demonstrate the "central role played by small-scale natural history cabinets," by guidebook authors and by mineral dealers—from the scientific entrepreneurs supplying the London trade to the rural poor, often women and children, selling souvenirs in situ—"in the advancement of natural knowledge as a whole."[38]

## Matlock Bath

The Devil's Cave narratives show that the "primitive" extends, as an aesthetic category, to Shakespeare and "immense Gothic churches." The category is also applied to rocks on the surface of the ground near Matlock. A different set of connotations attaches to the primitive in this landscape,

---

36. This account (from the *St. James Chronicle*) is quoted in Allen, "Wonders of Derbyshire," 59–60.

37. *Account of the Wonders*, 11.

38. Hamblyn, "Landscape and Contours of Knowledge," 23. In the quoted passage, Hamblyn refers to the vernacular guidebooks—mainly for the cabinet rather than the field—of J. R. Forster, John Lettsom, and John Hill. In a chapter on mineral dealers, he notes: "That a 'scenic' term [Derbyshire Diamond] should emerge in direct competition with an already established mineralogical term, indicates not only the strongly divergent quality of the various observational and naming practices that were taken to the landscape during the eighteenth century, but also the nature of the impact of tourism in the provinces" (90). In a chapter on mineral souvenirs, he relies on Faujas de Saint-Fond's *Journey through England and Scotland* as the only record of the activities of many of these "'low' cultural practitioners": "Most of his samples were bought from commercial dealers, but many of his contacts were guides and other workers—such as female souvenir-sellers—in the then fast-growing tourist industry" (126).

6.3   "Matlock High Tor, Derbyshire," from James Storer, *Antiquarian and Topographical Cabinet* (1807–11). Engraving by J. Greig from a drawing by G. H. Moore. Courtesy of Ellis Library, University of Missouri–Columbia.

partly because of the proximity of Arkwright's cotton mills. The chief industry in Matlock was the bath, as in Buxton, near the Devil's Cave. Murray Aust records the cost of meals in the "principal houses" of both, but notes that liquor "must be paid for besides, and procured by your own servant."[39] The "rude rocks" surrounding the bath—especially the gorge, Matlock Vale, which marks the course of the Derwent between Matlock and Cromford, with the sheer 120-yard cliff Matlock High Tor (featured in the pantomime's opening scene; see Fig. 6.3)—are thus forcefully juxtaposed not only against the technological sublime of the mills, but also against the polite milieu of the bath itself, with its servants and eating-houses. The verse and prose on Matlock register these contradictions in different ways. The poems especially tend to bring the "savage grandeur" of the rocks into concord with the social qualities of the landscape by placing the whole in a theological context. A number of prose accounts focus primarily on the rocks, but their relatively early dates reflect the site's accessibility, and the discourse they inhabit itself marks them as belonging to the polite culture of the region. Appropriately, the hot springs themselves—which supply the

39. Murray Aust, *Companion and Useful Guide*, 1:6–7.

baths, and had been known since Roman times—provide the common factor linking all the accounts, as the agency shaping both land and culture.

Several poems on Matlock share a religious tenor, connected in various
ways with the "medicinal rills" that draw tourists and patients to Matlock
Vale. The quoted phrase is the "Late Religious Lady's." This poet, while du-
bious about the aesthetic virtue of Cromford Mill's "gigging wheels" and
the moral virtue of visitors to the gorge who "wanton in nocturnal shade,"
takes the entire scene as a prompting to holy meditation, especially the Tor,
"which over-hangs the flood, / Haply the haunt of holy men and good." Her
somewhat morbid admonition to repent, addressed to the fashionable ladies
who frequent Matlock and read the *New Lady's Magazine*, derives additional
force from her presumed death, which gives this "Farewell to Matlock Bath"
its special poignancy.[40] In Thomas Lowndes' poem on Matlock, a bountiful
nature mimics the "medicinal rills" in "show'r[ing] . . . the mechanic powers" over Arkwright, whose "stately mills" teach us "the great Mechanic of
this World t'adore." The more optimistic religion accompanies a more extravagant aesthetic: "Such huge crags o'erhang the mountain's base, / . . . as
threat'n to crush whole townships in their fall." The sublime danger of Matlock High Tor is offset, again, by a merciful God who pours forth the
"min'ral waters . . . / That healing spring, which Heav'n has caused to
flow."[41] A third poem provides the theological context uniting this "healing
spring" and the uncouth-looking "rifted rocks"—the Deluge.[42] This anonymous poem transvalues such epithets as "savage," "rude," and "misshapen"
as signs of a divine dispensation, articulating the logic that joins "rude" geology and "polite" culture in all three poems. Whether sinister or salvific,
the striking forms of the limestone, sculpted by the Derwent, provide the
material basis of these adjectives.

The mechanist Deity in Whitehurst's *Inquiry* is amoral compared even to
the "great Mechanic" of the Lowndes poem, but Matlock Bath provides the
first concrete example of His operations, as well. The hot springs provide
important evidence for Whitehurst's actualist premise as well as the theory
of global volcanism built on that premise. Whitehurst is not uniformitarian
in the thoroughgoing sense of Hutton or Lyell, since he calls for a planetary
catastrophe, but he anticipates these thinkers by insisting that this catas-

40. The magazine's subtitle is worth citing here: *Polite, Entertaining, and Fashionable Companion for the Fair Sex* (2:663–64).
41. Lowndes, "Account of Matlock Bath," 156.
42. Both proceed, as this anonymous poet writes, from the "Primal Fount Divine": the
"clefts" in the rocks are a "corrective wound," producing healing waters that stand for the grace
made possible by the Flood, as a second fortunate Fall ("Lines, written at Matlock," 374–75).
Erasmus Darwin, by contrast, "burns" with "hopeless love" (rather than "chaste desires") at the
sight of Matlock High-Tor, "Nature's rudest child" ("Ode written on the River Darwent").

trophe can be deduced from observable processes and laws. In the first instance, Whitehurst states that it is not "the business of philosophy to inquire
into what the DEITY might have done, with respect to the formation of the
earth, but to discover if possible, by what mode he is pleased to act, in the
continuation and government of his works" (*Inquiry*, 28). Here in the early
stages of his geogony, Whitehurst argues that "the chaos was formed into an
habitable world" "by the same mode" active in ongoing "operations of nature." This logic makes the accretion of calcium around solid bodies in the
Matlock hot springs into a paradigm for the formation of solid matter
(28–29):

> The plants and fruits of the earth rise to maturity from their seeds, or first
> principles, in a regular uniform progression . . . for instance, the springs at
> Matlock-Bath in Derbyshire, tho' extremely pellucid and friendly to the
> human constitution, are nevertheless plentifully saturated with calcarious [*sic*]
> matter, which readily adheres to vegetables and other substances immersed in
> their streams, and thus, by a constant accretion, large masses of stone are
> gradually formed. The banks on which the bath-houses stand, and likewise
> the buildings themselves, are mostly composed of such materials.

Humphry Davy comments that Whitehurst's "sphere of observations was
unfortunately limited to a particular district" (*Lectures*, 53), but concedes
that "many interesting phenomena of the mineral kingdom . . . occur upon
a great scale in that county" (49). One of these phenomena is limestone: the
rock at Matlock is a reef limestone, formed of shells on a seafloor—somewhat as Whitehurst imagines—during the Lower Carboniferous. Another is
the famous "toadstone," now again recognized as lava. Geologists now believe that eruptions took place on the same seafloor, covering and penetrat-
ing the limestone strata.[43]

These two materials, both present at Matlock, represent the "rude" and
"polite" elements in Whitehurst's reading of the site. The limestone, of
which the bathhouses themselves are formed, is organic in origin, formed in
orderly and progressive fashion, while the toadstone is a product of the
"subterraneous convulsions" that disordered those original strata. Whitehurst infers a general catastrophe from the continued existence of earthquakes and volcanoes. He draws some of his local evidence from Matlock:
from a general association between hot springs and volcanism, he concludes
that "the springs at Bath, Buxton, and Matlock, may derive their temperatures from subterraneous fires, although there are no existing volcanos in
England" (*Inquiry*, 113). This move anticipates the second stage of the argument: "we shall endeavor to investigate from the same principles the cause

43. John Whittow, *Geology and Scenery in Britain*, 217.

of an universal flood, the origin of mountains, continents, craggy rocks, clifts, and other such disorderly appearances" (114). The analysis of the Derbyshire strata follows the same logic. In explicating his section representing Matlock High Tor (cf. Fig. 4.5), Whitehurst emphasizes the discontinuity observable by comparing the two banks of the river, which he adduces as a further proof of the catastrophe: "[this] serves to shew the effects produced by subterraneous convulsions; and likewise that mountains are not primary productions of nature, but of a very distant period of time from the creation of the world" (189). Toadstone, then, is not "primitive" in the sense of "original," but remains as a sign of the "confusion" producing disorderly topography like that of Matlock Vale, while the bathhouses officiate serenely as monuments of the original order. Arkwright's mills, as depicted in Joseph Wright's painting (1782–83), officiate in the same manner, and art historians have argued that this and other works by Wright betray Whitehurst's influence.[44] Wright's "glowing paintings of the gorge" and Whitehurst's "subterraneous fire" also point to a common origin in the local legend of Salmandore, portrayed in *The Wonders of Derbyshire* as a genius loci

> Whose mighty pow'r
> Directs the subterranean fire,
> And melts the solid rock to liquid ore.[45]

An earlier account of the hot springs brings out another sense of "primitive," a sense of the natural process that produces representative solid bodies under conditions like those of a laboratory or nursery. Moreton Gilks, a Fellow writing to the Royal Society in the late 1730s, observes that the springs give a "beautiful Representation, of the whole Business of Petrefaction."[46] Gilks considers only the orderly side of Matlock's geology, focusing in great detail on the first example given by Whitehurst, the buildup of material on plants immersed in the springs. Gilks first identifies the rock around the springs as an "Incrustation, formed upon the original Rock; composing a factitious Stone, of Earth, Vegetables, &c." Further investigation suggests a narrative of rock formation: "the farther you penetrate into

---

44. Drawing on the work of David Fraser, Stephen Daniels suggests that "the transverse view of the valley, the mills contained in a craterlike hollow, recalls Whitehurst's geological sections of the gorge at Matlock High Tor" (*Fields of Vision*, 62). Daniels notes that Whitehurst is studying this particular section in Wright's portrait of him (58; Fig. 4.5 above is a section by Whitehurst in the same style). The portrait also features a view of Vesuvius in eruption that for Daniels marks the geological connection, via Whitehurst, between Vesuvius and Wright's "glowing paintings" of Matlock (57).

45. *Account of the Wonders*, 22.

46. Gilks, "Letter," 354. Cf. the *St. James Chronicle* review of the 1779 pantomime: "Mr. de Loutherbourg has imitated Nature in the very process of Petrefaction" (quoted in Allen, "Wonders of Derbyshire," 64).

the Mountain, the closer and more compact the Stone appears." Gilks insists that "it is contrary to the Laws of Matter" to think of this as a metamorphosis, inferring instead that as a plant dissolves, it is replaced by "a uniform Stone in the Shape of the original Plant," and that in the end "a great many of these petrified Plants, and other Bodies united together, will compose large masses and whole *Strata* of Stone." Gilks' conclusion, that *all* limestone (and marble) is "such a Petrefaction as I have been describing, quite finished," leads to some weighty (and potentially heretical) reflections concerning the structure of matter and the age of the earth: "if we had as good Authority to suppose it 60,000 years old, as we have 6,000, it would be worth the while to trace the Origin and Source of these petrifying Exhalations a little deeper . . . and might produce . . . a more rational System of the Earth than has yet appeared."[47]

Matlock Vale appears to provide privileged access to planetary processes, whether understood as orderly artifice or rude sublimity, or through the theological lens of the periodical verse. All three interpretations respond to the obviously altered nature of the gorge as a landform, especially its variously shaped erosion-resistant pinnacles of limestone, as well as the visibility of ongoing processes in the hot springs and the downcutting of the river. For Gilks, the hot springs provide a model not only for the origin of rocks, but for the structure of matter itself, in the sense that this model confirms the empiricist premise of the integrity of bodies. A "petrifaction" can provide geological material only after it actually becomes a stone. Solid bodies are defined by the exclusion of other bodies, and petrifaction becomes possible only through the decay of the plant, which at the same time marks the difference between the vegetable and mineral kingdoms, the organic and inorganic, despite appearances to the contrary. Thomas Whately, writing about Matlock in 1770, shares this philosophical interest in form and structure, and to some extent the interest in geological process. His "laws of matter," though, are aesthetic categories: Matlock Vale illustrates "dignity," the first of three types of "character" displayed by rocks. Whately draws his prototypes for "terror" (Middleton Dale) and "fancy" (Dovedale) from the Peak District as well.

Whately's analysis distinguishes the aesthetic character of Matlock most sharply from the "primitive" associated with the Devil's Bottom and the Peak in general. Of the three characters, dignity, or "that which inspires

---

47. Gilks, "Letter," 353, 355, 356. Whitehurst stresses that the period of "three or four thousand years" succeeding the Flood is only a brief moment in the earth's history, and flexibility on this point is detectable much earlier than is generally recognized. Paolo Rossi has identified radical speculations on the age of the earth dating back as far as 1668 (*Dark Abyss of Time*, 12–13).

ideas of greatness, . . . has less wildness in it" than the other two (*OMG*, 99). His choice of Matlock as the prototype of this class thus confirms the associations of order, whether human, divine, or geological, established in the other texts. Whately uses Matlock Vale as a prototype because it is a wild or "unimproved" landscape, but it displays a dignified wildness, not incompatible with "some appearances of culture" (105). He introduces the gorge as "a vale near three miles long, shut up at one end by a rising moor, and at the other by vast cliffs of rock: the entrance into it is hewn through one of them, and is indeed a noble rude portal to a scene of romantic magnificence" (103). In keeping with his aesthetic category, Whately emphasizes the perpendicular lines of the gorge, which appears, especially from the top of Matlock High Tor, as "one vast precipice" (104). He is also attentive to variety in the "construction," which ranges from "irregularly jointed" to "more uniformly ribbed" to banks "composed of enormous masses of stone heaped upon each other." This variety most clearly evokes the properties of reef limestone, hinted at in the other accounts; because it lacks distinct horizontal bedding planes, it resists erosion, and massive pieces of it remain projecting into the landscape.[48] This variety, however, remains subordinate to scale in Whately's account, which draws on the classical prototype of Pelion upon Ossa as well as a more broadly literary formula: "height, breadth, solidity, boldness of idea, and unity of style, combine to form a character of greatness . . . unallayed with any extravagance."[49] Whately recommends judicious use of vegetation to accentuate this "character": "A thicket . . . makes the rocks which rise out of it seem larger than they are" and can also be used "to conceal the fragments and rubbish . . . which are often unsightly" (100–101). Geology itself becomes disruptive in the landscape designer's practical idiom, which appropriates to itself the process of imparting "delicacy" to "vast and rude" rocks. The "degree of wildness," in short, "runs sometimes to an excess that requires correction" (98). Nature must be made to conform to itself, and even rocks are subject to the pedagogy stated at the outset of the *Observations*: "to supply its defects, to correct its faults, and to improve its beauties" (1).

All these accounts distinguish Matlock through a unique combination of

48. Whittow, *Geology and Scenery in Britain*, 217. Whately's description is geologically specific inasmuch as his strong distinction between the splendid white colors of Matlock and the dark rocks in Middleton Dale corresponds to a geological difference between the denuded Carboniferous limestone of the White Peak and the younger, surrounding "Gritstone Edge," where Middleton Dale is located.

49. Whately's "dignity" carries the longest aesthetic pedigree of his three "characters," which further explains the orderly wildness of Matlock. Whately clearly echoes the ideas of Addison, as well as the "sedate" or composed sublime of John Baillie and Alexander Gerard, and the protomathematical sublime of Burke's "magnificence."

geological and cultural features. Whately's "dignity" provides one logic for resolving the apparent contradiction between rude rocks and polite culture. The remaining accounts seem to confirm this classification by drawing on theology and/or "laws of matter," stable and orderly systems that match the "character" of greatness. In the poems, appearances of chaos and of danger in the vale are offset by the divine dispensation of healing waters. The baths themselves are also a focus of scientific and of economic interest, apparently stimulating Whately's notions of magnificence as well. The mineral content of the springs, for Gilks and Whitehurst, images the orderly aspect of "original" processes of rock formation. At the same time, especially in Whitehurst, the rugged appearance of the gorge provides central evidence of the other "primitive" process, the disruption of this original order. This disruption is redeemed by the deist argument that it represents a "continuation" of the divine work, deducible from observable processes and laws. For Whitehurst as for Gilks, the geology here is not a "hidden" or magical property, but transparent and universal, a "beautiful representation." The wild appearances of Matlock, then, for all these writers, provide privileged access to some form of universal stability. Whately expresses this somewhat paradoxical synthesis in the dual function of Matlock Vale as both basis and object of an aesthetic protocol, a synthesis echoed in the range of perspectives on Matlock High Tor—frowning yet ultimately benign in the poems, and "disorderly" yet also an open book in Whitehurst's geology.

## Dovedale

Geologically, Dovedale resembles Matlock Vale: both consist of diversely weathered limestone features further exposed and sculpted by the downcutting river. The more diversified and fantastic shapes of Dovedale have to do with subterranean processes at work on the limestone strata long before they were exposed—a form of geological agency compatible with Whitehurst's and later theories. These "limestone pinnacle and other spectacular rock forms" were "sculptured [by] the existing network of underground drainage that had already been carved out of the massively jointed Mountain Limestone. In fact, the downcutting of the river Dove has merely opened up the maze of underground karstic features."[50] There is a strong sense of remoteness in Whately's account of Dovedale, in poems by Anthony Champion and Bernard Barton, and in John Farey's geology, where this remoteness is most explicitly figured as a descent into the bowels of the earth. The absence of spas, industry, or other development helps to account for the

50. Whittow, *Geology and Scenery in Britain*, 219.

seclusion of the place. Whately observes that "the only trace of men is a blind path, but lightly and seldom trodden, by those whom curiosity leads to see the wonders they have been told of Dovedale" (*OMG*, 114). Even by Sarah Murray Aust's time (about twenty years later) it is still necessary to go farther than a guide will take one in order to get to the heart of the place. Dovedale stands out as the most "romantic" of the Derbyshire sites, so called because the solitude, seclusion, and fantastic shapes of the rocks evoke the romance both of a mythical, literary past (as in the Champion poem quoted earlier) and a geological one. In several accounts, this romance also connotes individual liberty and imaginative freedom. The most famous romantic landscape in this sense is probably the "deep romantic chasm" in Coleridge's "Kubla Khan."[51]

Murray Aust's brief account emphasizes personal freedom of movement. The Peak comes up in the *Companion and Useful Guide* only as an optional side-trip along the route to Scotland, Murray Aust's chief subject and destination. Besides her extended account of the Devil's Cave and the discussion of accommodations, Murray Aust mentions only Dovedale and Chatsworth. A guide is de rigueur for all her excursions, especially those in Scotland, which frequently require a whole entourage. But she leaves the guide behind—at least briefly—in Dovedale, a move suggesting that this "very romantic place" is especially accessible to women and middle-class travelers: "pray walk entirely through it to the caves, for there are fine rocks near them. Do not suffer the guide to deter you from stepping from stone to stone, up a small part of the river, in order to get at the Caves; for, by the help of a stick, and a little attention not to slip off the stones, you will easily accomplish it; if you do not go so far, you will not see the most beautiful part of Dovedale."[52] This encouraging advice marks Dovedale as a site of individual adventure, uniquely associated with beauty, as against the overwhelming and sublime scenery that Murray Aust consistently prefers throughout her tour. Because of its accessibility, Dovedale has the unique ritual function, within Murray Aust's "cult of nature," of allowing the individual initiate to penetrate to the heart of the mystery.

In Whately's *Observations*, too, Dovedale is associated with alternative aesthetic categories: it embodies "fancy," a "character" entirely free of the techniques and man-made "accompaniments" typical of both "dignity" and "terror." A rhetorical sign of this freedom is that Whately keeps the abstract definition of "fancy" short, while actual description occupies the bulk of the

---

51. *Poems*, 297, l. 12. James Engell has argued that this description was inspired by a visit to Wookey Hole, a famous cavern in the Mendips not far from Nether Stowey ("A Yet Deeper Well"). See also James McKusick, "'Kubla Khan' and Theory."

52. Murray Aust, *Companion and Useful Guide*, 1:6.

chapter; description in the other two cases has the clearly subordinate role of illustration. The brief definition of fancy, however, is dense and suggestive: "Sometimes a spot, remarkable for nothing but its wildness, is highly romantic; and when this wildness rises to fancy, when the most singular, the most opposite forms and combinations are thrown together, then . . . [they] display the inexhaustible variety of nature" (111). While both dignity and terror are forms of the sublime, "fancy" seems more strongly associated with beauty, and with third categories such as novelty (Addison), variety (Hutcheson), and the picturesque.[53] Whately's "fancy" evokes both wildness and romance. Anthony Champion's poem ("To a Friend. The Scene Dovedale," 1756) shows that notions of enchantment were already attached to the place, and Whately's Dovedale is so wild and variegated as to seem magical—an explanation analogous to the early geology that invokes divine agency to explain the wild diversity of landforms. This logic also allows him to reassert the boundary between art and nature, to distinguish an authentic wildness that will not admit of improvements or even of imitation (116). The air of enchantment makes wilderness "romantic."

Whately's connoisseurship is inevitably artistic, but he limits the artist's role here to passive admiration. In the case of Dovedale, he devotes most of his efforts merely to describing and distinguishing the varied forms of the rocks. While in some places "many different shapes are confusedly tumbled together," elsewhere the forms achieve a striking level of definition: "Sometimes they are broken into slender sharp pinnacles, which rise upright, often two and three together. . . . Sometimes they stand out quite detached, heaving up in cumbrous piles, or starting into conical shapes, like vast spars, an hundred feet high . . . some, split and undermined, are wonderfully upheld by fragments apparently unequal to the weight they sustain" (*OMG*, 113–14). Like the speaker of "Resolution and Independence," Whately "wonders" at the unfathomable process by which the rocks acquired their present form and disposition. This phantasm of an obscure and spectacular past gives the whole its "air of enchantment": "the forms all around, grotesque as chance can cast, wild as nature can produce, and various as imagination can invent; the force which seems to have been exerted to place some of the rocks where they are now fixed immoveable; the magick by which others appear still to be suspended" (115)—all this serves to corroborate the impression of an alien agency, whether natural or supernatural.

53. Whately's brief discussion "Of Picturesque Beauty" (*OMG*, 146–49) is one of the earliest, preceding all but Arthur Young's use of the term, and seems to have influenced Gilpin if not Young. Whately's definition is quite literal, however, and the later sense of "picturesque"—as derived and adapted from Gilpin by Price and Knight—applies more accurately to this description of Dovedale.

Whately's choice of magic as a category for this agency, especially given the absence of any religious references, places it with phenomena that can have both natural and supernatural explanations, a slippage that Moreton Gilks also illustrates by demystifying the ostensibly "hidden properties" of petrifaction.

Though more reticent on the subject of improvements, Whately also studies Dovedale in terms of its "composition," and geological process becomes artistic through his personification of nature as a designer of landscapes. He remarks of the rocks that "the changes in their disposition are infinite," one of many phrases suggesting an intentional process of arrangement (*OMG*, 114). Another observation alludes specifically to pictorial composition: "through some are to be seen several more uncouth arches, and rude pillars, all detached, and retiring beyond each other, with the light shining in between them, till a rock far behind them closes the perspective." Whately seems to anticipate Gilpin, whose quest for landscapes that called for the frame of the landscape painter famously canonized a certain way of reading landscape, associated with the Claude glass and the tour. But Whately assertively distinguishes the art of gardening from painting: "it is as superior to landskip painting, as a reality to a representation" (1). The rocks in Dovedale have a significantly tactile as well as a visual dimension, and Whately—like the geologists—reads their "arrangement" in terms of physical processes. In the case of Dovedale, the physical reality of the scene does not call forth a mimetic intervention by the gardener. Whately excludes "marks of inhabitants" and "ornamental buildings" as "too artificial in a place so absolutely free from restraint. The only accompaniments proper for it are wood and water"—and these, indeed, would be difficult to eliminate (115).

What makes Dovedale so "absolutely free"? Whately limits the power of art in this description, but at the same time magnifies it by acknowledging that it is ordinarily capable of placing restraints on nature. His chapter "Of Rocks," and especially this reading of Dovedale, serve to establish the tremendous scale and scope, and the resistance—the *materiality*—of the "reality" that provides the medium of his art. It is this materiality that makes the gardener's landscape "superior" to "a representation," and gives to gardening the power of rearranging the earth's material. If, even from this vantage point, "art must almost despair of improving" such a scene, then it is "absolutely" wild and free, one of very few places that requires no "correction" in its natural state (116). Dovedale provides, instead, an ideal (unattainable) prototype for made landscapes. The limestone pinnacles enable Whately to imagine a place outside the order of enclosure and improvement that he represents. Both "nature" and "freedom" are co-opted by their use as aesthetic models for the landed gentry, but the use of these terms also depends on Dovedale's literal freedom from enclosure, just as its aesthetic value results from a lack of any except aesthetic resources. The landscape garden itself depends on a notion of material reality represented by these "mere rocks," the earth's material apparently in its original state, with no signs of human impact—a significant rarity in a landscape so long and so intensively cultivated as that of England.[54]

John Farey's massive survey of the "agriculture and minerals of Derbyshire" makes an odd context for a description of Dovedale, since it is of interest for neither minerals (in Farey's economic sense) nor agriculture. The description also stands out in the midst of this endless factual recitative for the enthusiasm that it sparks in Farey as a geologist. "The highly curious Dovedale" first appears as a relatively brief item in a list of "the principal narrow and rocky valleys . . . in Derbyshire," an appearance that marks it as an element of the tourist economy otherwise ignored by him. A more substantial geological description distinguishes Dovedale as "having the two surprising Hills of 4th Limestone . . . at the entrance of the Dale, and between which the Excavation is very narrow, and the Rocks precipitous; and so it continues . . . the Dove running, in great part of this distance, upon lower strata in this immense 4th Lime Rock, and in the general Series, than are perhaps anywhere else visible in the British islands."[55] Farey's precise stratigraphy is influenced by William Smith, and his sense of geological succession is more advanced than Whitehurst's (note the position of Derbyshire Limestone in Smith's tables, Figs. 4.1, 4.3). This modern scientific orientation—Farey is one of the "practical men" who gave geology its empirical focus—naturally influences the language used to express admiration. The river, aided by the erosion of younger rocks, and by the underground drainage patterns we now recognize, has penetrated farther beneath the surface of the earth than any other river or gorge in Britain.

Farey notices the same elements of this scene noticed by a range of writers over many decades, and these same elements—"precipitous" topography, "grand" forms, depth, wildness, seclusion—seem to provoke a similar aesthetic response, particular to Dovedale. Farey uses his more precise scientific terminology to locate an objective basis for his response: this does not only appear to be a deep valley; it actually penetrates to an earlier stage in the succession of strata than is visible anywhere else. This observation confirms the particular sense of "primitive" associated with Dovedale. Like Matlock, the site provides privileged access to geological processes and phe-

---

54. On the English landscape as a phenomenon of cultivation, see W. G. Hoskins, *Making of the English Landscape*. Hoskins argues that "the English landscape itself . . . is the richest historical record we possess" (14).

55. Farey, *General View*, 1:66, 478.

nomena. Unlike Matlock, however, this geological origin is displayed in a sphere that seems utterly remote from human ordering. The absolute depth and antiquity of the strata and the ancient patterns of weathering on the rocks are also what suggests to Whately that he is witnessing the earth's material—ordinarily a medium for the gardener's art—in its original, alien, inimitable state. The spectacular rock formations at Dovedale seem to imply the action of an utterly alien agency, an appearance compromised at Matlock by the presence of human ordering, reflected in turn by the theology and the study of natural laws characterizing the accounts of Matlock.

Dovedale is the most "romantic" place in the Peak because it seems a borderland between natural and supernatural agency, between art and nature. The evidence suggests that "romantic" does not mean "sublime" in this case. The descriptions of Matlock, by contrast, help to make clear that the sublime is a humanizing category, too bound up with grandeur and magnificence to address the particular mystery of Dovedale. Its wildness, according to both Whately and Champion, is so extreme as to be magical. This appearance of enchantment in turn makes it "romantic" in two senses. First, enchantment belongs to a literary past, to the genre of romance. Second, it belongs to the geological past, a time so remote that its vestiges can be read only as signs of obscure, titanic processes. The time of chivalry or that of the Druids provides an exact literary analogue for this period, for an enchanted past, invoked most obviously in Whately's accounts of the rocks' strange shapes and positions; these make Dovedale uniquely romantic from a geological, as well as literary, point of view. Farey's modern scientific explanation is quite the opposite, following the principle of "truth is stranger than fiction": he presents physical evidence of a period so distant that it *cannot* be imagined. The absence of development grounding these constructions of wildness may not be extensive enough by more recent American standards. De Loutherbourg's genre painting of Dovedale (Fig. 6.4) decisively foregrounds the cattle, though the dramatic reviewer who challenged his scenes in the pantomime might well have argued that this painting owes more to Claude Lorrain than to the Derbyshire landscape. The sheepfolds appearing in Bernard Barton's "Dove Dale. A Descriptive Sketch" (1809) might also seem to compromise the later idea of wilderness—though they do nothing in the poem itself to disturb an enchanted solitude that very much resembles the earlier accounts.[56] Nevertheless, the "made" character

---

56. "The sheep-folds' simple bound" somewhat incongruously accompanies the dale's geological "forms grotesque" in Barton's poem (*Metrical Effusions*, 180–81). Barton assures us that his sketch is a faithful record of his observations, but he must not have made it in the very steep and wooded central part of the dale described by Whately, Murray, Champion, and others. Gilpin, who was not favorably impressed by the Peak District, complained that "the axe had been at work" in Dovedale in his *Observations*, originally made in 1772 ("Explanations of the

6.4    *Dovedale in Derbyshire,* by Philippe-Jacques de Loutherbourg (1784). Oil on canvas. Reproduced by permission of York Museums Trust (York Art Gallery).

of the entire English landscape creates a context in which extremes of wildness must be remarkable, and in which the primitive materiality of romantic rocks might seem to authenticate ideas of nature and of freedom.

## Conclusion

The "romantic," as embodied by Peak District landforms, is a uniquely English category. Dovedale's indifference to the sublime, for example, calls to mind that this category is associated with an Alpine wildness implausible for the English countryside and perhaps, as Wordsworth insisted, irrelevant to it. Wordsworth, of course, attempts to redefine a local and national sub-

---

Prints," 1:xii; cf. his sketch on 2:227). There is currently pasture land along the Dove in Mill Dale, just above Dovedale, on the Derbyshire side—now itself under National Trust protection.

lime, "to reconcile a Briton to the scenery of his own country" (*Guide*, 106). Earlier experiments on Derbyshire as a representative national landscape and a site of productive disorder anticipate such a project but also inform Jane Austen's sketch of Pemberley. The most romantic landscapes in England seem to be those few that stand in some way outside human control or human impact. Such landscapes are typically imagined as a borderland between art and nature.[57] The examples of Staffa and the Giant's Causeway show how this aesthetic notion becomes naturalized in landscapes across the United Kingdom. These matching sets of prismatic basalt formations, on cither side of the Irish Sea, inhabit the same borderland as Dovedale, carrying an even stronger suggestion of art not made by human agency. These places are more remote (Staffa was not "discovered" until 1772), and so the literary traditions surrounding them are of a later date than those surrounding the Peak. But the writing on these places—including William Drummond's long poem and other narratives on the Giant's Causeway, Wordsworth's 1833 sonnets and Boswell's narrative on Staffa, and many other texts—shares important qualities with eighteenth-century writing on the Peak. These landscapes provoke similar reflections on the production of matter and the dynamics of order and disorder in the national landscape. But they are also less productive landscapes, more sparse in population and natural resources, anticipating the polarized discourse on wilderness that developed in the nineteenth century.

This chapter, then, represents one of many possible case studies of places generating a certain kind of literary culture, ranging from occasional poetry to earth science, and providing the ground on which these discourses and disciplines gradually became separate from each other. Such places provide an opportunity to observe the common origins of imaginative literature and other forms: if we recognize the gardener's landscape, or the geologist's, in the poet's, then it becomes clear that a literary landscape is always in some way physical and historical, never just conventional. The impact of the literary, in turn, caused the factual or practical study of land*forms* to be articulated in terms of land*scape*. It also becomes clear that cultural imaginings of landforms and landscape are unified around several fundamental issues: economic productivity, aesthetic experience, and the status of physical realities. The physical is increasingly identified with the particular partly because

---

57. The contemplation of natural phenomena as artworks is a staple of eighteenth-century aesthetics. My particular point here is that the perception of nature as art has a local significance and a local urgency in a landscape so thoroughly domesticated as that of England. If wilderness is an oddity—as it must be especially in a country intensively farmed and subject early on to industrialization—then its exceptional status strengthens the suggestion that it belongs to the aesthetic sphere.

unique landscapes such as those of the Peak provide increasingly significant resources for cultural endeavor. The early industrial dialectic of aesthetic and economic value, so powerfully at work in the period's representations of the earth's material, surfaces vividly in the diverse literature on the Peak District. Partly because its literary tradition predates the Act of Union, the Peak suggests itself as a representative English place, a national stage on which large-scale social and economic changes are rehearsed. The unique combination of physical and cultural features in the Peak District drew a re-markable level of attention in the eighteenth century, providing the basis for a tourist economy that is still booming long after the mines have been exhausted.

# Conclusion: Aesthetic Geology and Critical Discourse

Limestone exhibits a strength and boldness of character scarcely inferior to those of the primitive rocks, and the effects of the mixture of the disordered with the regular strata, and the relations of the curved lines to each other, are well calculated to awaken the strong emotions connected with the perception of natural beauty.
    —HUMPHRY DAVY, *Lectures on Geology* (1805)

**H**umphry Davy's paean to limestone covers, in a short space, much of the territory covered by this book. He introduces England's most visible rock in terms of "character," a concept first applied to the earth's material by theorists of the landscape garden such as Thomas Whately. Davy explains this character in relation to the "primitive," a category from early earth science that fascinated Blake, Wordsworth, Shelley, and other poets. Since Britain is essentially a chunk of limestone, Davy is rating the national character at the same time. Here we catch a glimpse of Wordsworth's concern to prove the Cumbrian mountains "scarcely inferior" to the Alps, those famous "primitive" mountains that, for Shelley, give the lie to English liberalism. For Blake, the anxiety about primitive rocks is a misguided quest for national origins, abandoning Albion to "stony Druids." "Strength and boldness" may characterize geological origins, but Davy's phrasing suggests a sense of cultural belatedness as well. The structural variety of limestone compensates for this belatedness by demonstrating the rock's enduring and universal aesthetic value. The "relations" of order and disorder and of curved lines in limestone all exemplify the axioms of aesthetics laid down by English theorists such as William Hogarth (the "line of beauty") and Francis Hutcheson (variety). The "strong emotions" aroused by limestone thus testify to the empirical validity of aesthetic theory as well as the timeless power of "natural beauty," as urged by Wordsworth, Coleridge, and poets less well known to Davy. By declining to name the agency that "calculated" the forms of limestone to arouse these emotions, Davy distances himself cautiously from the natural theology that both Romanticism and modern science helped to dislodge.

This short passage demonstrates how vital aesthetics remained to geology in 1805. Davy's Royal Institution lectures assert the disciplinary sovereignty of "science" in a sense already fairly close to modern usage, but he makes his case in a common cultural idiom. Readers of this book might also hear echoes of Erasmus Darwin in Davy's account of limestone. Darwin takes advantage of what I have called "specimen culture" to incorporate a long digression on sculpture into *The Botanic Garden*'s narrative of limestone and marble. Davy's limestone, with its "strength and boldness of character," also sounds like a classical or neoclassical sculpture. Davy's lecture hall provides an even better forum than Darwin's illustrated encyclopedic poem for capitalizing on the visual culture surrounding natural history. The manuscript of the geology lectures includes a sketch in Davy's hand titled "Sketch for a Large Painting for Next Thursday" (*Lectures*, xxxiii). We can deduce that the painter Thomas Webster rose to the occasion with a magnificent panorama, for the same image was later engraved for Davy's *Elements of Agricultural Chemistry* (see the frontispiece to this volume). The manuscript also indicates that Davy showed specimens frequently in the course of a lecture and encouraged the audience to study them with admiration. Martin Rudwick has demonstrated the fundamental importance of specimens for the development of modern earth science.[1] Specimen-collecting in its older, aristocratic guise formed the basis of the British Museum and other public collections, and by the early nineteenth century collecting in all branches of natural history had become popular across the class spectrum.[2] The visual emphasis of Davy's nonspecialist idiom is thus perhaps less surprising than his verbal

---

1. Rudwick, "Minerals, Strata, and Fossils," 266.
2. The collection of Sir Hans Sloane, president of the Royal Society from 1727 to 1741, became the core of the British Museum. Of the many recent works on museum culture, one specifically addresses geology while two address Romanticism. See, respectively, Simon J. Knell, *Culture of English Geology;* Eric Gidal, *Poetic Exhibitions;* and Gillen Darcy Wood, *Shock of the Real.*

commitment to aesthetic tropes. As a chemist, his attention to composition and other formal properties seems to encourage aestheticizing digressions in the midst of mineralogical discussion. Just as the specimens supplement and in some cases substitute for explanation, in the case of limestone an aesthetic response jostles against mineralogical analysis. Granite ("no rock is grander in form nor more sublime in structure") is the sublime to limestone's beautiful in Davy's schema of aesthetic geology. A specifically Wordsworthian commitment seems to suggest his emphasis on Jean-André de Luc's Alpine childhood in accounting for the Swiss naturalist's theory that Alpine summits are "part of the primordial matter of the globe" (*Lectures*, 53).

In my discussion of such geological texts, the term "aesthetic geology" designates a style comprehending spectacular description, catastrophic scenarios, literary allusions, appeals to aesthetic response and aesthetic theory, and hints concerning a vast realm beyond the pale of knowledge. Romantic poetry and early geology, as I have argued, form each other through the shared concepts of landscape aesthetics, and aesthetic geology also extends to the geological content of poetic figures. Aesthetic geology is literary from the start, and persists in poetry because it provides literary access to the valuable tool of geological description at a time when aesthetic experience is being constructed increasingly around localized landscapes. The geological sublime, in particular, is a major category for at least three of the canonical Romantic poets. Travel narratives and poems based on them, such as "Mont Blanc" and "Ode: The Pass of Kirkstone," share the local interests of aesthetic geology in its pervasive cultural sense, giving highly literary descriptions of closely observed physical features of their respective sites. These poems, no less than scientific travel and popular natural history, elaborate aesthetic categories constructed specifically around rocks. The geological content of these poems, as well as geographically less settled poems such as Blake's *Book of Urizen*, extends to primitive materiality, converging with geology's inquiry into the nature of "unorganized" matter. Aesthetically motivated inquiry into rocks generates widely shared speculation about the otherness of the physical. In *The Excursion* and *Prometheus Unbound*—as in Hutton and William Smith—geological "reading" reveals form as both a visual and a literary quality of geological objects. The rocky languages of nature range from nearly impenetrable codes of "rude" characters, suggesting violent and chaotic origins, to the highly transparent rock records of Shelley and Smith.

The geologists of the period (as they have come to be called)—Whitehurst, Hutton, Davy, Smith—were avid and self-consciously literary readers of rocks and generally strove for stylistic elegance in their writings. But they would probably be chagrined to find themselves the objects of strictly literary analysis. The Huttonian apologist John Playfair, in an 1811 essay, rejects key aspects of what I have called "aesthetic geology" as "a species of mental derangement, in which the patient raved continually of comets, deluges, volcanos and earthquakes; or talked of reclaiming the great wastes of the chaos, and converting them into a terraqueous and habitable globe."[3] The term "mental derangement" seems to refer to the quasi-poetic inspiration informing both the style and substance of early geology, which he dismisses as an "unreal mockery" that has nevertheless "continued even to the present day." Playfair seems especially concerned to reject the notion of catastrophic agency, so much at odds with the slow and gradual change posited by Hutton; yet the plutonic forces of Huttonian geology are equal in *magnitude* to the catastrophes of visionary geology, and Hutton himself had to borrow from the earlier texts to represent that magnitude. Playfair's great achievement as Hutton's popularizer was to render the theory in a more lucid and orderly style. But in the view of his contemporaries this very feat aligns him with the old fabulous geology, so that the charges quoted here rebound against him. John Murray, for example, argues that Playfair's style, because it is so mellifluous and reasonable, is simply more subtle and dangerous than that of the fabulists.[4]

The received narrative of the history of geology (questioned in the past fifteen years by scholars including Stephen Jay Gould, Rachel Laudan, and Simon Knell) depicts geology's struggle to free itself from the tradition of fanciful speculation and armchair theorizing in the process of becoming a modern science.[5] That struggle is apparent in the passages from Playfair and his critic John Murray. But it constitutes itself, especially in Murray, as a struggle against literary style. Playfair commits the same sort of imposture as Thomas Burnet, if his style, though superficially more sober and rational, makes unscientific theory aesthetically palatable. In retrospect, Murray's charge of "visionary" excess seems unfair and mistaken, since Huttonian uniformitarianism, especially as popularized by Playfair, eventually won the day after it was adopted by Charles Lyell. Notwithstanding this vindication of the theory's truth-claims, Murray's criticism remains as evidence of the importance of style in geological writing. Moreover, Murray's implicit accu-

3. This account of the history of geology (published in the *Edinburgh Review*) is quoted in Roy Porter's *Making of Geology*, 1.

4. Murray protests: "To the Author of the present treatise, the Huttonian doctrines, whatever may be their ingenuity and novelty, appear visionary and inconsistent with the phenomena of Geology; and to a defence of them so able, and so well calculated to give a favourable impression of the general system, he has been induced by that interest which every one feels in the opinions he believes to be just, to endeavour to reply"; *Comparative View*, iii–iv.

5. Martin Rudwick's *Scenes from Deep Time* questions "Whiggish" assumptions about scientific progress in geology by emphasizing the popular appeal of geological illustration. Knell's *Culture of English Geology* actively privileges factors traditionally ignored by history of geology, such as collecting and provincial societies.

sation, that Playfair has subordinated substance to style, indicates the power of aesthetic geology in the hands of a geologist who might be less concerned, or less equipped, with empirical fact. The Playfair-Murray debate characterizes a late stage in the transition from "letters" to "science" as the cultural rubric for study of the natural world.

Specialist discourses have relied, from the beginning, on the literary as a category against which the scientific may be set off. As George Levine puts it, "science established itself professionally in England in part by rejecting literature—at least those excesses of literature that seemed to the Royal Society to corrupt thought."[6] Writing in 1684, Thomas Burnet adopts such a rhetorical stance when he distinguishes the "Oratour's" view of the earth as "a beautiful and regular globe" from the "impartial" prose of the "Philosopher" who recognizes it as "a broken and confus'd heap of bodies."[7] While Playfair's charges against Burnet and his followers certainly have some merit—despite the shared rhetoric, geology never did become organized in the rigorous way that the physical sciences did under the Royal Society—it is striking that Playfair's complaint in 1811 participates in the same slow dialectic that was launched in the seventeenth century. Geology, too, must repudiate its literary affiliations in the process of discipline-formation. This process was certainly in motion when the Geological Society was formed in 1807, but it worked against a set of literary and cultural obstacles very different from those impinging on the Royal Society. Playfair's rejection of spectacular geology is typical; in the Romantic period some naturalists began to conflate earlier natural history with "literature" in Levine's sense. While Playfair attempted to distinguish his own self-consciously modern geology from older natural history, Erasmus Darwin had insisted only twenty years before on the importance of obsolete or fanciful theories even in the rigorous precincts of natural philosophy: "extravagant theories . . . in those parts of philosophy where our knowledge is yet imperfect, are not without their use; as they encourage the execution of laborious experiments, or the investigation of ingenious deduction, to confirm or refute them."[8] Literary qualities—whether embraced for their imaginative "extravagance" or discarded as "raving" excess—were thus essential, in one way or another, to specialization and professionalization in numerous areas we now see as strictly scientific. A generally informed readership was not the least of these literary requirements, as the same print market served the professionalization of both literature and science.

This study, being particularly concerned with geological poetry ("roman-

---

6. George Levine, ed., *One Culture*, 11.
7. Burnet, *Sacred Theory of the Earth*, 90–91.
8. "Advertisement," in *BG*, p. v.

tic rocks"), also emphasizes the stylistic investments of early earth science ("aesthetic geology"). Playfair's charge of "mental derangement" makes a useful test for such investments in the geologists considered here, especially Davy, Whitehurst, and even Playfair's own teacher, James Hutton. Whitehurst's geological style lies somewhere between the "raving" of Burnet and the persuasively lucid style of Playfair (which in turn was held to conceal a "visionary" agenda). His geological arguments also form an important link between these two extremes, for while Whitehurst's *Inquiry* is the last major theory of the earth to observe the conventions of physicotheology, its catastrophes have plausible empirical foundations, duly noted by Hutton.[9] But Whitehurst's catastrophist idiom and deliberately spectacular style make him doubly liable to Playfair's charge of "raving." Whitehurst's style may be interpreted as a geological poetics of the sublime. By 1778 the sublime provided a well-established interpretive framework for imposing rocks and mountains, partly because of the influence of Burnet. As a way of experiencing landscape, it had proved its capacity for prompting scientific inquiry. The style associated with this kind of experience had enjoyed a recent surge of popularity. Besides offering astonishing descriptions in that style, Whitehurst's theory also promises the historical legibility of sublime phenomena by means of the rock record. The basalt columns of the Giant's Causeway, for example, have their sublimity rather heightened than diminished by appearing as universally "intelligible characters" of a catastrophic past "anterior to history." The idea that they represent "columns of liquid fire" corroborating the disappearance of Atlantis contains both the geological "truth" of that formation's volcanic origin and the sort of ancient poetic "fable" that Whitehurst wants to see scientifically vindicated.

This geological method requires a combination of direct experience of landscape with literary allusion, "natural" with "rhetorical" sublimity. Whitehurst emphasizes, for example, that he has actually gone to see the Causeway (in spite of Dr. Johnson, perhaps, and unlike many other naturalists who discussed it) (*Inquiry*, 245–46). The presence of words from the standard lexicon used in descriptions of sublime mountains—"rude," "misshapen," "stupendous," and so forth—in the empirical descriptions of the Causeway paves the way for the more idealized sublimity of Atlantis and the

---

9. On Whitehurst and Hutton, see Dennis Dean, *James Hutton*, 13–14. Dean adopts Playfair's emphasis on Hutton's originality (e.g., 264) against Roy Porter's contention that Hutton was, like Whitehurst, a late adherent of the physicotheological tradition. (Cf. Porter, *Making of Geology*, chap. 8.) In my view, Whitehurst's theory of the present earth belongs more to Plutonism and modern geology than to physicotheology or Neptunism. Despite his initial emphasis on "primeval islands" rising from an original ocean, Whitehurst's Deluge becomes a mere figure for more plausible efficient causes of change in landforms, such as undersea volcanism and a substratum of molten rock.

imagined catastrophe. Similarly, Whitehurst's overarching theory of "subterraneous convulsions" is inspired by the rugged or "rude and romantic" countryside of his native Derbyshire. Since both landscapes are evidently in a quiescent state, the only available evidence for the "great alterations" they imply lies in the more recent catastrophes of Lisbon (1755) and Vesuvius (Pompeii was discovered in 1748); and for these Whitehurst must turn to second- and thirdhand accounts. This argument by analogy requires empirical correlates for the magnitude of the prehistoric convulsions; the literary portion of the method is to present these correlates as mere shadows of that magnitude. Accounts of historical earthquakes and eruptions present a magnitude that is not inconsiderable: Whitehurst cites accounts of whole cities "erased" by tidal waves or buried by volcanoes. He amplifies the "great revolutions anterior to history" (66) by harnessing the wonder generated by these real events.

Playfair's account of geological "raving" provides a good starting point for a discussion of aesthetic geology because it singles out cosmology as a field too large for modern scientific inquiry, and because it implies a connection between cosmology and poetry or myth. The "raving" geologist parodies the divine inspiration associated through Longinus with sublime composition. The scope of Whitehurst's narrative is more confined than those of Burnet and the physicotheologists (William Derham, John Ray, and others), but it is still clearly associated with cosmology. Such a geology operates by means of dark conceits, analogies and parables, borrowing the style of the Hebrew Bible whose authors Robert Lowth identified as the poet-prophets of their culture. The early geologists' emulation of Old Testament style and cosmology thus locates their accounts within a particular rhetorical sublime.[10] The figures of this rhetoric, especially the figures of catastrophic agency, amplify the aesthetic otherness derived from landscape aesthetics into an impressive natural sublime. Geological process, as inscribed in the rocks, itself remains obscure and allegorical. Aesthetic geology in the era of Whitehurst and Hutton registers the resistance of rocks to scientific inquiry even as it prosecutes that inquiry. It helps to maintain rock as the illegible primitive substance of nature, which shadows forth "sublime truths" by virtue of its indeterminacy and satisfies the aesthetic demand for an alien, primeval landscape. At the same time, aesthetic geology increas-

---

10. Lowth's *Lectures on the Sacred Poetry of the Hebrews* were first given shortly before they were published in Latin in 1759, but Joseph Johnson's publication of a first English edition in 1787, as critics have frequently noted, makes this book a very contemporary document for the Romantic period. Lowth himself at times shows a geological orientation (e.g., 1:134). Here is a classic example from Burnet: "The everlasting Hills, the Mountains and Rocks of the Earth, are melted as Wax before the Sun. . . . This huge mass of Stone is soften'd and dissolv'd as a tender Cloud into Rain . . . and swallowed up in a red Sea of Fire"; *Sacred Theory of the Earth*, 306.

ingly locates such landscapes in poems and narratives on specific sites with cultural dimensions, such as Dovedale and the Giant's Causeway.

Hutton's *Theory of the Earth* emphasizes systematic local description to the point of being famously difficult to read, but he shares Whitehurst's premise of prehistoric and interior, hence empirically unknowable, geological causes. Most of the *Theory* concentrates instead on a legible text of surface features, so that a chapter seeking to establish the "force and existence" of plutonic processes implied by these features suddenly requires an expanded aesthetic repertoire. Like Whitehurst, Hutton must provide a hint of the magnitude of these processes by which rock is molten and reconstituted. But he must establish this magnitude in four dimensions, since for him geological process is uniform rather than "convulsive." According to associationist psychology, physical magnitude produces the magnitude of the ideas associated with it. John Baillie's *Essay on the Sublime* (1747) consequently focuses explicitly on material objects, insisting repeatedly that "large Objects only constitute the *Sublime*."[11] Alexander Gerard, in his more influential treatises, draws heavily on Baillie, and Edmund Burke popularizes a similar idea of the sublime. Despite the psychologization of aesthetics in the latter part of the century, Wordsworth still feels obliged to protest in 1820 that "a notion of grandeur, as connected with magnitude, has seduced persons of taste into a general mistake on this subject" (*Guide*, 33). Hutton, when writing of a projected physical magnitude, can appeal implicitly to this tradition of aesthetic discourse in order to authenticate his claims (that there are subterranean forces sufficiently vast, both mechanically and in duration, to melt, reconsolidate, and elevate entire continents). He still needs earthquakes and volcanoes and the old prophetic style, borrowing the authority of aesthetic theory and even gaining plausibility from the association: while Baillie's "large Objects" are not the only sublime objects, Hutton's inquiry is a priori limited to physical science, and so the sublime he generates seems to carry a necessary association with real physical forces (as opposed to moral or intellectual greatness).[12] As I argued in chapter 3, the implicit appeal to aesthetics is powerfully at work in a typical rhetorical question that adduces aesthetic response as evidence for the existence of empirically unknowable forces: "Are those powerful operations of fire, or subterraneous heat, which

11. Baillie, *Essay on the Sublime*, 41; cf. 6–7.

12. When Alexander von Humboldt sets out to debunk the old style of geology, which "sought only the wonderful in volcanic appearances," he does so by means of measurements, focusing on physical magnitude as the locus of aesthetic geology. He presents his measurements of Vesuvius, taken systematically over a long period, in order to show that the size of the crater, the volume of material ejected, and so forth have been consistently exaggerated by "science" and "literature" alike (*Ansichten der Natur*, 371–72; see also the carefully prepared tables of measurements concluding the chapter).

so often have filled us with terror and astonishment, to be considered as having always been?" (*TE*, 1:143).

Early geology's reading public participated in the enthusiasm about the earth's material, but not uncritically. Geology became such an important arena for criticism and debate because it addressed its public not only as readers of text but also as readers of rocks, just like the authors themselves. Middle-class provincial intellectuals such as Whitehurst and Smith particularly stressed the earth's accessibility, inviting readers to interpret a "language and characters equally intelligible to all nations" (*Inquiry*, 257). Playfair would have cited Hutton's fidelity to this language, his abundant empirical evidence, to exempt his occasional appeals to "terror and astonishment" from the charge of "raving" about "volcanos and earthquakes." John Murray clearly did not agree, and neither did Goethe. Dismissing the arguments of Plutonism as "mad spouting" ("tolle Strudeleien"), Goethe uses it to restate the central Faustian theme of *Gelehrtensatire* (intellectual satire) in *Faust*, part II. A self-contained, quiescent, crustal earth served his literary and philosophical purposes, and as a trained mining engineer he supported this view with an informed defense of Werner's Neptunism, to which he remained committed until his death in 1832. The Plutonist-Neptunist debate snakes its way in seriocomic fashion through the phantasmagoric *klassische Walpurgisnacht* (act II), in which Goethe exposes the Plutonist (Huttonian) view as a set of fantastic hallucinations.[13] The debate returns in a context of much higher seriousness in act IV, in which Faust espouses a discreet Neptunism and Mephisto, naturally, a hotheaded Plutonism (10039–90). It is here that Faust asserts himself with finality by dismissing this view as "tolle Strudeleien" (10104), a charge that exposes the boiling, eddying content of Plutonism as an epiphenomenon of style. In this late passage, Goethe insists on the aesthetic dimension that had been dismissed from geology by that time, ultimately using geology to authenticate his own status as a poet of nature.

Charlotte Smith similarly uses her critical reading of geology to lay claim to a privileged poetic connection with nature. Though they are very different kinds of poems, both *Beachy Head* (1807) and *Faust* II (1831) rely on the critical ethos of public debate surrounding geology for their poetic ends.

13. *Faust*, part II, 87–99 (II.7495–7950). Thales and Anaxagoras, adopting the rhetoric of their modern counterparts (the Neptunists and Plutonists, respectively), debate the origin of an allegorical landscape on the river Peneus, a recurring scene in this portion of the drama. The rapidly changing geomorphology of this landscape and the fate of its main actor, Homunculus, suggest that Plutonism constructs a purely poetic landscape in an unreliable, phantasmagoric setting. According to Helmut Hölder, Goethe's geological activity began in 1776 during his service under Herzog Carl August, who assigned him the task of reviving a defunct copper mine in Ilmenau ("Goethe als Geologe," 233).

*Beachy Head* responds to an earlier phase of geological discipline-formation, as underscored by its posthumous appearance in the same year that George Bellas Greenough and others formed the Geological Society of London. Smith establishes a pattern of geological skepticism at the very beginning of *Beachy Head,* which offers an "extravagant theory" for the origin of the Sussex coast and the English Channel, but then questions this theory in her footnotes. She develops this ambivalent stance in a passage that culminates with a morally grounded rejection of geology in favor of botany. Remembering the fossils she found on the Sussex Downs in her youth, Smith's narrator entertains three possible explanations for her observation that these shells seemed to be made of the same chalk that surrounded them. A long note further details these youthful observations and remarks: "I have never read any of the late theories of the earth, nor was I ever satisfied with the attempts to explain many of the phenomena which call forth conjecture in those books I happened to have had access to on this subject." In the verse, she takes this insufficiency as evidence for the vanity of science in general:

> Ah! very vain is Science' proudest boast,
> And but a little light its flame yet lends
> To its most ardent votaries, since from whence
> These fossil forms are seen, is but conjecture,
> Food for vague theories, or vain dispute,
> While to his daily task the peasant goes,
> Unheeding such inquiry.[14]

Given the timing of this poem, it seems likely that Smith means to challenge the new geology's claims of expertise and professionalism. Her rejection sheds light both on geological discipline-formation and on the particular masculine associations of rocks and mountains so well exploited by the male Romantic poets. Smith identifies geology with "vague theories," "vain dispute," and "theories of the earth," one of the several names for earth science that a new generation of geologists was trying at the time to replace with the name "geology." The Geological Society reserved the term "theories of the earth" for old-fashioned armchair geology, including that of James Hutton, which it dismissed for political as well as scientific reasons. Smith seems aware of this pejorative sense, partly because she goes on to compare this sort of theorizing about fossils with antiquarianism, implicitly opposed to legitimate history as speculative and socially useless, just as geol-

---

14. Charlotte Smith, *Beachy Head* 5–10 and n., 368–89 and n., 372–89, 375n., 390–96 (*Poems,* 217, 232–33).

ogy is opposed to natural history and taxonomy.[15] As *Beachy Head* was being edited for publication, the infant Geological Society's investment in Neptunist theory led its members to minimize the importance of fossil evidence, though this was soon to change; younger poets such as Wordsworth and Blake larded their poetry with images of rocks and mountains devoid of fossil content; and Hutton, in 1795, insisted on theoretical grounds that all rocks were capable both of bearing fossils and of being transformed by heat. Smith, by contrast, celebrates the aesthetic experience of fossils, warning against the "vain" application of geological theory to such phenomena. When the Geological Society did turn to fossils, co-opting the fieldwork of William Smith and others, it was a matter of "taking something that formerly was not science and making it part of science itself" (in Paolo Rossi's words).[16] Smith's poem indicates why fossils might have been avoided by both poets and geologists seeking to establish themselves in terms of a new discipline: a masculine identification of nature with the earth's material accompanied professionalization in both literature and science. The geological idiom of Romanticism continued to skirt the issue of fossils because of an investment in illegibility strongly reinforced by recent generations of critics.

Geological issues provided an occasion for Smith and other women writers to negotiate an ambivalent authorial subjectivity within and against an increasingly masculinized discipline. She is one of several women writers whose ambivalence toward the objects and theory of geology takes shape as a conventional modesty often undercut by the assertion of a privileged identification with nature. Smith's diffidence, in using verbs like "hinting" and "alluding" to characterize the relationships between her images and scientific geology, resembles the professed geological modesty of other women writers, including Anna Seward, Ann Radcliffe, and Sarah Murray Aust. On the other hand, Smith's use of geological explanations as raw material for verse—just as she uses legend and fable—suggests a skepticism toward modern geology echoing that of Radcliffe and especially Murray Aust. In Radcliffe's German travels of 1794, a serious geological interest emerges clearly even as she apologizes for the absence of "proper and scientific denomination." Conventional modesty leads Murray Aust to defer to her (male) geological predecessors when speculating about the origins of Fingal's Cave, on Staffa, but her subsequent assertiveness on geological questions undermines and ironizes this modesty.[17] These geological asides are political gestures in

15. Ibid., lines 404–10.
16. Rossi, *Dark Abyss of Time*, xiv.
17. Ann Radcliffe, *Journey*, 259; Sarah Murray Aust, *Companion and Useful Guide*, 2:156; cf. 160, 177–79. The presence of numerous detailed and sometimes technical descriptions in Radcliffe's *Journey* suggests that her scientific modesty masks a legitimate confidence in her talent for geological description. This pattern is illuminated by Elizabeth Bohls' suggestion that

the sense that they challenge the bifurcation of public and domestic spheres tacitly embraced by women writing in what became "permissible" scientific genres, mainly didactic ones. This criticism of the politics of discipline-formation takes an explicitly scientific form in the 1822 debate between Maria Callcott and Greenough, then president of the Geological Society. Callcott published her account of an earthquake in Chile in the society's *Transactions*, only to have her observations and her science attacked by Greenough in a presidential lecture. Callcott responded defiantly to these criticisms: "as to ignorance of the science of Geology, Mrs. Callcott confesses it; and perhaps, that circumstance, and her consequent indifference to all theories connected with it, render her unbiassed testimony of the more value."[18]

This refusal of theory testifies to the complexity of what I have called "aesthetic materialism." The most modern geologists and their fiercest critics alike subscribed to a general fascination with geological objects, a fascination predicated on new perceptions of the materiality of nature. For Playfair, Goethe, and Smith, geology's aesthetic affiliations indicate theoretical overreaching of various kinds. Goethe and Smith reject geology's materialism even as they celebrate the aesthetic experience of landscape. William Smith, also a geological outsider, disavows geological theory to signify his aesthetic and moral sensibility as well as his fidelity to the earth's text: "My observations . . . are entirely original, and unincumbered with theories, for I have none to support: nor do I refer my reader to foreign countries . . . [but have] described the face of a country whose internal contents are more deeply explored than any other part of the earth's surface; and in which everyone . . . is a critic."[19] Like Humphry Davy's paean to limestone, Smith's ingenuous "observations" represent rocks as the national body or the "face of a country" whose landscapes provided material for an unprecedented explosion of literary and economic productivity during the Romantic period. W. G. Hoskins[20] has argued that "the English landscape itself . . .

women's travel writing eschews the generic subject position of landscape aesthetics and restores particularity as a means of working through exclusion from landownership (*Women Travel Writers*, 18, 66–68).

18. Quoted in Michele L. Aldrich, "Women in Geology," 51. See further Mary R. S. Creese and Thomas M. Creese, "British Women," 29–30.

19. William Smith, *Stratigraphical System of Organized Fossils*, vi–vii. Thomas Tredgold interprets Smith's refusal of theory as a patriotic gesture in his "Remarks on the geological principles of Werner, and those of Mr. Smith." Richard Hamblyn locates such skeptical gestures within the tradition of Dissenting science: "it is evident that the ingenuous dismissals of theory which often characterised the tone of such publications stemmed not always from the need to provide simplified notions to untutored audiences, but often from a stated position of impatience and rivalry with the officially sanctioned scientific institutions" ("Landscape and Contours of Knowledge," 52).

20. Hoskins, *Making of the English Landscape*, 14, 166.

is the richest historical record we possess," reactivating the social dimension of the rock record. One of Hoskins' examples, the "romantically beautiful valley" of Coalbrook-dale, attracted the interest of the burgeoning Shropshire iron industry and of Wordsworth and Anna Seward, who described "the other side of this romantic scene." Seen in this light, Romantic poems are the index fossils of a relatively recent stratum of English history dominated by the romance of landscape. Cultivated by an expanding critical public, the geological romance of that landscape's origins shaped and was shaped by the romance of an original national literature.

# Works Cited

ABBREVIATIONS

BIQ   *Blake: An Illustrated Quarterly*
BJHS  *British Journal for the History of Science*
SiR   *Studies in Romanticism*
TWC   *The Wordsworth Circle*


PRIMARY SOURCES

*An Account of the Wonders of Derbyshire, as Introduced in the Pantomime Entertainment at the Theatre-Royal, Drury-Lane.* London: W. Randall, 1779.

Addison, Joseph, and Richard Steele. *The Spectator.* Ed. Donald F. Bond. 5 vols. Oxford: Clarendon, 1965.

Aikin, Arthur. *Journal of a Tour through North Wales and Part of Shropshire; with Observations in Mineralogy and Other Branches of Natural History.* London: J. Johnson, 1797.

Aikin, John. *Essay on the Application of Natural History to Poetry.* London: J. Johnson, 1777.

Aristotle. *The Poetics.* In *A New Aristotle Reader,* ed. J. L. Ackrill, 540–56. Princeton: Princeton University Press, 1987.

Austen, Jane. *The Complete Novels of Jane Austen.* New York: Modern Library, n.d.

Baillie, John. *Essay on the Sublime.* London: R. Dodsley, 1747.

Barbauld, Anna Letitia. *Poems.* London: J. Johnson, 1773.

Barton, Bernard. *Metrical Effusions and The Triumph of the Orwell and The Convict's Appeal.* 1809. Ed. Donald H. Reiman. New York: Garland, 1977.

Beattie, James. *The Minstrel, or, The Progress of Genius.* 1771–74. London: C. Dilly and W. Creech, 1797.

Bertuch, Friedrich Justin. *Bilderbuch für Kinder.* 12 vols. Weimar: Verlag des Landes-Industrie-Comptoirs, 1798–1830.

Blake, William. *The Book of Urizen* (Copy G). Ed. Kay P. Easson and Roger R. Easson. Boulder and New York: Shambhala/Random House, 1978.

——. *The Complete Poetry and Prose of William Blake.* Ed. David V. Erdman. Rev. ed. New York: Doubleday, 1988.

——. *The Illuminated Blake.* Ed. David V. Erdman. Garden City, N.Y.: Anchor, 1974.

——. *Jerusalem. Blake's Illuminated Books.* Vol. 1. Ed. Morton Paley. Princeton: Princeton University Press, 1991.

——. *The Urizen Books. Blake's Illuminated Books.* Vol. 6. Ed. David Worrall. Princeton: Princeton University Press, 1995.

Boswell, James. *Life of Johnson.* Ed. R. W. Chapman. Oxford: Oxford University Press, 1980.

Brooke, Henry. *Universal Beauty.* 1735. In *Works of the English Poets*, ed. Chalmers. Vol. 17:337–65.

Burke, Edmund. *A Philosophical Enquiry into the Origin of Our Ideas of the Sublime and Beautiful.* 1757. Ed. Adam Phillips. New York: Oxford University Press, 1990.

Burnet, Thomas. *The Sacred Theory of the Earth.* 1690–91. London: Centaur, 1965.

Carroll, Lewis. *The Annotated Alice.* Ed. Martin Gardner. Cleveland: World Publishing, 1960.

Chalmers, Alexander, ed. *The Works of the English Poets from Chaucer to Cowper.* 24 vols. London: J. Johnson, 1810.

Chambers, William. *A Dissertation on Oriental Gardening.* 1772. Westmead: Gregg International, 1972.

——. *An Explanatory Discourse by Tan Chet-Qua, of Quang-Chew-Fu, Gent.* 1773. Ed. Richard E. Quaintance Jr. Los Angeles: Augustan Reprint Society, 1978.

Champion, Anthony. "To a Friend. The Scene Dovedale. 1756." In *Miscellanies, in Verse and Prose*, 48–52. London: Dent, 1801.

Coleridge, Samuel Taylor. *The Poems of Samuel Taylor Coleridge.* Ed. Ernest Hartley Coleridge. Oxford: Oxford University Press, 1912.

*The Compact Edition of the Oxford English Dictionary.* 2 vols. Oxford: Oxford University Press, 1971.

Cotton, Charles. *The Wonders of the Peake.* London: Joanna Brome, 1681.

Cumberland, George. *An Attempt to Describe Hafod.* London: T. Egerton, 1796.

——. *The Captive of the Castle of Sennaar.* 1798. Ed. G. E. Bentley Jr. Montreal: McGill-Queen's University Press, 1991.

——. "Mr. Cumberland in Defence of the Mosaic System." *Monthly Magazine* 40 (1815): 18–20.

——. "Mr. Cumberland on Proper Objects of Geology." *Monthly Magazine* 40 (1815): 130–33.

——. *Reliquiae Conservatae, from the Primitive Materials of Our Present Globe, with Popular Descriptions of the Prominent Characters of Some Remarkable Fossil Encrinites.* Bristol: J. Gutch, 1826.

Darwin, Erasmus. *The Botanic Garden. A Poem, in Two Parts . . . with Philosophical Notes.* 1791. London: Jones, 1824.

——. *The Letters of Erasmus Darwin.* Ed. Desmond King-Hele. Cambridge: Cambridge University Press, 1981.

——. *The Loves of the Plants.* 1789. Ed. Jonathan Wordsworth. New York: Woodstock, 1991.

——. "Ode written on the River Darwent, in a romantic Valley near its Source. By Dr. D——, of Derby." *Gentleman's Magazine* 55.2 (1785): 640.

——. *The Temple of Nature; or, The Origin of Society. A Poem with Philosophical Notes.* 1802. Ed. Donald Reiman. New York: Garland, 1978.

Davy, Humphry. *The Collected Works of Sir Humphry Davy.* Ed. John Davy. 9 vols. London: Smith, Elder, 1839.

——. *Fragmentary Remains, Literary and Scientific, of Sir Humphry Davy.* Ed. John Davy. London: J. Churchill, 1858.

——. *Humphry Davy on Geology: The 1805 Lectures for the General Audience.* Ed. Robert Siegfried and Robert Dott. Madison: University of Wisconsin Press, 1980.

Dennis, John. *The Critical Works of John Dennis.* Ed. Edward Niles Hooker. 2 vols. Baltimore: Johns Hopkins Press, 1943.

Drayton, Michael. *The Poly-Olbion: A Chorographical Description of Great Britain.* 1613–22. 2 vols. London: Spenser Society, 1899–90.

Drummond, William. *Clontarf, a Poem.* Dublin: Archer, 1822.

——. *The Giants' Causeway.* Belfast: Archer and Wirling, 1811.

Dyer, George. *Poetics: or, A Series of Poems, and Disquisitions on Poetry.* 2 vols. London: J. Johnson, 1812.

Dyer, John. *The Ruins of Rome.* 1740. In *The Poets of Great Britain,* ed. Samuel Johnson. Vol. 52:19–38. London: Cadell and Davies, 1807.

Farey, John. *General View of the Agriculture and Minerals of Derbyshire, with Observations on the Means of Their Improvement.* 3 vols. London: Sherwood, Neely, and Jones, 1811–17.

——. "Mr. William Smith's Discoveries in Geology." *Annals of Philosophy* 11 (1818): 359–64.

Faujas de Saint-Fond, Barthélemy. *A Journey through England and Scotland to the Hebrides in 1784.* 1797. Ed. and trans. Archibald Geikie. 2 vols. Glasgow: Hugh Hopkins, 1907.

Ferguson, James. "Mr. Ferguson's Description of the Devil's Cave, at Castleton, in the Peak of Derbyshire." *Gentleman's Magazine* 42 (1772): 518–19.

Forster, Johann Reinhold. *An Introduction to Mineralogy.* London: J. Johnson, 1768.

Gerard, Alexander. *An Essay on Taste.* 1759. Gainesville, Fla.: Scholars' Facsimiles and Reprints, 1963.

Gilks, Moreton. "A Letter . . . giving some account of the Petrefactions near Matlock Baths in Derbyshire." *Philosophical Transactions of the Royal Society of London* 41 (1740): 353–56.

Gilpin, William. *Observations on Several Parts of England, Particularly the Mountains and Lakes of Cumberland and Westmoreland, Relative Chiefly to Picturesque Beauty.* 3d ed. 2 vols. London: Cadell and Davies, 1808.

——. *Three Essays: On Picturesque Beauty, On Picturesque Travel, and On Landscape Painting.* 1792. Westmead: Gregg International, 1972.

Goethe, Johann Wolfgang von. *Faust. Der Tragödie zweiter Teil.* 1831. Ed. Lothar Scheithauer. Stuttgart: Reclam, 1971.

——. *Geologische und mineralogische Schriften. Gesamtausgabe der Werke und Schriften in zweiundzwanzig Bänden.* Vol. 20. Ed. Helmut Hölder and Eugen Wolf. Stuttgart: Cotta, 1960.

——. *Goethe's Collected Works.* Vol. 12. Ed. and trans. Douglas Miller. New York: Suhrkamp/Insel, 1988.

Gray, Thomas, Horace Walpole, Richard West, and Thomas Ashton. *The Correspondence of Gray, Walpole, West, and Ashton.* Ed. Paget Toynbee. 2 vols. Oxford: Clarendon, 1915.

Hamilton, Sir William. *Campi Phlegraei.* Naples, 1776.

Hazen, Robert M., ed. *The Poetry of Geology.* London: Allen and Unwin, 1982.

Hegel, G. W. F. *Phänomenologie des Geistes.* 1807. Ed. Hans-Friedrich Wessels and Heinrich Clairmont. Hamburg: Felix Meiner, 1988.

Hobbes, Thomas. *De Mirabilibus Pecci. Being the Wonders of the Peak in Darby-shire.* 1636 (Latin). 1678 (English). 5th ed. London: William Crook, 1683.

Hoffmann, E. T. A. *E. T. A. Hoffmann's Ausgewählte Werke.* 6 vols. Stuttgart: J. G. Cotta, n.d.

Hogarth, William. *The Analysis of Beauty.* 1753. Ed. Ronald Paulson. New Haven: Yale University Press, 1997.

Humboldt, Alexander von. *Ansichten der Natur.* Ed. Wilhelm Bölsche. 1823. Leipzig: Reclam, 1920.

——. *Personal Narrative of Travels to the Equinoctial Regions of America during the Years 1799–1804.* Trans. Thomasina Ross. 3 vols. London: Routledge, 1851.

——. *Political Essay on the Kingdom of New Spain.* Trans. John Black. 4 vols. London: Longman, 1811.

Hunt, John Dixon, and Peter Willis, eds. *The Genius of the Place.* New York: Harper and Row, 1975.

Hutcheson, Francis. *An Inquiry into the Original of Our Ideas of Beauty and Virtue.* 2d ed. London: J. Darby, 1726.

Hutton, James. *A Dissertation upon the Philosophy of Light, Heat, and Fire.* Edinburgh: Cadell, Junior, and Davies, 1794.

——. *Theory of the Earth.* 2 vols. 1795. Weinheim: Engelmann, 1959.

Jago, Richard. *Edge-hill.* 1767. In *The Works of the British Poets,* ed. Thomas Park. Vol. 27. London: J. Sharpe, 1808.

Kant, Immanuel. *The Critique of Judgment.* 1790. Trans. Werner S. Pluhar. Indianapolis: Hackett, 1987.

——. *The Critique of Pure Reason.* 1781. Trans. Norman Kemp Smith. 1929. New York: St. Martin's, 1965.

——. *Kritik der reinen Vernunft. Werke.* Vol. 3. Ed. Wilhelm Weischedel. Darmstadt: Wissenschaftliche Buchgesellschaft, 1983.

——. *Kritik der Urteilskraft. Werke.* Vol. 8. Ed. Wilhelm Weischedel. Darmstadt: Wissenschaftliche Buchgesellschaft, 1983.

Keate, George. *The Alps, A Poem.* London: R. and J. Dodsley, 1763.

Keats, John. *Complete Poems.* Ed. Jack Stillinger. Cambridge, Mass.: Belknap Press of Harvard University Press, 1982.

"A Late Religious Lady's Farewell to Matlock Bath." *New Lady's Magazine* 2 (1787): 663–64.

"Lines, written at Matlock, the latter end of June, 1787." *European Magazine* 45 (1804): 374–75.

Locke, John. *An Essay Concerning Human Understanding.* 1690. Ed. Peter H. Nidditch. Oxford: Clarendon, 1975.

Lowndes, Thomas. *Tracts in Prose and Verse.* Dover: William Bonython, 1825.

Lowth, Robert. *Lectures on the Sacred Poetry of the Hebrews.* 1787. New York: G. Olms, 1969.

Lyell, Charles. *Principles of Geology.* 3 vols. London: John Murray, 1830.

Mallet, David. *The Excursion.* 1728. In *Works of the English Poets*, ed. Chalmers. Vol. 14:16–26.

Maton, William. *Observations Relative Chiefly to the Natural History, Picturesque Scenery, and Antiquities, of the Western Counties of England.* Salisbury: J. Easton, 1797.

Montgomery, James. *Greenland.* 1819. Reprinted in *The Poetry of Geology*, ed. R. Hazen, 46–49.

——. *The Pelican Island.* 1827. Reprinted in *The Poetry of Geology*, ed. R. Hazen, 40–46.

Moule, Thomas. *The Landscape Album: or, Great Britain Illustrated in a series of sixty views by W. Westall.* London: Charles Tilt, 1832.

Murray, John. *A Comparative View of the Huttonian and Neptunian Systems of Geology, in Answer to the Illustrations of the Huttonian Theory of the Earth, by Professor Playfair.* 1802. New York: Arno, 1978.

Murray Aust, Sarah. *A Companion and Useful Guide to the Beauties of Scotland.* 1799. 3d ed. 2 vols. London: G. and W. Nicol, 1810.

Nicholson, William. *An Introduction to Natural Philosophy.* 2 vols. London: Joseph Johnson, 1782.

Novalis (Friedrich von Hardenberg). *The Novices of Sais.* 1802. Trans. Ralph Manheim. New York: Curt Valentin, 1949.

——. *Schriften.* Ed. Paul Kluckhohn and Richard Samuel. 6 vols. 3d ed. Stuttgart: Kohlhammer, 1977.

Paris, John Ayrton. *The Life of Sir Humphry Davy.* London: H. Colburn and R. Bentley, 1831.

Parkinson, James. *Organic Remains of a Former World: An Examination of the Mineral Remains of Vegetables and Animals of the Antediluvian World.* 3 vols. London: Nornaville and Fell et al., 1808–11.

Pope, Alexander. *The Poems of Alexander Pope.* Ed. John Butt. New Haven: Yale University Press, 1963.

Price, Uvedale. *Sir Uvedale Price on the Picturesque.* Ed. Thomas Dick Lauder. Edinburgh: Caldwell, Lloyd, 1842.

Radcliffe, Ann. *A Journey Made in the Summer of 1794.* London: G. G. and J. Robinson, 1795.

——. *The Mysteries of Udolpho.* 1794. Ed. Bonamy Dobrée. 1966; reprint, New York: Oxford University Press, 1980.

Raffles, Thomas. *Letters, during a Tour through some parts of France, Savoy, Switzerland, Germany, and the Netherlands, in the Summer of 1817.* 2d ed. Liverpool: Thomas Taylor, 1819.

Rogers, Charles. *A Collection of Prints in Imitation of Drawings.* 2 vols. London: Boydell, White, and Molini, 1778.

Schlegel, Friedrich. *Kritische und theoretische Schriften.* Ed. Andreas Huyssen. Stuttgart: Reclam, 1978.

Seward, Anna. *The Letters of Anna Seward.* 3 vols. Edinburgh: Constable, 1811.

———. *The Poetical Works of Anna Seward.* Ed. Walter Scott. 3 vols. Edinburgh: Ballantyne, 1810.

Shelley, Mary. *Frankenstein; or, The Modern Prometheus.* Ed. D. L. McDonald and Kathleen Scherf. 1818. Peterborough, Ont.: Broadview, 1994.

———. *The Journals of Mary Shelley, 1814–1844.* Ed. Paula Feldman and Diana Scott-Kilvert. 2 vols. Oxford: Clarendon, 1987.

Shelley, Percy Bysshe. *The Complete Works of Percy Bysshe Shelley.* Ed. Roger Ingpen and Walter E. Peck. 10 vols. London: Julian Editions, 1929.

———. *Shelley's Poetry and Prose.* Ed. Neil Fraistat and Donald H. Reiman. 2d ed. New York: Norton, 2002.

Sheppard, Thomas. *William Smith: His Maps and Memoirs.* Hull: A. Brown, 1920.

Simond, Louis. *Switzerland; or, a Journal of a Tour.* 2 vols. London: John Murray, 1822.

Smith, Charlotte. *The Poems of Charlotte Smith.* Ed. Stuart Curran. Oxford: Oxford University Press, 1993.

Smith, William. *A Memoir to the Map and Delineation of the Strata of England and Wales, with Part of Scotland.* London: John Cary, 1815.

———. *Strata Identified by Organized Fossils, containing Prints on Colored Paper of the Most Characteristic Specimen in each Stratum.* London: W. Arding, 1816.

———. *Stratigraphical System of Organized Fossils, with Reference to the Specimens of the Original Collection in the British Museum: Explaining Their State of Preservation and Their Use in Identifying the British Strata.* London: E. Williams, 1817.

Storer, James. *Antiquarian and Topographical Cabinet, Containing a Series of Elegant Views of the Most Interesting Objects of Curiosity in Great Britain.* 10 vols. London: W. Clarke, 1807–11.

Stukeley, William. *Abury, a Temple of the British Druids.* 1743. New York: Garland, 1984.

———. *Stonehenge: A Temple Restored to the British Druids.* 1740. New York: Garland, 1984.

Taylor, Thomas. *A Vindication of the Rights of Brutes.* London: Edward Jeffery, 1792.

Thomson, James. *The Seasons.* 1746. In *The Complete Poetical Works of James Thomson,* ed. J. Logie Robertson, 1–225. London: Oxford University Press, 1908.

*Transactions of the Royal Geological Society of Cornwall.* Vol. 1. London: William Philips, 1818.

Tredgold, Thomas. "Remarks on the geological Principles of Werner, and those of Mr. Smith." *Philosophical Magazine* 51 (1818): 36–38.

Werner, Abraham Gottlob. *Kurze Klassifikation der verschiedenen Gebirgsarten.* 1786. Trans. A. M. Ospovat. New York: Hafner, 1971.

Whately, Thomas. *Observations on Modern Gardening.* 2d ed. 1770. New York: Garland, 1978.

Whitehurst, John. *Inquiry into the Original State and Formation of the Earth.* London: J. Cooper, 1778.

——. *Inquiry into the Original State and Formation of the Earth.* 2d ed. 1786. New York: Arno, 1978.

Wollstonecraft, Mary. *Historical and Moral View of the Origin and Progress of the French Revolution.* 2d ed. London: Joseph Johnson, 1795.

Wordsworth, Dorothy. "Grasmere—a Fragment." In *English Romantic Writers,* ed. David Perkins, 496–97. Rev. ed. Fort Worth: Harcourt Brace, 1995.

——. *The Journals of Dorothy Wordsworth.* Ed. Ernest de Selincourt. 2 vols. New York: Macmillan, 1941.

Wordsworth, Dorothy, and William Wordsworth. *The Letters of William and Dorothy Wordsworth: The Early Years, 1787–1805.* Ed. Ernest de Selincourt. Oxford: Clarendon, 1967.

Wordsworth, William. *Lyrical Ballads.* Ed. R. L. Brett and A. R. Jones. London: Methuen, 1965.

——. *Poems, in Two Volumes, and Other Poems, 1800–1807.* Ed. Jared Curtis. Ithaca: Cornell University Press, 1983.

——. *The Poetical Works of William Wordsworth.* Ed. Ernest de Selincourt. 5 vols. Oxford: Clarendon, 1940–49.

——. *The Prelude: 1799, 1805, 1850.* Ed. Jonathan Wordsworth, M. H. Abrams, and Stephen Gill. New York: Norton, 1979.

——. *The Prose Works of William Wordsworth.* Ed. W. J. B. Owen and Jane Smyser. 3 vols. Oxford: Clarendon, 1974.

——. *Wordsworth's Guide to the Lakes.* 5th ed. 1835. Ed. Ernest de Selincourt. 1906. Oxford: Oxford University Press, 1977.

——. *Wordsworth's Literary Criticism.* Ed. Nowell C. Smith. London: Henry Frowde, 1905.

SECONDARY SOURCES

Abrams, M. H. *The Mirror and the Lamp.* 1953. New York: Oxford University Press, 1971.

Adorno, Theodor, and Max Horkheimer. *Dialectic of Enlightenment: Philosophical Fragments.* Trans. Edmund Jephcott. Stanford: Stanford University Press, 2002.

——. *Dialektik der Aufklärung. Philosophische Fragmente.* 1944. Frankfurt: S. Fischer, 1969.

Aldrich, Michele L. "Women in Geology." In *Women of Science: Righting the Record,* ed. G. Kass-Simon and P. Farnes, 42–71. Bloomington: Indiana University Press, 1990.

Allen, David. *The Naturalist in Britain: A Social History.* 1976; reprint, Princeton: Princeton University Press, 1994.

Allen, Ralph G. "The Wonders of Derbyshire: A Spectacular Eighteenth-Century Travelogue." *Theatre Survey* 2 (1961): 54–66.

Altick, Richard D. *The English Common Reader: A Social History of the Mass Reading Public, 1800–1900.* Chicago: University of Chicago Press, 1957.

——. *The Shows of London.* Cambridge, Mass.: Belknap Press of Harvard University Press, 1978.

Andrews, Malcolm. *The Search for the Picturesque: Landscape Aesthetics and Tourism in Britain, 1760–1800.* Stanford: Stanford University Press, 1989.

Aubin, Robert A. "Grottoes, Geology, and the Gothic Revival." *Studies in Philology* 31 (1934): 408–16.

——. *Topographical Poetry in XVIII-Century England.* New York: Modern Language Association, 1936.

Ault, Donald. *Visionary Physics: Blake's Response to Newton.* Chicago: University of Chicago Press, 1974.

Barrell, John. *The Idea of Landscape and the Sense of Place, 1730–1840: An Approach to the Poetry of John Clare.* Cambridge: Cambridge University Press, 1972.

Bate, Jonathan. *Romantic Ecology.* London: Routledge, 1991.

Bedell, Rebecca. *The Anatomy of Nature: Geology and American Landscape Painting, 1825–1875.* Princeton: Princeton University Press, 2001.

Beer, Gillian. *Darwin's Plots: Evolutionary Narrative in Darwin, George Eliot, and Nineteenth-Century Fiction.* London: Routledge, 1983.

——. "Has Nature a Future?" In *The Third Culture,* ed. Shaffer, 15–27.

Benjamin, Marina, ed. *Science and Sensibility: Gender and Scientific Inquiry, 1780–1945.* Oxford: Blackwell, 1991.

Bentley, G. E., Jr., ed. *A Bibliography of George Cumberland (1754–1848).* New York: Garland, 1975.

——. *Blake Records.* Oxford: Clarendon, 1969.

Berman, Morris. *Social Change and Scientific Organization: The Royal Institution, 1799–1844.* Ithaca: Cornell University Press, 1978.

Bewell, Alan. *Romanticism and Colonial Disease.* Baltimore: Johns Hopkins University Press, 1999.

——. *Wordsworth and the Enlightenment.* New Haven: Yale University Press, 1989.

Bode, Christoph. "A Kantian Sublime in Shelley: 'Respect for Our Own Vocation' in an Indifferent Universe." *1650–1850: Ideas, Aesthetics, and Inquiries in the Early Modern Era* 3 (1997): 329–58.

Boehme, Hartmut. "Das Steinerne: Anmerkungen zur Theorie des Erhabenen aus dem Blick des 'Menschenfremdesten.'" In *Das Erhabene: Zwischen Grenzerfahrung und Größenwahn,* ed. Christine Pries, 119–41. Darmstadt: VCH, 1989.

Bohls, Elizabeth. *Women Travel Writers and the Language of Aesthetics, 1716–1818.* Cambridge: Cambridge University Press, 1995.

Brinkley, Robert. "Spaces between Words: Writing *Mont Blanc.*" In *Romantic Revisions,* ed. Brinkley and Keith Hanley, 243–67. Cambridge: Cambridge University Press, 1992.

Butler, Judith. *Bodies That Matter: On the Discursive Limits of "Sex".* New York: Routledge, 1993.

Challinor, John. *The History of British Geology: A Bibliographical Study.* New York: Barnes and Noble, 1971.

Clark, William, Jan Golinski, and Simon Schaffer, eds. *The Sciences in Enlightened Europe.* Chicago: University of Chicago Press, 1999.

Copley, Stephen. "William Gilpin and the Black-Lead Mine." In *The Politics of the Picturesque,* ed. Copley and Peter Garside, 212–36. Cambridge: Cambridge University Press, 1994.

Creese, Mary R. S., and Thomas M. Creese. "British Women Who Contributed to Research in the Geological Sciences in the Nineteenth Century." *BJHS* 27 (1994): 23–54.

Cronon, William. "The Trouble with Wilderness; or, Getting Back to the Wrong Nature." In *Uncommon Ground: Rethinking the Human Place in Nature*, ed. Cronon, 69–90. New York: Norton, 1996.

Cunningham, Andrew, and Nicholas Jardine, eds. *Romanticism and the Sciences*. Cambridge: Cambridge University Press, 1990.

Daniels, Stephen. *Fields of Vision: Landscape Imagery and National Identity in England and the United States*. Princeton: Princeton University Press, 1993.

——. *Joseph Wright*. Princeton: Princeton University Press, 1999.

Davies, Gordon. *The Earth in Decay: A History of British Geomorphology, 1578–1878*. London: MacDonald, 1968.

Dean, Dennis R. "Geology and English Literature: Crosscurrents, 1770–1830." Ph.D. diss., University of Wisconsin, 1968.

——. *Gideon Mantell and the Discovery of Dinosaurs*. Cambridge: Cambridge University Press, 1999.

——. *James Hutton and the History of Geology*. Ithaca: Cornell University Press, 1992.

De Bolla, Peter. *The Discourse of the Sublime: Readings in History, Aesthetics, and the Subject*. Oxford: B. Blackwell, 1989.

De Luca, Vincent A. "Blake's Wall of Words." In *Unnam'd Forms*, ed. Nelson Hilton and T. Vogler, 218–41. Berkeley: University of California Press, 1986.

——. *Words of Eternity: Blake and the Poetics of the Sublime*. Princeton: Princeton University Press, 1991.

De Man, Paul. "Phenomenality and Materiality in Kant." In *Hermeneutics: Questions and Prospects*, ed. G. Shapiro and A. Sica, 121–44. Amherst: University of Massachusetts Press, 1984.

——. *The Rhetoric of Romanticism*. New York: Columbia University Press, 1984.

Engell, James. "A Yet Deeper Well: 'Kubla Khan,' Wookey Hole, Cain." *TWC* 26 (1995): 3.

Erdman, David V. *Blake: Prophet against Empire*. 3d ed. Princeton: Princeton University Press, 1977.

——. *A Concordance to the Poetry of William Blake*. 2 vols. Ithaca: Cornell University Press, 1967.

Essick, Robert. "Wordsworth and Leech-Lore." *TWC* 12 (1981): 100–102.

Essick, Robert, and Morton Paley. "'Dear Generous Cumberland': A Newly Discovered Letter and Poem by William Blake." *BIQ* 32 (1998): 4–13.

Esterhammer, Angela. "Calling into Existence: *The Book of Urizen*." In *Blake in the Nineties*, ed. Steve Clark and David Worrall, 114–32. New York: St. Martin's, 1999.

Eyles, Joan M. "William Smith: Some Aspects of His Life and Work." In *Toward a History of Geology*, ed. Cecil J. Schneer, 142–58. Cambridge, Mass.: MIT Press, 1969.

Ferguson, Frances. "Shelley's *Mont Blanc*: What the Mountain Said." In *Romanticism and Language*, ed. Arden Reed, 202–14. Ithaca: Cornell University Press, 1984.

——. *Solitude and the Sublime: Romanticism and the Aesthetics of Individuation*. New York: Routledge, 1992.

Force, James. *William Whiston: Honest Newtonian.* Cambridge: Cambridge University Press, 1985.

Foreman, Amanda. *Georgiana, Duchess of Devonshire.* London: HarperCollins, 1998.

Foucault, Michel. *The Order of Things: An Archaeology of the Human Sciences.* 1966. New York: Random House, 1970.

Fulford, Tim. "Coleridge, Darwin, Linnaeus: The Sexual Politics of Botany." *TWC* 28 (1997): 124–30.

Frey, Heather. "Defining the Self, Defiling the Countryside: Travel Writing and Romantic Ecology." *TWC* 28 (1997): 162–66.

Fry, Paul H. *A Defense of Poetry: Reflections on the Occasion of Writing.* Stanford: Stanford University Press, 1995.

Gandhi, Leela. *Postcolonial Theory: A Critical Introduction.* New York: Columbia University Press, 1998.

Gates, Barbara, and Ann Shteir, eds. *Natural Eloquence: Women Reinscribe Science.* Madison: University of Wisconsin Press, 1997.

Gidal, Eric. *Poetic Exhibitions: Romantic Aesthetics and the Pleasures of the British Museum.* Lewisburg, Pa.: Bucknell University Press, 2001.

Gilmartin, Kevin. "Popular Radicalism and the Public Sphere." In *Romanticism and Its Publics,* ed. Klancher, 549–57.

Gold, Helmut. *Erkenntnisse unter Tage: Bergbaumotive in der Literatur der Romantik.* Opladen: Westdeutscher Verlag, 1990.

Golinski, Jan. *Science as Public Culture: Chemistry and Enlightenment in Britain, 1760–1820.* Cambridge: Cambridge University Press, 1992.

Gould, Stephen Jay. "The Man Who Set the Clock Back." *New York Review of Books* 48.15 (October 4, 2001): 51–56.

———. *Time's Arrow, Time's Cycle: Myth and Metaphor in the Discovery of Geological Time.* Cambridge, Mass.: Harvard University Press, 1987.

Grabo, Carl. *A Newton among Poets: Shelley's Use of Science in Prometheus Unbound.* Chapel Hill: University of North Carolina Press, 1930.

Greenberg, Mark. "Blake's 'Science.'" *Studies in Eighteenth-Century Culture* 12 (1983): 115–30.

Guillory, John. "The English Common Place: Lineages of the Topographical Genre." *Critical Quarterly* 33 (1991): 3–27.

Habermas, Jürgen. *The Structural Transformation of the Public Sphere.* Trans. Thomas Burger. Cambridge, Mass.: MIT Press, 1989.

———. *Strukturwandel der Öffentlichkeit.* 1962. 2d ed. Frankfurt: Suhrkamp, 1990.

Hamblyn, Richard. "Landscape and the Contours of Knowledge: The Literature of Travel and the Sciences of the Earth in Eighteenth-Century Britain." Ph.D. diss., Cambridge University, 1994.

———. "Private Cabinets and Popular Geology: The British Audiences for Volcanoes in the Eighteenth Century." In *Transports: Travel, Pleasure, and Imaginative Geography, 1600–1830,* ed. Chloe Chard and Helen Langdon, 179–205. New Haven: Yale University Press, 1996.

Harré, Rom. "Davy and Coleridge: Romanticism in Science and the Arts." In *Common Denominators in Art and Science,* ed. Pollock, 53–64.

——. "What Is the *Zeitgeist*?" In *Common Denominators in Art and Science*, ed. Pollock, 1–8.

Harris, John, and Michael Snodin, eds. *Sir William Chambers*. New Haven: Yale University Press, 1996.

Hartman, Geoffrey. *The Unmediated Vision*. 1954; reprint, New York: Harcourt, Brace and World, 1966.

Heppner, Christopher. "Another 'New' Blake Engraving: More about Blake and William Nicholson." *BIQ* 12 (1978): 193–97.

Heringman, Noah. "Introduction: The Commerce of Literature and Natural History." In *Romantic Science: The Literary Forms of Natural History*, ed. Heringman, 1–19. Albany: SUNY Press, 2003.

——. "Stones So Wonderous Cheap." *SiR* 37 (spring 1998): 43–62.

——. "The Style of Natural Catastrophes." *Huntington Library Quarterly* 66 (fall 2003): 97–133.

Hertz, Neil. *The End of the Line: Psychoanalysis and the Sublime*. New York: Columbia University Press, 1985.

Hilton, Nelson. "Blake and the Perception of Science." *Annals of Scholarship* 4 (1986): 54–68.

——. *Literal Imagination: Blake's Vision of Words*. Berkeley: University of California Press, 1983.

——. "An Original Story." In *Unnam'd Forms: Blake and Textuality*, ed. Hilton and T. Vogler, 69–104. Berkeley: University of California Press, 1986.

——. "The Spectre of Darwin." *BIQ* 15 (1981): 36–48.

——. "The Sweet Science of Atmospheres in *The Four Zoas*." *BIQ* 12 (1978): 80–86.

Hitt, Christopher. "Toward an Ecological Sublime." *New Literary History* 30.3 (1999): 603–23.

Hogan, Charles Beecher, ed. *The London Stage, 1660–1800*. 12 vols. Carbondale: Southern Illinois University Press, 1968.

Hogle, Jerrold. *Shelley's Process: Radical Transference and the Development of His Major Works*. New York: Oxford University Press, 1988.

Hölder, Helmut. "Goethe als Geologe." *Goethe-Jahrbuch* 111 (1994): 231–45.

Hoskins, W. G. *The Making of the English Landscape*. 1955; reprint, Harmondsworth: Penguin, 1970.

Hughes, D. J. "Potentiality in *Prometheus Unbound*." In *Shelley's Poetry and Prose*, ed. Sharon B. Powers and Donald H. Reiman, 603–20. New York: Norton, 1977.

Jacob, Margaret C. *The Cultural Meaning of the Scientific Revolution*. Philadelphia: Temple University Press, 1988.

——. *Scientific Culture and the Making of the Industrial West*. New York: Oxford University Press, 1997.

Jardine, Nicholas, James Secord, and E. C. Spary, eds. *Cultures of Natural History*. Cambridge: Cambridge University Press, 1996.

Johnson, Mary Lynn. "Blake, Democritus, and the 'Fluxions of the Atom': Some Contexts for Materialist Critiques." In *Historicizing Blake*, ed. Steve Clark and David Worrall, 105–24. New York: St. Martin's, 1994.

Jordanova, Ludmilla, and Roy Porter, eds. *Images of the Earth: Essays in the History of the Environmental Sciences.* Chalfont St. Giles: British Society for the History of Science, 1979.

Kelley, Theresa. *Wordsworth's Revisionary Aesthetics.* Cambridge: Cambridge University Press, 1988.

Keynes, Geoffrey. *Blake Studies.* Rev. ed. Oxford: Clarendon, 1971.

King, Amy Mae. *Bloom: The Botanical Vernacular in the English Novel.* New York: Oxford University Press, 2003.

King-Hele, Desmond. *Erasmus Darwin: A Life of Unequalled Achievement.* London: Giles de la Mare, 1999.

———. *Erasmus Darwin and the Romantic Poets.* Houndmills: Macmillan, 1986.

Klancher, Jon. *The Making of English Reading Audiences, 1790–1832.* Madison: University of Wisconsin Press, 1987.

———, ed. *Romanticism and Its Publics: A Forum. SiR* 33 (winter 1994).

Knapp, Steven. "The Sublime, Self-Reference, and Wordsworth's 'Resolution and Independence.'" *Modern Language Notes* 99 (1984): 1007–22.

Knell, Simon J. *The Culture of English Geology, 1815–51: A Science Revealed through Its Collecting.* Aldershot: Ashgate, 2000.

Knight, David M. "Commissions, Creativity, Esteem: What Made Humphry Davy Tick?" In *Common Denominators,* ed. Pollock, 46–52.

———. *Humphry Davy: Science and Power.* Oxford: Blackwell, 1992.

Kuhn, Thomas. *The Structure of Scientific Revolutions.* 1962. Rev. ed. Chicago: University of Chicago Press, 1970.

Lang, W. D. "Mary Anning, of Lyme, Collector and Vendor of Fossils, 1799–1847." *Natural History Magazine* 5 (1935): 66–81.

Laudan, Rachel. *From Mineralogy to Geology: The Foundations of a Science, 1650–1830.* Chicago: University of Chicago Press, 1987.

Lawrence, Christopher. "The Power and the Glory: Humphry Davy and Romanticism." In *Romanticism and the Sciences,* ed. Cunningham and Jardine, 213–27.

Leask, Nigel. "Mont Blanc's Mysterious Voice: Shelley and Huttonian Earth Science." In *The Third Culture,* ed. Shaffer, 182–203.

Leonard, D. C. "Erasmus Darwin and William Blake." *Eighteenth-Century Life* 4 (1978): 78–81.

Levine, George. *One Culture: Essays in Science and Literature.* Madison: University of Wisconsin Press, 1987.

Levinson, Marjorie. *Wordsworth's Great Period Poems.* Cambridge: Cambridge University Press, 1986.

Liu, Alan. *Wordsworth: The Sense of History.* Stanford: Stanford University Press, 1989.

Liu, Lydia. "Robinson Crusoe's Earthenware Pot: Science, Aesthetics, and the Metaphysics of True Porcelain." In *Romantic Science,* ed. Heringman, 137–69.

Logan, James V. *The Poetry and Aesthetics of Erasmus Darwin.* Princeton: Princeton University Press, 1936.

Lyotard, Jean-François. "The Interest of the Sublime." 1984. In *Of the Sublime: Presence in Question,* ed. Jean-Luc Nancy, 109–32. Trans. Jeffrey Librett. Albany: SUNY Press, 1993.

Mahoney, Dennis. "Human History as Natural History in *Die Lehrlinge zu Sais* and

*Heinrich von Ofterdingen.*" In *Subversive Sublimities: Undercurrents of the German Enlightenment,* ed. Eitel Timm, 1–11. Columbia, S.C.: Camden House, 1992.

Matthews, G. M. "A Volcano's Voice in Shelley." 1957. In *Shelley's Poetry and Prose,* ed. Fraistat and Reiman, 550–68.

Matthews, Susan. "*Jerusalem* and Nationalism." 1992. In *William Blake,* ed. John Lucas, 80–100. New York: Addison Wesley Longman, 1998.

McKibben, Bill. *The End of Nature.* New York: Random House, 1989.

McKusick, James. *Green Writing: Romanticism and Ecology.* New York: St. Martin's, 2000.

——. " 'Kubla Khan' and the Theory of the Earth." In *Samuel Taylor Coleridge and the Sciences of Life,* ed. N. Roe, 134–151. Oxford: Oxford University Press, 2001.

McNeil, Maureen. "The Scientific Muse: The Poetry of Erasmus Darwin." In *Languages of Nature: Critical Essays on Science and Literature,* ed. Ludmilla Jordanova, 164–203. London: Free Association Books, 1986.

——. *Under the Banner of Science: Erasmus Darwin and His Age.* Manchester: Manchester University Press, 1987.

Mee, Jon. *Dangerous Enthusiasm: William Blake and the Culture of Radicalism in the 1790s.* Oxford: Clarendon, 1992.

Mellor, Anne. "*Frankenstein*: A Feminist Critique of Science." In *One Culture,* ed. Levine, 287–312.

——. *Mothers of the Nation: Women's Political Writing in England, 1780–1830.* Bloomington: Indiana University Press, 2000.

Monastersky, Richard. "The Marriage of Art and Science." *Chronicle of Higher Education* 47.38 (June 1, 2001): A12–15.

Monk, Samuel H. *The Sublime: A Study of Critical Theories in XVIII-Century England.* New York: Modern Language Association, 1935.

Nicolson, Marjorie. *Mountain Gloom and Mountain Glory: The Development of the Aesthetics of the Infinite.* 1959. Seattle: University of Washington Press, 1997.

——. *Newton Demands the Muse: Newton's Opticks and the Eighteenth-Century Poets.* 1946; reprint, Westport, Conn.: Greenwood, 1979.

Nurmi, Martin K. "Negative Sources in Blake." In *William Blake: Essays for S. Foster Damon,* ed. A. Rosenfeld, 303–318. Providence: Brown University Press, 1969.

Oerlemans, Onno. *Romanticism and the Materiality of Nature.* Toronto: University of Toronto Press, 2002.

Oldroyd, David. *Thinking about the Earth: A History of Ideas in Geology.* Cambridge, Mass.: Harvard University Press, 1996.

Paley, Morton. "Blake and Thomas Burnet's *Sacred Theory of the Earth.*" *BIQ* 25 (1991): 75–78.

——. *The Continuing City: William Blake's Jerusalem.* Oxford: Clarendon, 1983.

——. "Thomas Johnes, 'Ancient Guardian of Wales.' " *BIQ* 2 (1967): 65–67.

Peterfreund, Stuart. *William Blake in a Newtonian World: Essays on Literature as Art and Science.* Norman: University of Oklahoma Press, 1998.

Pollock, Martin, ed. *Common Denominators in Art and Science.* Aberdeen: Aberdeen University Press, 1983.

Porter, Roy. "The Industrial Revolution and the Rise of Geology." In *Changing Per-*

*spectives in the History of Science*, ed. M. Teich and R. Young, 320–43. Dordrecht: D. Reidel, 1973.

——. *The Making of Geology: Earth Science in Britain 1660–1815*. Cambridge: Cambridge University Press, 1977.

Pratt, Mary Louise. *Imperial Eyes: Travel Writing and Transculturation*. New York: Routledge, 1992.

Pyle, Forest. *The Ideology of Imagination: Subject and Society in the Discourse of Romanticism*. Stanford: Stanford University Press, 1995.

Rajan, Tilottama. *The Supplement of Reading: Figures of Understanding in Romantic Theory and Practice*. Ithaca: Cornell University Press, 1990.

Randel, Fred V. "*Frankenstein*, Feminism, and the Intertextuality of Mountains." *SiR* 23 (1984): 515–32.

Rappaport, Rhoda. *When Geologists Were Historians, 1665–1750*. Ithaca: Cornell University Press, 1997.

Richardson, Alan. *Literature, Education, and Romanticism: Reading as Social Practice, 1780–1832*. Cambridge: Cambridge University Press, 1995.

Ross, Catherine E. " 'Twin Labourers and Heirs of the Same Hopes': The Professional Rivalry of Humphry Davy and William Wordsworth." In *Romantic Science*, ed. Heringman, 23–52.

Rossi, Paolo. *The Dark Abyss of Time: The History of the Earth and the History of Nations from Hooke to Vico*. 1979. Trans. Lydia Cochrane. Chicago: University of Chicago Press, 1984.

Rudwick, Martin. *Georges Cuvier, Fossil Bones, and Geological Catastrophes*. Chicago: University of Chicago Press, 1997.

——. *The Great Devonian Controversy: The Shaping of Scientific Knowledge among Gentlemanly Specialists*. Chicago: University of Chicago Press, 1985.

——. *The Meaning of Fossils: Episodes in the History of Palaeontology*. Rev. ed. New York: Science History Publications, 1976.

——. "Minerals, Strata, and Fossils." In *Cultures of Natural History*, ed. Jardine, Secord, and Spary, 266–86.

——. *Scenes from Deep Time*. Chicago: University of Chicago Press, 1994.

Rupke, Nicolaas. "The Apocalyptic Denominator in English Culture of the Early Nineteenth Century." In *Common Denominators in Art and Science*, ed. Pollock, 30–41.

——. "Caves, Fossils, and the History of the Earth." In *Romanticism and the Sciences*, ed. Cunningham and Jardine, 241–59.

Schiebinger, Londa. "The Private Life of Plants: Sexual Politics in Carl Linnaeus and Erasmus Darwin." In *Science and Sensibility*, ed. Benjamin, 121–43.

Schulman, Samuel. "The Spenserian Enchantments of Wordsworth's 'Resolution and Independence.' " *Modern Philology* 79 (1981): 24–44.

Secord, James. *Controversy in Victorian Geology: The Cambrian-Silurian Dispute*. Princeton: Princeton University Press, 1986.

Shaffer, Elinor S., ed. *The Third Culture: Literature and Science*. Berlin: De Gruyter, 1998.

Sharrock, Roger. "The Poet and the Chemist." *Notes and Records of the Royal Society* 17 (1962): 57–76.

Shortland, Michael. "Darkness Visible: Underground Culture in the Golden Age of Geology." *History of Science* 32 (1994): 1–61.

Shteir, Ann. *Cultivating Women, Cultivating Science: Flora's Daughters and Botany in England, 1760–1860.* Baltimore: Johns Hopkins University Press, 1996.

Siskin, Clifford. *The Work of Writing: Literature and Social Change in Britain, 1700–1830.* Baltimore: Johns Hopkins University Press, 1998.

Stafford, Barbara. *Voyage into Substance: Art, Science, Nature, and the Illustrated Travel Account, 1760–1840.* Cambridge, Mass.: MIT Press, 1984.

Stafford, Barbara, and Frances Terpak. *Devices of Wonder.* Los Angeles: Getty Research Institute, 2001.

Thomas, Keith. "Coleridge, Wordsworth, and the New Historicism: 'Chamouny; The Hour before Sunrise, A Hymn' and Book 6 of *The Prelude.*" *SiR* 33 (spring 1994): 81–117.

Todd, Ruthven. *Tracks in the Snow: Studies in English Science and Art.* New York: Scribner's, 1947.

Torrens, Hugh. "Arthur Aikin's Mineralogical Survey of Shropshire, 1796–1816, and the Contemporary Audience for Geological Publications." *BJHS* 16 (1983): 111–53.

———. "Geological Communication in the Bath Area in the Last Half of the Eighteenth Century." In *Images of the Earth*, ed. Jordanova and Porter, 215–47.

———. "Mary Anning (1799–1847) of Lyme; 'The Greatest Fossilist the World Ever Knew.'" *BJHS* 28 (1995): 257–84.

Toulmin, Stephen. *The Return to Cosmology: Postmodern Science and the Theology of Nature.* Berkeley: University of California Press, 1987.

Vesely, Suzanne. "The Daughters of Eighteenth-Century Science: A Rationalist and Materialist Context for William Blake's Female Figures." *Colby Quarterly* 34.1 (1998): 5–24.

Vincent, Patrick. "What the Mountain Should Have Said." Paper presented at the annual convention of the North American Society for the Study of Romanticism, Phoenix, Ariz., September 2000.

Walker, Eric. "The Pass of Kirkstone." *TWC* 19 (1989): 116–21.

Wallace, Anne D. "Picturesque Fossils, Sublime Geology? The Crisis of Authority in *Beachy Head.*" *European Romantic Review* 13 (winter 2002): 77–93.

Wasserman, Earl. *Shelley: A Critical Reading.* Baltimore: Johns Hopkins Press, 1971.

Weindling, Paul Julian. "Geological Controversy and Its Historiography: The Prehistory of the Geological Society of London." In *Images of the Earth*, ed. Jordanova and Porter, 248–71.

Weiskel, Thomas. *The Romantic Sublime: Studies in the Structure and Psychology of Transcendence.* Baltimore: Johns Hopkins University Press, 1976.

Whittow, John. *Geology and Scenery in Britain.* London: Chapman and Hall, 1992.

Williams, Raymond. *The Country and the City.* New York: Oxford University Press, 1973.

———. *Marxism and Literature.* Oxford: Oxford University Press, 1977.

Winchester, Simon. *The Map That Changed the World: William Smith and the Birth of Modern Geology.* New York: HarperCollins, 2001.

Wood, Andy. *The Politics of Social Conflict: The Peak Country, 1520–1770.* Cambridge: Cambridge University Press, 1999.

Wood, Gillen Darcy. *The Shock of the Real: Romanticism and Visual Culture, 1760–1860*. New York: Palgrave, 2001.

Worrall, David. "Blake's Derbyshire: A Visionary Locale in *Jerusalem*." *BIQ* 11 (1977): 34–35.

———. "Blake's *Jerusalem* and the Visionary History of Britain." *SiR* 16 (spring 1977): 189–216.

———. "William Blake and Erasmus Darwin's *Botanic Garden*." *Bulletin of the New York Public Library* 79 (1975): 397–417.

Wyatt, John. *Wordsworth and the Geologists*. Cambridge: Cambridge University Press, 1995.

Ziolkowski, Theodore. *German Romanticism and Its Institutions*. Princeton: Princeton University Press, 1991.

Žižek, Slavoj. *The Sublime Object of Ideology*. London: Verso, 1989.

# Index

Milton Keynes UK
Ingram Content Group UK Ltd.
UKHW022234130624
444101UK00001BA/62